Birkhäuser

Frontiers in Mathematics

Mikhail Borsuk

Transmission Problems for Elliptic Second-Order Equations in Non-Smooth Domains

Mikhail Borsuk
Department of Mathematics and Informatics
University of Warmia and Mazury
M. Oczapowskiego str. 2
10-719 Olsztyn
Poland
e-mail: borsuk@uwm.edu.pl

2010 Mathematics Subject Classification: 35J25, 35J60, 35J65, 35J70, 35J85, 35B05, 35B45, 35B65

ISBN 978-3-0346-0476-5 e-ISBN 978-3-0346-0477-2
DOI 10.1007/978-3-0346-0477-2

Library of Congress Control Number: 2010930426

Cover design: deblik, Berlin

Printed on acid-free paper

Springer Basel AG is part of Springer Science+Business Media

www.birkhauser-science.com

To my wife Tatiana

Contents

Preface

The goal of this book is to investigate the behaviour of weak solutions to the elliptic transmisssion problem in a neighborhood of boundary singularities: angular and conic points or edges. We consider this problem both for linear and quasi-linear (very little studied) equations. In style and methods of research, this book is close to our monograph [14] together with Prof. V. Kondratiev.

The book consists of an Introduction, seven chapters, a Bibliography and Indexes. Chapter 1 is of auxiliary character. We recall the basic definitions and properties of Sobolev spaces and weighted Sobolev-Kondratiev spaces. Here we recall also the well-known Stampacchia's Lemma and derive a generalization for the solution of the Cauchy problem – the Gronwall-Chaplygin type inequality.

Chapter 2 deals with the eigenvalue problem for m-Laplace-Beltrami operator. By the variational principle we prove a new integro-differential Friedrichs-Wirtinger type inequality. This inequality is the basis for obtaining of precise exponents of the decreasing rate of the solution near boundary singularities.

Chapter 3 deals with the investigation of the transmission problem for linear elliptic second order equations in the domains with boundary conic point.

Chapter 4 is devoted to the transmission problem in conic domains with N different media for an equation with the Laplace operator in the principal part.

Chapters 5, 6 and 7 deal with the investigation of the transmission problem for quasi-linear elliptic second order equations in the domains with boundary conic point (Chapters 5–6) or with an edge at the boundary of a domain.

All results are given in the book with complete proofs. The book is based on the author's research he had made over the past years (see [8, 9, 10, 11, 12, 13]).

I would like to express my gratitude to Dr. Mikhail Kolev from University of Warmia and Mazury in Olsztyn who improved my English and Dr. Mykhaylo Plesha from Lviv (Ukraine) who executed figures in TEX. It also should be mentioned that the work on the book was made possible by the support of the Polish Ministry of Science and Higher Education through Grant Nr N201 381834. This help is gratefully acknowledged.

Olsztyn, Poland

Spring 2010

Introduction

Transmission problems appear frequently in various fields of physics and technics. For instance, one of the important problems of the electrodynamics of solid media is the research on electromagnetic processes in ferromagnetic media with different dielectric constants. This type of problem appears as well in solid mechanics if a body consists of composite materials. Let us mention also vibrating folded membranes, composite plates, folded plates, junctions in elastic multi-structures etc.

In this work we obtain estimates of the weak solutions to the elliptic transmission problem near singularities on the boundary (conical boundary point or edge). Analogous results were established in [14] for the Dirichlet, mixed and Robin problems in domains with singularities on the boundary without interfaces.

Many mathematicians have considered transmission problems for *linear* elliptic equations. For the first time, M. Schechter [65], O.A. Oleinik [61], V.A. Il'in [31], O.A. Ladyzhenskaya and N.N. Ural'tseva [43], Z.G. Sheftel [66], M.V. Borsuk [6] studied general interface problems for linear second-order elliptic operators in *smooth* domains. V.A. Il'in and I.A. Shishmarev [31, 32] investigated the classical solvability of the Dirichlet and Neumann problems for general (non-divergence) linear elliptic second-order equations with discontinuous coefficients. They used the potential method. Van Tun [69] extended their results and established Schauder estimates. M. Schechter [65], O.A. Oleinik [61] as well as O.A. Ladyzhenskaya and N.N. Ural'tseva [43] (see also [44]) investigated the weak solvability of the Dirichlet transmission problem for divergence linear elliptic second-order equations and established the smoothness of weak solutions near the interface. Z.G. Sheftel [66] extended the S. Agmon, A. Douglis and L. Nirenberg results [1] to general boundary value problems for elliptic equations of any order. M.V. Borsuk [6] derived exact Schauder estimates of solutions to the transmission problem for general linear elliptic second-order equations with the third boundary condition. The exactness of these estimates consists in their explicit dependence on the smoothness of coefficients in estimating constants. A. Lorenzi [50] proved an existence and uniqueness theorem for solutions in $W^{2,p}(\mathbb{R}^n)$ of second-order linear elliptic equations, whose coefficients are constant-valued in the half-spaces $\mathbb{R}^n_+$ and $\mathbb{R}^n_-$. M. Costabel and E. Stephan [21] applied a direct boundary integral equation method for transmission problems.

M. Borsuk, *Transmission Problems for Elliptic Second-Order Equations in Non-Smooth Domains*, Frontiers in Mathematics, DOI 10.1007/978-3-0346-0477-2_1, © Springer Basel AG 2010

Starting in the 1970s many mathematicians have studied linear transmission problems in *non-smooth* domains in some particular linear cases (see for example [15, 16, 17, 19, 51, 53, 54, 55, 56] and references cited in [58, 59]). T. Petersdorf [62] proved an existence and uniqueness theorem for solutions in $W^1(G)$, where G is the number of adjacent Lipschitz domains, of the transmission problem for the homogeneous Helmholz equation: he reduced the transmission problem to a system of boundary integral equations. R. Kellogg [35, 36, 37, 38], Ben M'Barek and M. Merigot [3], K. Lemrabet [45], M. Dobrowolski [24] investigated the behavior of solutions of the transmission Dirichlet and Neumann problems for the Laplace operator in a neighborhood of an angular boundary point of a plane-bounded domain. K. Lemrabet [46] studied also the case of bounded three-dimensional domains. H. Blumenfeld [5] studied the regularity of solutions of the mixed transmission problem in a polygon and derived the approximate solution by the finite element method. M. Dauge and S. Nicaise [22, 23, 58] extended some of the V. Kondratiev–V. Maz'ya–P. Grisvard classical results, concerning the singular behavior of the weak solution of a boundary value problem in a non-smooth domain, to the transmission problems for the Laplace operator with oblique derivative boundary and interface condition on a two-dimensional polygonal topological network: they obtained index formulae, a calculation of the dimension of the kernel, an expansion of weak solutions into regular and singular parts and established higher regularity results. L. Escauriaza, E. Fabes, G. Verchota [27] investigated the regularity for weak solutions to transmission problems in plane domains with internal Lipschitz boundaries. The $H^{1+\frac{1}{4}}$-regularity result for the Laplace interface problem in two dimensions was obtained by M. Petzoldt [63]. General *linear* two- or three-dimensional interface problems in polygonal and polyhedral domains were considered by S. Nicaise, A.-M. Sändig in [59]: they studied the solvability, the regularity and the solution asymptotics in weighted Sobolev spaces. They investigated also [60] the regularity and boundary integral equations for two- or three-dimensional transmission problems with the Laplace and elasticity operators (see also the overview by D. Knees–A.-M. Sändig [40]). Regularity results in terms of weighted Sobolev-Kondratiev spaces were obtained as well by W. Chikouche, D. Mercier and S. Nicaise [19] for two- and three-dimensional transmission problems for the Laplace operator using a penalization technique. D. Kapanadze and B.-W. Schulze studied boundary-contact problems with conical [33] singularities and edge [34] singularities at the interfaces for general linear, any order, elliptic equations (as well as systems). They constructed parametrices and showed the regularity with asymptotics of solutions in weighted Sobolev-Kondratiev spaces.

Concerning the transmission problem for quasi-linear elliptic equations we know a few works. At first the quasi-linear transmission problem was investigated only in *smooth* domains. M.V. Borsuk [6] and then V.Ja. Rivkind–N.N. Ural'tseva [64] proved the classical solvability of the transmission problem for quasi-linear elliptic second-order equations with co-normal derivative boundary and interface conditions. Later N. Kutev–P.L. Lions [41] investigated the transmission problem for general uniform elliptic second-order equations with Dirichlet boundary condition. They proved by the regularization method the existence and the uniqueness of $\mathbf{C}^1(G)\cap C^0(\overline{G})-$ solutions with appropriate conditions on the nonlinearities and the regularity on each side of the interface.

We know only two papers, namely that by D. Knees [39] and by C. Ebmeyer, J. Frehse and M. Kassmann [26], which are concerned with the study of the regularity of weak solutions of special quasi-linear transmission problems on polyhedral domains. In particular, the regularity in the Nikolskii spaces up to the transmission surface and the boundary for the Dirichlet transmission problems are proved in [26].

A principal new feature of our work is the consideration of our estimates of weak solutions to the transmission problem for *linear* elliptic equations with *minimal smooth coefficients in n-dimensional conic domains.* Our examples demonstrate this fact. Moreover we consider the estimates of weak solutions for *general divergence quasi-linear* elliptic second-order equations in *n-dimensional conic* domains or in domains with *edges.* We shall study the following elliptic transmission problems:

- for the linear transmission problem

$$\begin{cases} \mathcal{L}[u] \equiv \frac{\partial}{\partial x_i}\left(a^{ij}(x)u_{x_j}\right) + a^i(x)u_{x_i} + a(x)u = f(x), & x \in G \setminus \Sigma_0; \\ [u]_{\overline{\Sigma_0}} = 0, \quad \mathcal{S}[u] \equiv \left[\frac{\partial u}{\partial \nu}\right]_{\Sigma_0} + \frac{1}{|x|}\sigma\left(\frac{x}{|x|}\right)u(x) = h(x), & x \in \Sigma_0; \\ \mathcal{B}[u] \equiv \frac{\partial u}{\partial \nu} + \frac{1}{|x|}\gamma\left(\frac{x}{|x|}\right)u = g(x), & x \in \partial G \setminus \{\Sigma_0 \cup \mathcal{O}\}; \end{cases} \tag{L}$$

- for the Laplace operator with N different media and mixed boundary condition

$$\begin{cases} a\triangle u - pu(x) = f(x), & x \in G\setminus \bigcup\limits_{k=1}^{N-1} \Sigma_k; \\ [u]_{\overline{\Sigma_k}} = 0, & x \in \Sigma_k, \\ \mathcal{S}_k[u] \equiv \left[a\frac{\partial u}{\partial n_k}\right]_{\Sigma_k} + \frac{1}{|x|}\beta_k(\omega)u(x) = h_k(x), & k = 1, \dots, N-1; \\ \mathcal{B}[u] \equiv \alpha(x)\cdot a\frac{\partial u}{\partial \vec{n}} + \frac{1}{|x|}\gamma(\omega)u(x) = g(x), & x \in \partial G \setminus \left\{\bigcup\limits_{k=1}^{N-1} \Sigma_k \cup \mathcal{O}\right\}, \end{cases} \tag{LN}$$

where $a > 0,\ p \geq 0$; $\alpha(x) = \begin{cases} 0, & \text{if } x \in \mathcal{D} \\ 1, & \text{if } x \notin \mathcal{D} \end{cases}$ and $\mathcal{D} \subseteq \partial G$ is the part of the boundary ∂G where we consider the Dirichlet boundary condition;

- for weak nonlinear equations

$$\begin{cases} -\dfrac{d}{dx_i}\left(|u|^q a^{ij}(x)u_{x_j}\right) + b(x,u,\nabla u) = 0, \quad q \geq 0, & x \in G \setminus \Sigma_0; \\ [u]_{\overline{\Sigma_0}} = 0, & \\ \mathcal{S}[u] \equiv \left[\frac{\partial u}{\partial \nu}\right]_{\Sigma_0} + \frac{1}{|x|}\sigma\left(\frac{x}{|x|}\right) u\cdot|u|^q = h(x,u), & x \in \Sigma_0; \\ \mathcal{B}[u] \equiv \frac{\partial u}{\partial \nu} + \frac{1}{|x|}\gamma\left(\frac{x}{|x|}\right) u\cdot|u|^q = g(x,u), & x \in \partial G \setminus \{\Sigma_0 \cup \mathcal{O}\}; \end{cases} \qquad (WQL)$$

here $q \geq 0$; if $q = 0$ and $b(x,u,\nabla u) = a^i(x)u_{x_i} + a(x)u - f(x)$ then we have linear transmission problem (L);

- for general elliptic divergence quasi-linear equations

$$\begin{cases} -\dfrac{d}{dx_i}a_i(x,u,\nabla u) + b(x,u,\nabla u) = 0, & x \in G \setminus \Sigma_0; \\ [u]_{\overline{\Sigma_0}} = 0, & \\ \mathcal{S}[u] \equiv \left[\frac{\partial u}{\partial \nu}\right]_{\Sigma_0} + \frac{1}{|x|^{m-1}}\sigma\left(\frac{x}{|x|}\right) u\cdot|u|^{q+m-2} = h(x,u), & x \in \Sigma_0; \\ \mathcal{B}[u] \equiv \frac{\partial u}{\partial \nu} + \frac{1}{|x|^{m-1}}\gamma\left(\frac{x}{|x|}\right) u\cdot|u|^{q+m-2} = g(x,u), & x \in \partial G \setminus \{\Sigma_0 \cup \mathcal{O}\}; \end{cases} \qquad (QL)$$

- and for elliptic divergence quasi-linear equations with triple degeneration

$$\begin{cases} -\frac{d}{dx_i}a^i(x,u,\nabla u) + b(x,u,\nabla u) + a_0 c(x,u) = f(x), & x \in G \setminus \Sigma_0; \\ [u]_{\Sigma_0} = 0, & x \in \Sigma_0; \\ \mathcal{S}[u] \equiv \left[a a^i(x,u,\nabla u)n_i\right]_{\Sigma_0} & \\ \qquad + \beta_0 r^{\tau-m+1}u|u|^{q+m-2} = h(x,u), & x \in \Sigma_0; \\ \mathcal{B}[u] \equiv a a^i(x,u,\nabla u)n_i & \\ \qquad + \gamma(\omega) r^{\tau-m+1}u|u|^{q+m-2} = g(x,u), & x \in \partial G \setminus \{\Sigma_0 \cup \Gamma_0\}, \end{cases} \qquad (TDQL)$$

where $q \geq 0,\ m > 1,\ a > 0,\ a_0 \geq 0,\ \beta_0 > 0,\ \tau \geq m-2$ are given numbers and $\overline{\Sigma_0}$ contains an edge Γ_0.

Summation over repeated indices from 1 to n is understood; here everywhere $\frac{\partial}{\partial \nu}$ denotes the conormal derivative operator.

Chapter 1

Preliminaries

1.1 List of symbols

Let $G \subset \mathbb{R}^n$, $n \geq 2$ be a bounded domain with boundary ∂G that is a smooth surface everywhere except at the origin $\mathcal{O} \in \partial G$ and near the point $\mathcal{O}$ it is a conical surface with vertex at $\mathcal{O}$. We assume that $G = G_+ \cup G_- \cup \Sigma_0$ is divided into two subdomains G_+ and G_- by a $\Sigma_0 = G \cap \{x_n = 0\}$, where $\mathcal{O} \in \overline{\Sigma_0}$. Let us fix some notation used throughout the work:

- $[l]$ – the integral part of l (if l is not integer);
- $\mathbb{R}$ – the set of real numbers;
- $\mathbb{R}_+$ – the set of positive numbers;
- $\mathbb{R}^n$ – the n-dimensional Euclidean space, $n \geq 2$;
- $\mathbb{N}$ – the set of natural numbers;
- $\mathbb{N}_0 = \mathbb{N} \cup \{0\}$ – the set of non-negative integers;
- S^{n-1} – a unit sphere in $\mathbb{R}^n$ centered at $\mathcal{O}$;
- $(r, \omega), \omega = (\omega_1, \omega_2, \ldots, \omega_{n-1})$ – spherical coordinates of $x \in \mathbb{R}^n$ with pole $\mathcal{O}$: $x_1 = r\cos\omega_1$, $x_2 = r\cos\omega_2\sin\omega_1, \ldots, x_{n-1} = r\cos\omega_{n-1}\sin\omega_{n-2}\ldots\sin\omega_1$, $x_n = r\sin\omega_{n-1}\sin\omega_{n-2}\ldots\sin\omega_1$;
- $\mathcal{C}$ – a rotational cone $\{x_1 > r\cos\frac{\omega_0}{2}\}$ with the vertex at $\mathcal{O}$;
- $\partial\mathcal{C}$ – the lateral surface of $\mathcal{C} : \{x_1 = r\cos\frac{\omega_0}{2}\}$;
- Ω – a domain on the unit sphere S^{n-1} with smooth boundary $\partial\Omega$ obtained by the intersection of the cone $\mathcal{C}$ with the sphere S^{n-1};
- $\partial\Omega = \partial\mathcal{C} \cap S^{n-1}$;
- $G_a^b = \{(r, \omega) \mid 0 \leq a < r < b; \omega \in \Omega\} \cap G$ – a layer in $\mathbb{R}^n$;

M. Borsuk, *Transmission Problems for Elliptic Second-Order Equations in Non-Smooth Domains*, Frontiers in Mathematics, DOI 10.1007/978-3-0346-0477-2_2, © Springer Basel AG 2010

- $\Gamma_a^b = \{(r,\omega) \mid 0 \le a < r < b; \omega \in \partial\Omega\} \cap \partial G$ – the lateral surface of layer G_a^b;
- $G_d = G \setminus G_0^d;\quad \Gamma_d = \partial G \setminus \Gamma_0^d,\ d > 0;$
- $\Sigma_a^b = G_a^b \cap \{x_n = 0\} \subset \Sigma_0;\quad \Sigma_d = \Sigma_0 \setminus \Sigma_0^d,\ d > 0;$
- $\Omega_\rho = G_0^d \cap \{|x| = \rho\};\quad 0 < \rho < d;$
- $\Omega_+ = \Omega \cap \{x_n > 0\},\ \Omega_- = \Omega \cap \{x_n < 0\} \Longrightarrow \Omega = \Omega_+ \cup \Omega_- \cup \sigma_0;$
- $\sigma_0 = \Sigma_0 \cap \Omega;\ \partial_\pm\Omega = \overline{\Omega_\pm} \cap \partial\mathcal{C};\ \partial\Omega_\pm = \partial_\pm\Omega \cup \sigma_0;$
- $G^{(k)} := G_{2^{-(k+1)}d}^{2^{-k}d},\quad k = 0, 1, 2, \ldots;$
- $u(x) = \begin{cases} u_+(x), & x \in G_+, \\ u_-(x), & x \in G_-; \end{cases} \quad f(x) = \begin{cases} f_+(x), & x \in G_+, \\ f_-(x), & x \in G_- \end{cases}$ etc.;
- $[u]_{\Sigma_0}$ denotes the saltus of the function $u(x)$ on crossing Σ_0, i.e., $[u]_{\Sigma_0} = u_+(x)\Big|_{\Sigma_0} - u_-(x)\Big|_{\Sigma_0}$, where $u_\pm(x)\Big|_{\Sigma_0} = \lim\limits_{G_\pm \ni y \to x \in \Sigma_0} u_\pm(y);$
- $n_i = \cos(\overrightarrow{n}, x_i)$, $i = 1, \ldots, n$, where $\overrightarrow{n}$ denotes the unit outward vector with respect to G_+ (or G) normal to Σ_0 (respectively $\partial G \setminus \mathcal{O}$).

1.2 Operators and formulae related to spherical coordinates

Let us recall first some well-known formulae related to the spherical coordinates $(r, \omega_1, \ldots, \omega_{n-1})$ centered at the conical point $\mathcal{O}$:

$$dx = r^{n-1} dr d\Omega, \quad d\Omega_\rho = \rho^{n-1} d\Omega, \tag{1.2.1}$$

$d\Omega = J(\omega)d\omega$ denotes the $(n-1)$-dimensional area element of the unit sphere, where

$$J(\omega) = \sin^{n-2}\omega_1 \sin^{n-3}\omega_2 \ldots \sin\omega_{n-2}, \tag{1.2.2}$$

$$d\omega = d\omega_1 \ldots d\omega_{n-1}, \tag{1.2.3}$$

$ds = r^{n-2} dr d\sigma$ denotes the $(n-1)$-dimensional area element on $\partial\mathcal{C}$ and $d\sigma$ denotes the $(n-2)$-dimensional area element on $\partial\Omega$;

$$|\nabla u|^2 = \left(\frac{\partial u}{\partial r}\right)^2 + \frac{1}{r^2} |\nabla_\omega u|^2, \tag{1.2.4}$$

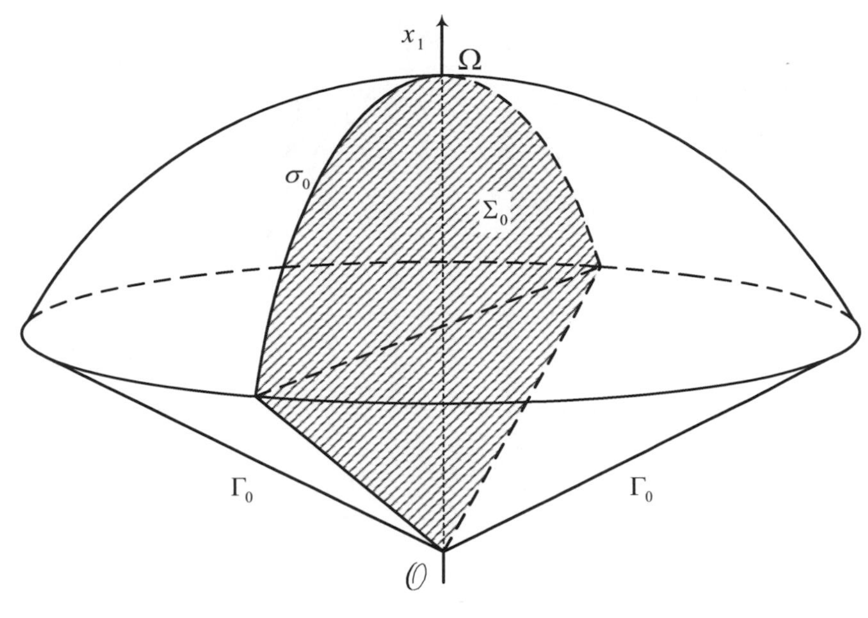

Figure 1

where $|\nabla_\omega u|$ denotes the projection of the vector ∇u onto the tangent plane to the unit sphere at the point ω:

$$\nabla_\omega u = \left\{ \frac{1}{\sqrt{q_1}} \frac{\partial u}{\partial \omega_1}, \ldots, \frac{1}{\sqrt{q_{n-1}}} \frac{\partial u}{\partial \omega_{n-1}} \right\}, \tag{1.2.5}$$

$$|\nabla_\omega u|^2 = \sum_{i=1}^{n-1} \frac{1}{q_i} \left(\frac{\partial u}{\partial \omega_i} \right)^2, \quad \text{where } q_1 = 1,\ q_i = (\sin\omega_1 \cdots \sin\omega_{i-1})^2,\ i \geq 2,$$

$$\Delta u = \frac{\partial^2 u}{\partial r^2} + \frac{n-1}{r} \frac{\partial u}{\partial r} + \frac{1}{r^2} \Delta_\omega u, \text{ denotes the Laplace operator,}$$

$$\Delta_\omega u = \frac{1}{J(\omega)} \sum_{i=1}^{n-1} \frac{\partial}{\partial \omega_i} \left(\frac{J(\omega)}{q_i} \cdot \frac{\partial u}{\partial \omega_i} \right) = \sum_{i=1}^{n-1} \frac{1}{q_j \sin^{n-i-1} \omega_i} \frac{\partial}{\partial \omega_i} \left(\sin^{n-i-1} \omega_i \frac{\partial u}{\partial \omega_i} \right) \tag{1.2.6}$$

denotes the Beltrami-Laplace operator,

$$\mathrm{div}_\omega u = \frac{1}{J(\omega)} \sum_{i=1}^{n-1} \frac{\partial}{\partial \omega_i} \left(\frac{J(\omega)}{\sqrt{q_i}} u_i \right), \quad u = (u_1, \ldots, u_{n-1}). \tag{1.2.7}$$

Lemma 1.1. *Assume for some $d > 0$ that G_0^d is the rotational cone with the vertex at $\mathcal{O}$ and the aperture ω_0, and let*

$$\Gamma_0^d = \left\{ (r,\omega) \Big| x_1^2 = \cot^2 \frac{\omega_0}{2} \sum_{i=2}^{n} x_i^2;\ |\omega_1| = \frac{\omega_0}{2},\ \omega_0 \in (0, 2\pi) \right\}. \tag{1.2.8}$$

Then

$$x_i \cos(\vec{n}, x_i)|_{\Gamma_0^d} = 0, \text{ and } \cos(\vec{n}, x_1)|_{\Gamma_0^d} = -\sin \frac{\omega_0}{2}. \tag{1.2.9}$$

Proof. By virtue of (1.2.8) we can rewrite the equation of Γ_0^d in the form $F(x) \equiv x_1^2 - \cot^2 \frac{\omega_0}{2} \sum\limits_{i=2}^{n} x_i^2 = 0$. We use the formula $\cos(\vec{n}, x_i) = \frac{\frac{\partial F}{\partial x_i}}{|\nabla F|}$, $\forall i = 1, \ldots, n$. Because of $\frac{\partial F}{\partial x_1} = 2x_1$, $\frac{\partial F}{\partial x_i} = -2\cot^2 \frac{\omega_0}{2} x_i$, $\forall i = 2, \ldots, n$, we obtain

$$x_i \cos(\vec{n}, x_i)\Big|_{\Gamma_0^d} = \frac{1}{|\nabla F|} x_i \frac{\partial F}{\partial x_i}\Big|_{\Gamma_0^d} = \frac{2}{|\nabla F|} \left(x_1^2 - \cot^2 \frac{\omega_0}{2} \sum_{i=2}^{n} x_i^2 \right) \Bigg|_{\Gamma_0^d} = 0.$$

Furthermore from

$$|\nabla F|^2 = \left(\frac{\partial F}{\partial x_1} \right)^2 + \sum_{i=2}^{n} \left(\frac{\partial F}{\partial x_i} \right)^2 = 4\left(x_1^2 + \cot^4 \frac{\omega_0}{2} \sum_{i=2}^{n} x_i^2 \right)$$

$$\Rightarrow |\nabla F|^2 \Big|_{\Gamma_0^d} = 4x_1^2 \left(1 + \frac{\cos^2 \frac{\omega_0}{2}}{\sin^2 \frac{\omega_0}{2}} \right) = \frac{4x_1^2}{\sin^2 \frac{\omega_0}{2}},$$

we get

$$\cos(\vec{n}, x_1)\Big|_{\Gamma_0^d} = -2x_1 \frac{\sin \frac{\omega_0}{2}}{2x_1} = -\sin \frac{\omega_0}{2}, \qquad \text{since} \quad \angle(\vec{n}, x_1) > \frac{\pi}{2}. \qquad \square$$

1.3 The quasi-distance function r_ε and its properties

Let us assume that the cone K is contained in a rotational cone $\mathcal{C}$ with the opening angle ω_0. Furthermore, let us suppose that the axis of $\mathcal{C}$ coincides with $\{(x_1, 0, \ldots, 0) : x_1 > 0\}$. In this case we define **the quasi-distance** $r_\varepsilon(x)$ as follows. We fix the point $Q = (-1, 0, \ldots, 0) \in S^{n-1} \setminus \bar{\Omega}$ and consider the unit radius-vector $\vec{l} = \mathcal{O}Q = \{-1, 0, \ldots, 0\}$. We denote by $\vec{r}$ the radius-vector of the point $x \in \bar{G}$ and introduce the vector $\vec{r_\varepsilon} = \vec{r} - \varepsilon\vec{l}$ for each $\varepsilon > 0$. Since $\varepsilon\vec{l} \notin G_0^d$ for all $\varepsilon \in]0, d[$, it follows that $r_\varepsilon(x) = |\vec{r} - \varepsilon\vec{l}| \neq 0$ for all $x \in \bar{G}$. It is easy to verify that $r_\varepsilon(x)$ has the following properties (see in detail §1.4 [14]):

1. *there exists $h > 0$ such that: $r_\varepsilon(x) \geq hr$ and $r_\varepsilon(x) \geq h\varepsilon$, $\forall x \in \overline{G}$, where* $h = \begin{cases} 1, & \textit{if } x_1 \geq 0, \\ \sin \frac{\omega_0}{2}, & \textit{if } x_1 < 0; \end{cases}$

2. *if* $x \in G_d$*, then* $r_\varepsilon(x) \geq \frac{d}{2}$ *for all* $\varepsilon \in]0, \frac{d}{2}[$;
3. $\lim_{\varepsilon \to 0+} r_\varepsilon(x) = r$*, for all* $x \in \bar{G}$;
4. $|\nabla r_\varepsilon|^2 = 1$*, and* $\triangle r_\varepsilon = \frac{n-1}{r_\varepsilon}$.

1.4 Function spaces

We use the standard function spaces:

- $C^k(\overline{G_\pm})$ with the norm $|u_\pm|_{k,G_\pm}$;
- the Lebesgue space $L_p(G_\pm), p \geq 1$ with the norm $\|u_\pm\|_{p,G_\pm}$;
- the Sobolev space $W^{k,p}(G_\pm)$ with the norm $\|u_\pm\|_{p,k;G_\pm}$

and introduce their direct sums

- $\mathbf{C}^k(\overline{G}) = C^k(\overline{G_+}) \dotplus C^k(\overline{G_-})$ with the norm $|u|_{k,G} = |u_+|_{k,G_+} + |u_-|_{k,G_-}$;
- $\mathbf{L}_p(G) = L_p(G_+) \dotplus L_p(G_-)$ with the norm

$$\|u\|_{\mathbf{L}_p(G)} = \left(\int\limits_{G_+} |u_+|^p dx\right)^{\frac{1}{p}} + \left(\int\limits_{G_-} |u_-|^p dx\right)^{\frac{1}{p}};$$

- $\mathbf{W}^{k,p}(G) = W^{k,p}(G_+) \dotplus W^{k,p}(G_-)$ with the norm

$$\|u\|_{p,k;G} = \left(\int\limits_{G_+} \sum_{|\beta|=0}^{k} |D^\beta u_+|^p dx\right)^{\frac{1}{p}} + \left(\int\limits_{G_-} \sum_{|\beta|=0}^{k} |D^\beta u_-|^p dx\right)^{\frac{1}{p}}.$$

For any integer $k \geq 0$ and real α we define the weighted Sobolev space $\mathbf{V}^k_{p,\alpha}(G)$ as the space of distributions $u \in \mathcal{D}'(G)$ with the finite norm

$$\|u\|_{\mathbf{V}^k_{p,\alpha}(G)} = \left(\int\limits_{G_+} \sum_{|\beta|=0}^{k} r^{\alpha+p(|\beta|-k)}|D^\beta u_+|^p \, dx\right)^{\frac{1}{p}} + \left(\int\limits_{G_-} \sum_{|\beta|=0}^{k} r^{\alpha+p(|\beta|-k)}|D^\beta u_-|^p \, dx\right)^{\frac{1}{p}}$$

and $\mathbf{V}^{k-\frac{1}{p}}_{p,\alpha}(\partial G)$ as the space of functions φ, given on ∂G, with the norm $\|\varphi\|_{\mathbf{V}^{k-\frac{1}{p}}_{p,\alpha}(\partial G)} = \inf \|\Phi\|_{\mathbf{V}^k_{p,\alpha}(G)}$, where the infimum is taken over all functions Φ such that $\Phi\big|_{\partial G} = \varphi$ in the sense of traces. We write

$$\mathbf{W}^k(G) \equiv \mathbf{W}^{k,2}(G), \quad \overset{\circ}{\mathbf{W}}{}^k_\alpha(G) \equiv \mathbf{V}^k_{2,\alpha}(G), \quad \overset{\circ}{\mathbf{W}}{}^{k-\frac{1}{2}}_\alpha(\partial G) \equiv \mathbf{V}^{k-\frac{1}{2}}_{2,\alpha}(\partial G).$$

1.5 Some inequalities

In this section we recall some elementary inequalities (see e.g. [2, 30]) which will be frequently used throughout this book.

Lemma 1.2 (Cauchy's Inequality). *For any $a, b \geq 0$ and $\varepsilon > 0$, we have*

$$ab \leq \frac{\varepsilon}{2}a^2 + \frac{1}{2\varepsilon}b^2. \tag{1.5.1}$$

Lemma 1.3 (Young's Inequality). *For any $a, b \geq 0$, $\varepsilon > 0$ and $p, q > 1$ with $\frac{1}{p} + \frac{1}{q} = 1$, we have*

$$ab \leq \frac{1}{p}(\varepsilon a)^p + \frac{1}{q}\left(\frac{b}{\varepsilon}\right)^q. \tag{1.5.2}$$

Lemma 1.4 (Hölder's Inequality). *For any non-negative real numbers a_i, b_i, $i = 1, \dots, n$, and $p, q \in \mathbb{R}$ with $\frac{1}{p} + \frac{1}{q} = 1$, we have*

$$\sum_{i=1}^{n} a_i b_i \leq \left(\sum_{i=1}^{n} a_i^p\right)^{1/p} \left(\sum_{i=1}^{n} b_i^q\right)^{1/q}. \tag{1.5.3}$$

Lemma 1.5. (Theorem 41 [30]). *For any non-negative real numbers a, b and $m \geq 1$ we have*

$$ma^{m-1}(a-b) \geq a^m - b^m \geq mb^{m-1}(a-b). \tag{1.5.4}$$

Lemma 1.6 (Jensen's Inequality (Theorem 65 [30])**).** *Let a_i, $i = 1, \dots, n$, be any non-negative real numbers and $p > 0$. Then*

$$\lambda \sum_{i=1}^{n} a_i^p \leq \left(\sum_{i=1}^{n} a_i\right)^p \leq \Lambda \sum_{i=1}^{n} a_i^p, \tag{1.5.5}$$

where $\lambda = \min(1, n^{p-1})$ and $\Lambda = \max(1, n^{p-1})$.

Lemma 1.7. *For any a, $b \in \mathbb{R}$ and $m > 1$ we have*

$$|b|^m \geq |a|^m + m|a|^{m-2}a(b-a). \tag{1.5.6}$$

Proof. By the Young inequality (1.5.2) with $\varepsilon = 1$, $p = m$, $q = \frac{m}{m-1}$, we obtain

$$m|a|^{m-2}ab \leq m|b| \cdot |a|^{m-1} \leq |b|^m + (m-1)|a|^m \Longrightarrow (1.5.6). \qquad \square$$

Theorem 1.8 (Hölder's Inequality, see Theorem 189 [30]**).**
Let $p, q > 1$ with $\frac{1}{p} + \frac{1}{q} = 1$ and $u \in L^p(G)$, $v \in L^q(G)$. Then

$$\int_G |uv| dx \leq \|u\|_{L^p(G)} \|v\|_{L^q(G)}. \tag{1.5.7}$$

If $p = 1$, then (1.5.7) is valid with $q = \infty$.

Corollary 1.9. *Let $1 \le p \le q$ and $u \in L^q(G)$. Then*

$$\|u\|_{L^p(G)} \le (\operatorname{meas} G)^{1/p-1/q}\|u\|_{L^q(G)}. \tag{1.5.8}$$

Corollary 1.10 (Interpolation inequality).
Let $1 < p \le q \le r$ and $1/q = \lambda/p + (1-\lambda)/r$. Then the inequality

$$\|u\|_{L^q(G)} \le \|u\|^{\lambda}_{L^p(G)}\|u\|^{1-\lambda}_{L^r(G)} \tag{1.5.9}$$

holds for all $u \in L^r(G)$.

Theorem 1.11 (Clarkson's Inequality, see §3.2, Chapter I [68]). *Let $u, v \in L^p(G)$. Then*

$$\left\|\frac{u+v}{2}\right\|^p_{L^p(G)} + \left\|\frac{u-v}{2}\right\|^p_{L^p(G)} \le \frac{1}{2}\Big(\|u\|^p_{L^p(G)} + \|v\|^p_{L^p(G)}\Big),\ 2 \le p < \infty;$$

$$\left\|\frac{u+v}{2}\right\|^{\frac{p}{p-1}}_{L^p(G)} + \left\|\frac{u-v}{2}\right\|^{\frac{p}{p-1}}_{L^p(G)} \le \Big(\frac{1}{2}\|u\|^p_{L^p(G)} + \frac{1}{2}\|v\|^p_{L^p(G)}\Big)^{\frac{1}{p-1}},\ 1 \le p \le 2.$$

Theorem 1.12 (Fatou's Theorem, see Theorem 19 §6, Chapter III [25]). *Let $\{f_k\} \in L^1(G)$, $k \subset \mathbb{N}$, be a sequence of non-negative functions which is convergent almost everywhere in G to the function f. Then*

$$\int\limits_G f dx \le \sup \int\limits_G f_k dx. \tag{1.5.10}$$

We need also the following well-known inequalities:

Theorem 1.13. (See e.g. (6.23), (6.24) Chapter I [42] or Lemma 6.36 [49]). *Let ∂G be piecewise smooth and $u \in W^{1,1}(G)$. Then there is a constant $c > 0$ that depends only on G such that*

$$\int\limits_\Gamma |u| ds \le c\int\limits_G (|u| + |\nabla u|)\, dx \quad \textit{for each}\ \Gamma \subseteq \partial\Omega; \tag{1.5.11}$$

$$\int\limits_{\partial G} v^2 ds \le \int\limits_G (\delta|\nabla v|^2 + c_\delta v^2)dx,\ \forall v(x) \in W^{1,2}(G), \forall\delta > 0. \tag{1.5.12}$$

1.6 Sobolev embedding theorems

We now recall the well-known *Sobolev inequalities* and *Kondrashov compactness results* which are frequently referred as the *embedding theorems* (see [68], §§1.4.5–1.4.6 [52], §7.7[29]).

Theorem 1.14 (Sobolev inequalities (see e.g. [70, Theorem 2.4.1], [29, Theorem 7.10])**).** *Let G be a bounded open domain in $\mathbb{R}^n$ and $p > 1$. Then*

$$W_0^{1,p}(G) \hookrightarrow \begin{cases} L_{\frac{np}{n-p}}(G) & \text{for } p < n, \\ C^0(\overline{G}) & \text{for } p > n. \end{cases} \tag{1.6.1}$$

Furthermore, there exists a constant $c = c(n,p)$ such that for all $u \in W_0^{1,p}(G)$ we have

$$\|u\|_{L_{np/(n-p)}(G)} \leq c\|\nabla u\|_{L_p(G)} \tag{1.6.2}$$

for $p < n$ and

$$\sup_G |u| \leq c(\operatorname{meas} G)^{1/n-1/p}\|\nabla u\|_{L_p(G)} \tag{1.6.3}$$

for $p > n$.

The following Embedding Theorems 1.15–1.20 were proved first by Sobolev [68] and can be found with complete proofs in [52, Section 1.4]. Let G be a $C^{0,1}$ bounded domain in $\mathbb{R}^n$.

Theorem 1.15. *Let $k \in \mathbb{N}$ and $p \in \mathbb{R}$ with $p \geq 1$, $kp < n$. Then the embedding*

$$W^{k,p}(G) \hookrightarrow L_q(G) \tag{1.6.4}$$

is continuous for $1 \leq q \leq np/(n-kp)$ and compact for $1 \leq q < np/(n-kp)$. If $kp = n$, then the embedding (1.6.4) is continuous and compact for any $q \geq 1$.

Theorem 1.16. (For the proof see, for example, (2.19) §2, chapter II [43]). *Let $u \in W^1(G)$. Then*

$$\|u\|^2_{\mathbf{L}_{\frac{2p}{p-2}}(G)} \leq \delta\|\nabla u\|^2_{\mathbf{L}_2(G)} + c(\delta, p, n, meas\, G)\|u\|^2_{\mathbf{L}_2(G)}, \quad p > n, \ \forall \delta > 0. \tag{1.6.5}$$

Theorem 1.17. *Let $k \in \mathbb{N}_0$, $m \in \mathbb{N}$ and let $p, q \in \mathbb{R}$ with $p, q \geq 1$. If $kp < n$, then the embedding*

$$W^{m+k,p}(G) \hookrightarrow W^{m,q}(G) \tag{1.6.6}$$

is continuous for any $q \in \mathbb{R}$ satisfying $1 \leq q \leq np/(n-kp)$. If $k = np$, then the embedding (1.6.6) is continuous for any $q \geq 1$.

Theorem 1.18. *Let $k, m \in \mathbb{N}_0$ and $p > 1$. Then the embedding $W^{k,p}(G) \hookrightarrow C^{m+\beta}(G)$ is continuous if*

$$(k-m-1)p < n < (k-m)p \quad \text{and} \quad 0 < \beta \leq k - m - n/p, \tag{1.6.7}$$

and compact if the inequality for β in (1.6.7) is sharp. If $(k-m-1)p = n$, then the embedding is continuous for any $\beta \in (0,1)$.

Theorem 1.19. *Let $u \in W^{k,p}(G)$ with $k \in \mathbb{N}$, $p \in \mathbb{R}$, $kp > n$ and $p > 1$. Then $u \in C^m(G)$ for $0 \le m < k - n/p$ and there exists a constant c, independent of u, such that*

$$\sup_{x \in G} |D^\alpha u(x)| \le c\|u\|_{W^{k,p}(G)}$$

for all $|\alpha| < k - n/p$.

Theorem 1.20. *Let G be a lipschitzian domain and $T_s \subset \overline{G}$ be a piecewise C^k-smooth s-dimensional manifold. Let $k \ge 1$, $p > 1$, $kp \le n$, $n - kp < s \le n$, $1 \le q \le q* = sp/(n - kp)$. Then the embedding $W^{k,p}(G) \hookrightarrow L_q(T_s)$ is continuous and the inequality*

$$\|u\|_{L_q(T_s)} \le c\|u\|_{W^{k,p}(G)} \tag{1.6.8}$$

holds. If $q < q*$, then the above embedding is compact.

1.7 The Cauchy problem for a differential inequality

Theorem 1.21. (See Theorem 1.57 [14]). *Let $V(\varrho)$ be a monotonically increasing, non-negative differentiable function defined on $[0, 2d]$ and satisfying the problem*

$$\begin{cases} V'(\rho) - \mathcal{P}(\varrho)V(\varrho) + \mathcal{N}(\rho)V(2\rho) + \mathcal{Q}(\rho) \ge 0, & 0 < \rho < d, \\ V(d) \le V_0, \end{cases} \tag{CP}$$

where $\mathcal{P}(\varrho), \mathcal{N}(\varrho), \mathcal{Q}(\varrho)$ are non-negative continuous functions defined on $[0, 2d]$ and V_0 is a constant. Then we have

$$V(\varrho) \le \exp\Big(\int_\varrho^d \mathcal{B}(\tau)d\tau\Big)\Big\{V_0 \exp\Big(-\int_\varrho^d \mathcal{P}(\tau)d\tau\Big) + \int_\varrho^d \mathcal{Q}(\tau)\exp\Big(-\int_\varrho^\tau \mathcal{P}(\sigma)d\sigma\Big)d\tau\Big\} \tag{1.7.1}$$

with

$$\mathcal{B}(\varrho) = \mathcal{N}(\varrho)\exp\Big(\int_\varrho^{2\varrho} \mathcal{P}(\sigma)d\sigma\Big). \tag{1.7.2}$$

1.8 Additional auxiliary results

1.8.1 Stampacchia's Lemma

Lemma 1.22. (See Lemma 3.11 of [57]). *Let $\varphi : [k_0, \infty) \to \mathbb{R}$ be a non-negative and non-increasing function which satisfies*

$$\varphi(h) \le \frac{C}{(h-k)^\alpha}[\varphi(k)]^\beta \quad \text{for} \quad h > k > k_0, \tag{1.8.1}$$

where C, α, β are positive constants with $\beta > 1$. Then $\varphi(k_0 + d) = 0$, where $d^\alpha = C\,|\varphi(k_0)|^{\beta-1}\,2^{\alpha\beta/(\beta-1)}$.

Proof. For each $s = 1, 2, \ldots$ we let $k_s = k_0 + d - \frac{d}{2^s}$ and consider the sequence $\{\varphi(k_s)\}$. From (1.8.1) we obtain

$$\varphi(k_{s+1}) \leq \frac{C2^{(s+1)\alpha}}{d^\alpha}[\varphi(k_s)]^\beta, \quad s = 1, 2, \ldots. \tag{1.8.2}$$

We now prove by induction that

$$\varphi(k_s) \leq \frac{\varphi(k_0)}{2^{-s\mu}}, \text{ where } \mu = \frac{\alpha}{1-\beta} < 0. \tag{1.8.3}$$

For $s = 0$ the claim is evident. Let us suppose that (1.8.3) is valid up to s. By (1.8.2) and the definition of d^α if follows that

$$\varphi(k_{s+1}) \leq C\frac{2^{(s+1)\alpha}}{d^\alpha}\frac{[\varphi(k_0]^\beta}{2^{-s\beta\mu}} \leq \frac{\varphi(k_0)}{2^{-(s+1)\mu}}. \tag{1.8.4}$$

Since the right-hand side of (1.8.4) tends to zero as $s \to \infty$, we obtain $0 \leq \varphi(k_0 + d) \leq \varphi(k_s) \to 0$. □

1.8.2 Other assertions

Lemma 1.23. (See Lemma 2.1 [20]). *Let us consider the function*

$$\eta(x) = \begin{cases} e^{\varkappa x} - 1, & x \geq 0, \\ -e^{-\varkappa x} + 1, & x \leq 0, \end{cases}$$

where $\varkappa > 0$. Let a, b be positive constants, $m > 1$. If $\varkappa > (2b/a) + m$, then we have

$$a\eta'(x) - b|\eta(x)| \geq \frac{a}{2}e^{\varkappa x}, \quad \forall x \geq 0; \tag{1.8.5}$$

$$\eta(x) \geq \left[\eta\left(\frac{x}{m}\right)\right]^m, \quad \forall x \geq 0. \tag{1.8.6}$$

Moreover, there exist some $d \geq 0$ and $M > 0$ such that

$$\eta(x) \leq M\left[\eta\left(\frac{x}{m}\right)\right]^m, \quad \eta'(x) \leq M\left[\eta\left(\frac{x}{m}\right)\right]^m, \quad \forall x \geq d, \tag{1.8.7}$$

$$|\eta(x)| \geq x, \quad \forall x \in \mathbb{R}. \tag{1.8.8}$$

Proof. We refer to Lemma 1.60 [14] for the proof. □

Lemma 1.24. (See Lemma 4.1 in Chapter 2 [18]). *Let $\psi(s)$ be a bounded non-negative function defined on the interval $[0, \varrho]$. Suppose that for any $0 \le \sigma < s \le \varrho$ the function $\psi(s)$ satisfies*

$$\psi(\sigma) \le \delta\psi(s) + \frac{A}{(s-\sigma)^\alpha} + B,$$

where $\delta \in (0,1)$, A, B and α are non-negative constants. Then

$$\psi(r) \le C\left[\frac{A}{(R-r)^\alpha} + B\right], \quad 0 \le r < R \le \varrho, \tag{1.8.9}$$

where C depends only on α, δ.

Theorem 1.25 (S. Bernstein [4]). *If y is a bounded analytic solution of the equation*

$$y'' = f(x, y, y'), \ a_0 < x < b_0, \quad \text{where } |f(x,y,y')| < Ay'^2 + B,$$

then y' is also bounded:

$$|y'| < \sqrt{\frac{B}{2A}} e^{4AM}, \quad M = \sup_{x \in (a_0, b_0)} |y(x)|.$$

Chapter 2

Eigenvalue problem and integro-differential inequalities

2.1 Eigenvalue problem for the m-Laplacian in a bounded domain on the unit sphere

Let $\Omega \subset S^{n-1}$ with a smooth boundary $\partial\Omega$ be the intersection of the cone $\mathcal{C}$ with the unit sphere S^{n-1}. Let $\overrightarrow{\nu}$ be the exterior normal to $\partial\mathcal{C}$ at points of $\partial\Omega$ and $\overrightarrow{\tau}$ be the exterior with respect to Ω_+ normal to σ_0 (lying in the tangent to Ω plane). Let $\gamma(\omega)$ be *a positive* bounded piecewise smooth function on $\partial\Omega$ and $\sigma(\omega)$ be *a positive* continuous function on σ_0. We consider the eigenvalue problem for the m-Laplace-Beltrami operator on the unit sphere:

$$\begin{cases} a\left(\operatorname{div}_\omega(|\nabla_\omega\psi|^{m-2}\nabla_\omega\psi) + \vartheta|\psi|^{m-2}\psi\right) = 0, & \omega \in \Omega, \\ [\psi]_{\sigma_0} = 0, \quad \left[a|\nabla_\omega\psi|^{m-2}\frac{\partial\psi}{\partial\overrightarrow{\tau}}\right]_{\sigma_0} + \sigma(\omega)|\psi|^{m-2}\psi\Big|_{\sigma_0} = 0, & \\ a|\nabla_\omega\psi|^{m-2}\frac{\partial\psi}{\partial\overrightarrow{\nu}} + \gamma(\omega)|\psi|^{m-2}\psi\Big|_{\partial\Omega} = 0, & \end{cases} \qquad (NEVP)$$

which consists in the determination of all values ϑ (eigenvalues) for which $(NEVP)$ has non-zero weak solutions (eigenfunctions); here $a = \begin{cases} a_+, & x \in \Omega_+, \\ a_-, & x \in \Omega_-, \end{cases}$ $a_\pm$ are positive constants.

M. Borsuk, *Transmission Problems for Elliptic Second-Order Equations in Non-Smooth Domains*, Frontiers in Mathematics, DOI 10.1007/978-3-0346-0477-2_3, © Springer Basel AG 2010

Definition 2.1. A function ψ is called a **weak** eigenfunction of problem $(NEVP)$ provided that $\psi \in \mathbf{C}^0(\overline{\Omega}) \cap \mathbf{W}^{1,m}(\Omega)$ and satisfies the integral identity

$$\int_\Omega a\left\{|\nabla_\omega \psi|^{m-2}\frac{1}{q_i}\frac{\partial\psi}{\partial\omega_i}\frac{\partial\eta}{\partial\omega_i} - \vartheta|\psi|^{m-2}\psi\eta\right\}d\Omega + \int_{\sigma_0}\sigma(\omega)|\psi|^{m-2}\psi\eta d\sigma + \int_{\partial\Omega}\gamma(\omega)|\psi|^{m-2}\psi\eta d\sigma = 0$$

for all $\eta(x) \in \mathbf{C}^0(\overline{\Omega}) \cap \mathbf{W}^{1,m}(\Omega)$.

Remark 1. We observe that $\vartheta = 0$ is **not** an eigenvalue of $(NEVP)$. In fact, setting $\eta = \psi$ and $\vartheta = 0$ we have

$$\int_\Omega a|\nabla_\omega\psi|^m d\Omega + \int_{\sigma_0}\sigma(\omega)|\psi|^m d\sigma + \int_{\partial\Omega}\gamma(\omega)|\psi|^m d\sigma = 0 \quad \Longrightarrow \quad \psi \equiv 0,$$

since $a > 0$, $\sigma(\omega) > 0$, $\gamma(\omega) > 0$.

We characterize the first eigenvalue $\vartheta(m)$ of the eigenvalue problem for the m-Laplacian by

$$\vartheta(m) = \inf_{\substack{\psi\in W^{1,m}(\Omega)\\ \psi\neq 0}} \frac{\int_\Omega a|\nabla_\omega\psi|^m d\Omega + \int_{\sigma_0}\sigma(\omega)|\psi|^m d\sigma + \int_{\partial\Omega}\gamma(\omega)|\psi|^m d\sigma}{\int_\Omega a|\psi|^m d\Omega}. \tag{2.1.1}$$

Let us introduce the following functionals on $\mathbf{C}^0(\overline{\Omega}) \cap \mathbf{W}^{1,m}(\Omega)$:

$$F[\psi] = \int_\Omega a|\nabla_\omega\psi|^m d\Omega + \int_{\sigma_0}\sigma(\omega)|\psi|^m d\sigma + \int_{\partial\Omega}\gamma(\omega)|\psi|^m d\sigma,$$

$$G[\psi] = \int_\Omega a|\psi|^m d\Omega,$$

$$H[\psi] = \int_\Omega a\Big\langle|\nabla_\omega\psi|^m - \vartheta|\psi|^m\Big\rangle d\Omega + \int_{\sigma_0}\sigma(\omega)|\psi|^m d\sigma + \int_{\partial\Omega}\gamma(\omega)|\psi|^m d\sigma$$

and the corresponding forms

$$\mathcal{F}(\psi,\eta) = \int_\Omega a|\nabla_\omega\psi|^{m-2}\frac{1}{q_i}\frac{\partial\psi}{\partial\omega_i}\frac{\partial\eta}{\partial\omega_i}d\Omega + \int_{\sigma_0}\sigma(\omega)|\psi|^{m-2}\psi\eta d\sigma + \int_{\partial\Omega}\gamma(\omega)|\psi|^{m-2}\psi\eta d\sigma,$$

$$\mathcal{G}(\psi,\eta) = \int_\Omega a|\psi|^{m-2}\psi\eta d\Omega.$$

Further we define the set $K = \{\psi \in \mathbf{C}^0(\overline{\Omega}) \cap \mathbf{W}^{1,m}(\Omega) \big| \ G[\psi] = 1\}$. Since $K \subset \mathbf{C}^0(\overline{\Omega}) \cap \mathbf{W}^{1,m}(\Omega)$, $F[\psi]$ is bounded from below for $\psi \in K$. The greatest lower bound of $F[\psi]$ for this family we denote by ϑ: $\inf_{\psi \in K} F[\psi] = \vartheta$. Now our aim is to establish

Theorem 2.2. *Let $\Omega \subset S^{n-1}$ be a domain with smooth boundary $\partial\Omega$ and $\gamma(\omega)$ be a positive bounded piecewise smooth function on $\partial\Omega$, $\sigma(\omega)$ be a positive continuous function on σ_0. There exist $\vartheta > 0$ and a function $\psi \in K$ such that $\mathcal{F}(\psi, \eta) - \vartheta\mathcal{G}(\psi, \eta) = 0$ for arbitrary $\eta \in \mathbf{C}^0(\overline{\Omega}) \cap \mathbf{W}^{1,m}(\Omega)$. In particular $F[\psi] = \vartheta$.*

Proof. Because $F[v]$ is bounded from below for $v \in K$, there is $\vartheta = \inf_{v \in K} F[v]$. Let us consider a sequence $\{v_k\} \subset K$ such that $\lim_{k\to\infty} F[v_k] = \vartheta$ (such a sequence exists by the definition of the infimum). By $K \subset \mathbf{W}^{1,m}(\Omega)$, sequence $\{v_k\}$ is bounded in $\mathbf{W}^{1,m}(\Omega)$ and therefore compact in $\mathbf{L}^m(\Omega)$. Choosing a subsequence, if needed, we can assume that $\{v_k\}$ is converging in $\mathbf{L}^m(\Omega)$. As a result we obtain the following property of the functional G: given any $\varepsilon > 0$ we can find $N(\varepsilon)$ such that

$$G[v_k - v_l] \leq a^* \cdot \|v_k - v_l\|^m_{\mathbf{L}^m(\Omega)} < \varepsilon, \quad \forall \varepsilon > 0 \tag{2.1.2}$$

for all $k, l > N(\varepsilon)$. Now we use the inequality (1.5.6):

$$\left|\frac{v_k + v_l}{2}\right|^m \geq |v_k|^m + \frac{m}{2}|v_k|^{m-2}v_k(v_l - v_k), \ m > 1.$$

By integrating this inequality over Ω:

$$\int_\Omega a\left|\frac{v_k + v_l}{2}\right|^m d\Omega \geq \int_\Omega a|v_k|^m d\Omega + \frac{m}{2}\int_\Omega a|v_k|^{m-2}v_k(v_l - v_k)d\Omega$$

and applying the Young inequality with $p = \frac{m}{m-1}$, $q = m$, we get

$$\left|\frac{m}{2}v_k|v_k|^{m-2}(v_l - v_k)\right| \leq \frac{m}{2}|v_k|^{m-1}|v_l - v_k| \leq \frac{m-1}{2}\delta^{\frac{m}{m-1}}|v_k|^m + \frac{1}{2\delta^m}|v_l - v_k|^m,$$

for all $\delta > 0$. This fact yields that

$$\int_\Omega a\left|\frac{v_k + v_l}{2}\right|^m d\Omega \geq \left(1 - \frac{m-1}{2}\delta^{\frac{m}{m-1}}\right)\int_\Omega a|v_k|^m d\Omega - \frac{1}{2\delta^m}\int_\Omega a|v_l - v_k|^m d\Omega,$$

for all $\delta > 0$. This implies that $G\left[\frac{v_k+v_l}{2}\right] \geq \left(1 - \frac{m-1}{2}\delta^{\frac{m}{m-1}}\right)G[v_k] - \frac{1}{2\delta^m}G[v_l - v_k]$, for all $\delta > 0$. By using $G[v_k] = G[v_l] = 1$ and $G[v_l - v_k] < \varepsilon_1$, we obtain $G\left[\frac{v_k+v_l}{2}\right] > 1 - \frac{m-1}{2}\delta^{\frac{m}{m-1}} - \frac{\varepsilon_1}{2\delta^m}$, $\forall \delta, \varepsilon_1 > 0$ for large k, l. Now, by choosing $\delta^m = \varepsilon_1^{\frac{m-1}{m}}$ and setting $\varepsilon = \frac{m\vartheta}{2}\varepsilon_1^{\frac{1}{m}}$, we get

$$G\left[\frac{v_k + v_l}{2}\right] > 1 - \frac{\varepsilon}{\vartheta} \tag{2.1.3}$$

for large k, l. The functionals $F[v]$ and $G[v]$ are homogeneous functionals and therefore their ratio $\frac{F[v]}{G[v]}$ does not change under the passage from v to cv ($c = const \neq 0$). Hence

$$\inf_{v \in \mathbf{W}^{1,m}(\Omega)} \frac{F[v]}{G[v]} = \inf_{v \in K} F[v] = \vartheta.$$

Therefore $F[v] \geq \vartheta G[v]$ for all $v \in \mathbf{W}^{1,m}(\Omega)$. Since $\frac{v_k + v_l}{2} \in \mathbf{W}^{1,m}(\Omega)$ together with $v_k, v_l \in K$, then

$$F\left[\frac{v_k + v_l}{2}\right] \geq \vartheta G\left[\frac{v_k + v_l}{2}\right] > \vartheta\left(1 - \frac{\varepsilon}{\vartheta}\right) = \vartheta - \varepsilon, \quad k, l > N(\varepsilon).$$

Let us take k and l large enough so that $F[v_k] < \vartheta + \varepsilon$ and $F[v_l] < \vartheta + \varepsilon$. We apply the Clarkson inequalities (see Theorem 1.11):

1) $m \geq 2$:

$$\begin{aligned} F\left[\frac{v_l - v_k}{2}\right] &\leq \frac{1}{2}F[v_l] + \frac{1}{2}F[v_k] - F\left[\frac{v_l + v_k}{2}\right] \\ &< \vartheta + \varepsilon - (\vartheta - \varepsilon) = 2\varepsilon \end{aligned}$$

2) $1 < m \leq 2$:

$$\begin{aligned} F^{\frac{1}{m-1}}\left[\frac{v_l - v_k}{2}\right] &\leq \left(\frac{1}{2}F[v_k] + \frac{1}{2}F[v_l]\right)^{\frac{1}{m-1}} - F^{\frac{1}{m-1}}\left[\frac{v_k + v_l}{2}\right] \\ &< (\vartheta + \varepsilon)^{\frac{1}{m-1}} - (\vartheta - \varepsilon)^{\frac{1}{m-1}} < \frac{2\varepsilon}{m-1}(\theta + \varepsilon)^{\frac{2-m}{m-1}}, \end{aligned}$$

by inequality (1.5.4). Consequently,

$$F[v_k - v_l] \to 0, \quad \text{as } k, l \to \infty. \tag{2.1.4}$$

From (2.1.2), (2.1.4) it follows that $\|v_k - v_l\|_{\mathbf{W}^{1,m}(\Omega)} \to 0$, as $k, l \to \infty$. Thus, $\{v_k\}$ is a Cauchy sequence in $\mathbf{W}^{1,m}(\Omega)$ and hence by the completeness of $\mathbf{W}^{1,m}(\Omega)$ there exists a function $\psi \in \mathbf{W}^{1,m}(\Omega)$ such that $\|v_k - \psi\|_{\mathbf{W}^{1,m}(\Omega)} \to 0$, as $k \to \infty$. Moreover,

$$\begin{aligned} F[v_k] - F[\psi] &= \int_\Omega a\left(|\nabla_\omega v_k|^m - |\nabla_\omega \psi|^m\right) d\Omega + \int_{\sigma_0} \sigma(\omega)\left(|v_k|^m - |\psi|^m\right) d\sigma \\ &+ \int_{\partial\Omega} \gamma(\omega)\left(|v_k|^m - |\psi|^m\right) d\sigma. \end{aligned}$$

Now, by (1.5.4) and the Hölder inequality, we have

$$\left|\int_\Omega a\left(|\nabla_\omega v_k|^m - |\nabla_\omega \psi|^m\right) d\Omega\right| \leq m \int_\Omega a|\nabla v_k|^{m-1}|\nabla_\omega(v_k - \psi)|d\Omega$$

$$\leq m\left(\int_\Omega a|\nabla_\omega(v_k-\psi)|^m d\Omega\right)^{1/m} \cdot \left(\int_\Omega a|\nabla_\omega v_k|^m d\Omega\right)^{(m-1)/m} \to 0, \text{ as } k \to \infty,$$

since $v_k \in \mathbf{W}^{1,m}$. Furthermore, by using (1.5.6), the Hölder inequality for integrals and (1.5.11), we obtain

$$\int_{\sigma_0} \sigma(\omega)\left(|v_k|^m - |\psi|^m\right) d\sigma \leq m \int_{\sigma_0} \sigma(\omega)|v_k|^{m-1} \cdot |v_k - \psi| d\sigma$$
$$\leq m \max_{\sigma_0} |\sigma(\omega)| \cdot \|v_k - \psi\|_{L_m(\sigma_0)} \cdot \|v_k\|_{L_m(\sigma_0)}$$
$$\leq c\|v_k - \psi\|_{\mathbf{W}^{1,m}(\Omega)} \cdot \|v_k\|_{\mathbf{W}^{1,m}(\Omega)} \to 0, \text{ as } k \to \infty,$$

and also

$$\int_{\partial\Omega} \gamma(\omega)\left(|v_k|^m - |\psi|^m\right) d\sigma \leq m \int_{\partial\Omega} \gamma(\omega)|v_k|^{m-1} \cdot |v_k - \psi| d\sigma$$
$$\leq m \max_{\partial\Omega} |\gamma(\omega)| \cdot \|v_k - \psi\|_{L_m(\partial\Omega)} \cdot \|v_k\|_{L_m(\partial\Omega)}$$
$$\leq c\|v_k - \psi\|_{\mathbf{W}^{1,m}(\Omega)} \cdot \|v_k\|_{\mathbf{W}^{1,m}(\Omega)} \to 0, \text{ as } k \to \infty.$$

Therefore we get $F[\psi] = \lim_{k\to\infty} F[v_k] = \vartheta$. Analogously it can be seen that $G[\psi] = 1$.

Suppose now that η is some function from $\mathbf{C}^0(\overline{\Omega}) \cap \mathbf{W}^{1,m}(\Omega)$. Consider the ratio $\frac{F[\psi+\varepsilon\eta]}{G[\psi+\varepsilon\eta]}$. It is a continuously differentiable function of ε on some interval around the point $\varepsilon = 0$. This ratio has a minimum at $\varepsilon = 0$ equal to ϑ and therefore, by the Fermat Theorem, we have

$$\left(\frac{F[\psi+\varepsilon\eta]}{G[\psi+\varepsilon\eta]}\right)'_{\varepsilon=0} = m\frac{\mathcal{F}(\psi,\eta)G[\psi] - F[\psi]\mathcal{G}(\psi,\eta)}{G^2[\psi]} = 0,$$

which because of $F[\psi] = \vartheta$, $G[\psi] = 1$, gives $\mathcal{F}(\psi,\eta) - \vartheta\mathcal{G}(\psi,\eta) = 0$, for all $\eta \in \mathbf{C}^0(\overline{\Omega}) \cap \mathbf{W}^{1,m}(\Omega)$. □

2.2 The Friedrichs-Wirtinger type inequality

Now from the variational principle we obtain

The Friedrichs-Wirtinger type inequality. *Let $\vartheta(m)$ be the least positive eigenvalue of the problem $(NEVP)$ (it exists according to Theorem 2.2). Let $\Omega \subset S^{n-1}$ and*

$\psi \in \mathbf{W}^{1,m}(\Omega)$, $\gamma(\omega)$ be a positive bounded piecewise smooth function on $\partial\Omega$, $\sigma(\omega)$ be a positive continuous function on σ_0. Then

$$\int\limits_{\Omega} a|\psi|^m \, d\Omega \le \frac{1}{\vartheta(m)} \left\{ \int\limits_{\Omega} a|\nabla_\omega \psi|^m \, d\Omega + \int\limits_{\sigma_0} \sigma(\omega)|\psi|^m d\sigma + \int\limits_{\partial\Omega} \gamma(\omega)|\psi|^m d\sigma \right\} \quad (W)_m$$

with the sharp constant $\frac{1}{\vartheta(m)}$.

Proof. By approximation arguments, it is clearly that we only need to consider the above described functionals $F[\psi], G[\psi], H[\psi]$ on $\mathbf{C}^1(\overline{\Omega}) \cap \mathbf{W}^{2,m}(\Omega)$. We will find the minimum of the functional $F[\psi]$ on the set K. For this we investigate the minimization of the functional $H[\psi]$ on all functions $\psi(\omega) \in \mathbf{C}^1(\overline{\Omega}) \cap \mathbf{W}^{2,m}(\Omega)$ which satisfy the boundary conditions from $(NEVP)$. We use formally the Lagrange multipliers and get the Euler equation from the condition $\delta H[\psi] = 0$. Calculating the first variation δH, we have

$$\begin{aligned}
\delta H[\psi] &= \delta \Bigg(\int\limits_{\Omega} a \left\{ \left\langle \sum_{i=1}^{N-1} \frac{1}{q_i} \left(\frac{\partial \psi}{\partial \omega_i} \right)^2 \right\rangle^{\frac{m}{2}} - \vartheta(\psi^2)^{\frac{m}{2}} \right\} d\Omega + \int\limits_{\sigma_0} \sigma(\omega)(\psi^2)^{\frac{m}{2}} d\sigma \\
&\quad + \int\limits_{\partial\Omega} \gamma(\omega)(\psi^2)^{\frac{m}{2}} d\sigma \Bigg) \\
&= -m \int\limits_{\Omega} a \sum_{i=1}^{N-1} \frac{\partial}{\partial \omega_i} \left(\frac{J(\omega)}{q_i} \cdot |\nabla_\omega \psi|^{m-2} \frac{\partial \psi}{\partial \omega_i} \right) \cdot \delta\psi d\omega \\
&\quad - m\vartheta \int\limits_{\Omega} a|\psi|^{m-2}\psi \cdot \delta\psi d\Omega + m \int\limits_{\partial\Omega} a|\psi|^{m-2} \frac{\partial \psi}{\partial \overrightarrow{\nu}} \cdot \delta\psi d\sigma \\
&\quad + m \int\limits_{\sigma_0} \left[a|\psi|^{m-2} \frac{\partial \psi}{\partial \overrightarrow{\tau}} \right] \cdot \delta\psi d\sigma + m \int\limits_{\sigma_0} \sigma(\omega)|\psi|^{m-2}\psi \cdot \delta\psi d\sigma \\
&\quad + m \int\limits_{\partial\Omega} \gamma(\omega)|\psi|^{m-2}\psi \cdot \delta\psi d\sigma \\
&= -m \int\limits_{\Omega} a \left\{ \mathrm{div}_\omega(|\nabla_\omega \psi|^{m-2} \nabla_\omega \psi) + \vartheta|\psi|^{m-2}\psi \right\} \cdot \delta\psi d\Omega \\
&\quad + m \int\limits_{\sigma_0} \left\{ \left[a|\psi|^{m-2} \frac{\partial \psi}{\partial \overrightarrow{\tau}} \right] + \sigma(\omega)|\psi|^{m-2}\psi \right\} \cdot \delta\psi d\sigma \\
&\quad + m \int\limits_{\partial\Omega} \left\{ a|\psi|^{m-2} \frac{\partial \psi}{\partial \overrightarrow{\nu}} + \gamma(\omega)|\psi|^{m-2}\psi \right\} \cdot \delta\psi d\sigma.
\end{aligned}$$

Hence, because of $\delta H[\psi] = 0$, $\forall \delta\psi \in \mathbf{C}^0(\overline{\Omega}) \cap \mathbf{W}^{1,m}(\Omega)$, we get the eigenvalue problem $(NEVP)$ for the m-Laplacian . Conversely, let $\vartheta, \psi(\omega)$ be a weak solution of the eigenvalue problem for the m-Laplacian. From the definition of the weak eigenfunction, by setting $\eta = \psi(\omega)$, we obtain

$$0 = F[\psi] - \vartheta G[\psi] \underset{(\text{by } K)}{=} F[\psi] - \vartheta \Rightarrow \vartheta = F[u],$$

consequently, the required minimum is the least eigenvalue of the eigenvalue problem for the m-Laplacian. The existence of a function $\psi \in K$ such that $F[\psi] \le F[v]$ for all $v \in K$ has been proved above. □

Remark 2. For $m = 2$ the eigenvalue problem $(NEVP)$ takes the form

$$\begin{cases} a_\pm\left(\triangle_\omega \psi_\pm + \vartheta\psi_\pm\right) = 0, \quad a_\pm \text{ are positive constants}; \quad \omega \in \Omega_\pm, \\ [\psi]_{\sigma_0} = 0, \quad \left[a\frac{\partial\psi}{\partial\vec{\tau}}\right]_{\sigma_0} + \sigma(\omega)\psi\Big|_{\sigma_0} = 0, \\ a_\pm\frac{\partial\psi_\pm}{\partial\vec{\nu}} + \gamma_\pm(\omega)\psi_\pm\Big|_{\partial_\pm\Omega} = 0, \end{cases} \tag{EVP}$$

and we set

$$\lambda = \frac{2 - n + \sqrt{(n-2)^2 + 4\vartheta}}{2}. \tag{2.2.1}$$

Then $(W)_m$ takes the form

$$\int\limits_\Omega a\psi^2(\omega)d\Omega$$
$$\le \frac{1}{\lambda(\lambda+n-2)}\left\{\int\limits_\Omega a|\nabla_\omega\psi|^2 d\Omega + \int\limits_{\sigma_0}\sigma(\omega)\psi^2(\omega)d\sigma + \int\limits_{\partial\Omega}\gamma(\omega)\psi^2(\omega)d\sigma\right\}, \tag{W)$_2$ ($}$$

for all $\psi(\omega) \in \mathbf{W}^1(\Omega)$.

Remark 3. The constants in $(W)_m$ and $(W)_2$ are the best possible ones.

Corollary 2.3. *Let $\vartheta(m)$ be the least positive eigenvalue of the problem $(NEVP)$. Let $v(x) \in \mathbf{W}^{1,m}(G_0^d)$, $\gamma(\omega)$ be a positive bounded piecewise smooth function on $\partial\Omega$, $\sigma(\omega)$ be a positive continuous function on σ_0. Then for any $\varrho \in (0,d)$ and for all α,*

$$\int\limits_{G_0^\varrho} ar^\alpha|v|^m dx$$
$$\le \frac{1}{\vartheta(m)}\Big\{\int\limits_{G_0^\varrho} ar^{\alpha+m}|\nabla v|^m dx + \int\limits_{\Sigma_0^\varrho} r^{\alpha+m}\frac{\sigma(\omega)}{r^{m-1}}|v|^m ds + \int\limits_{\Gamma_0^\varrho} r^{\alpha+m}\frac{\gamma(\omega)}{r^{m-1}}|v|^m ds\Big\}, \tag{2.2.2}$$

provided that integrals on the right are finite. In particular,

$$\int\limits_{G_0^\varrho} a|v|^m dx \le \frac{\varrho^m}{\vartheta(m)} \left\{ \int\limits_{G_0^\varrho} a|\nabla v|^m dx + \int\limits_{\Sigma_0^\varrho} \frac{\sigma(\omega)}{r^{m-1}} |v|^m ds + \int\limits_{\Gamma_0^\varrho} \frac{\gamma(\omega)}{r^{m-1}} |v|^m ds \right\}. \tag{$(H-W)_m$}$$

Proof. Consider the inequality $(W)_m$ for the function $v(r,\omega)$. By multiplying it by $r^{\alpha+n-1}$ and integrating over $r \in (0,\varrho)$, we obtain the desired inequality (2.2.2). Setting in it $\alpha = 0$ we get $(H-W)_m$. □

Similarly we have

Corollary 2.4. *Let* $u \in \mathbf{C}^0(\overline{G}) \cap \overset{\circ}{\mathbf{W}}{}^1_{\alpha-2}(G)$, *and* λ *be as above in* (2.2.1). *Let* $\sigma(\omega)$, $\omega \in \sigma_0$; $\gamma(\omega)$, $\omega \in \partial\Omega$ *be non-negative bounded piecewise smooth functions. Then*

$$\int\limits_{G_0^d} ar^{\alpha-4}u^2 dx \le \frac{1}{\lambda(\lambda+n-2)} \Big\{ \int\limits_{G_0^d} ar^{\alpha-2}|\nabla u|^2 dx + \int\limits_{\Sigma_0^d} r^{\alpha-3}\sigma(\omega)u^2(x) ds + \int\limits_{\Gamma_0^d} r^{\alpha-3}\gamma(\omega)u^2(x) ds \Big\}, \quad \forall\alpha. \tag{2.2.3}$$

Lemma 2.5. *Let* $v \in \mathbf{C}^0(\overline{G}) \cap \overset{\circ}{\mathbf{W}}{}^1_0(G)$ *and* $\sigma(\omega) \ge 0$, $\gamma(\omega) \ge 0$. *Then for any* $\varepsilon > 0$ *and for all* α,

$$\int\limits_{G_0^d} ar_\varepsilon^{\alpha-2} r^{-2} v^2 dx \le \widetilde{c} \cdot \left\{ \int\limits_{G_0^d} ar_\varepsilon^{\alpha-2}|\nabla v|^2 dx + \int\limits_{\Sigma_0^d} r^{-1} r_\varepsilon^{\alpha-2} \sigma(\omega) v^2(x) ds + \int\limits_{\Gamma_0^d} r^{-1} r_\varepsilon^{\alpha-2} \gamma(\omega) v^2 ds \right\},$$

$$\widetilde{c} = \frac{1}{\lambda(\lambda+n-2)} \left(\frac{2}{h}\right)^{(\alpha-2)sgn(\alpha-2)}, \tag{2.2.4}$$

where h *is defined in Subsection* 1.3.

Proof. Multiplying both sides of the Friedrichs-Wirtinger inequality $(W)_2$ by $(r+\varepsilon)^{\alpha-2} r^{n-3}$ and integrating over $r \in (\frac{\varrho}{2}, \varrho)$, we obtain

$$\int\limits_{G^{\varrho}_{\varrho/2}} a(r+\varepsilon)^{\alpha-2}r^{-2}v^2dx$$
$$\leq \frac{1}{\lambda(\lambda+n-2)}\Big\{\int\limits_{G^{\varrho}_{\varrho/2}} a(r+\varepsilon)^{\alpha-2}|\nabla v|^2dx + \int\limits_{\Gamma^{\varrho}_{\varrho/2}} r^{-1}(r+\varepsilon)^{\alpha-2}\gamma(\omega)v^2ds$$
$$+ \int\limits_{\Sigma^{\varrho}_{\varrho/2}} r^{-1}(r+\varepsilon)^{\alpha-2}\sigma(\omega)v^2ds\Big\},$$

for all $\varepsilon > 0$ or, because of the property 1) of r_ε §1.3, $r_\varepsilon \leq r+\varepsilon \leq \frac{2}{h}r_\varepsilon$ we find

$$\int\limits_{G^{\varrho}_{\varrho/2}} ar_\varepsilon^{\alpha-2}r^{-2}v^2dx$$
$$\leq \widetilde{c}\cdot\left\{\int\limits_{G^{\varrho}_{\varrho/2}} ar_\varepsilon^{\alpha-2}|\nabla v|^2dx + \int\limits_{\Sigma^{\varrho}_{\varrho/2}} r^{-1}r_\varepsilon^{\alpha-2}\sigma(\omega)v^2ds + \int\limits_{\Gamma^{\varrho}_{\varrho/2}} r^{-1}r_\varepsilon^{\alpha-2}\gamma(\omega)v^2ds\right\},$$

for all $\varepsilon > 0$. Letting $\rho = 2^{-k}d$, $(k = 0, 1, 2, \dots)$ and summing the obtained inequalities over all k we get the desired inequality (2.2.4). □

2.3 The Hardy and Hardy-Friedrichs-Wirtinger type inequalities

We recall first the classical Hardy inequality (see Theorem 330 [30]):

Theorem 2.6. *Let $d > 0$ and $v \in C^0[0,d] \cap W^{1,2}(0,d)$ with $v(0) = 0$. Then*

$$\int\limits_0^d r^{n-5+\alpha}v^2(r)dr \leq \frac{4}{(4-n-\alpha)^2}\int\limits_0^d r^{n-3+\alpha}\left(\frac{\partial v}{\partial r}\right)^2 dr \tag{2.3.1}$$

for $\alpha < 4-n$ provided that the integral on the right-hand side is finite.

Corollary 2.7. *Let $0 < \varepsilon < d$ and $v \in C^0[\varepsilon,d] \cap W^{1,2}(\varepsilon,d)$, with $v(\varepsilon) = 0$. Then*

$$\int\limits_\varepsilon^d r^{n-5+\alpha}v^2(r)dr \leq \frac{4}{(4-n-\alpha)^2}\int\limits_\varepsilon^d r^{n-3+\alpha}\left(\frac{\partial v}{\partial r}\right)^2 dr, \quad \alpha < 4-n. \tag{2.3.2}$$

Proof. We apply the inequality (2.3.1) to the function $v(r)$ extended by zero into $[0,\varepsilon)$. □

Now using the Hardy inequality we establish:

Proposition 2.8 (The Hardy-Friedrichs-Wirtinger type inequality).

Let $u \in \mathbf{C}^0(\overline{G}) \cap \overset{\circ}{\mathbf{W}}{}^1_{\alpha-2}(G)$ *and* $\sigma(\omega)$, $\gamma(\omega)$ *be positive bounded piecewise smooth functions. Then*

$$\int\limits_{G_0^d} ar^{\alpha-4}u^2dx \leq H(\lambda, n, \alpha)\Bigg\{\int\limits_{G_0^d} ar^{\alpha-2}|\nabla u|^2dx + \int\limits_{\Sigma_0^d} r^{\alpha-3}\sigma(\omega)u^2(x)ds.$$

$$+ \int\limits_{\Gamma_0^d} r^{\alpha-3}\gamma(\omega)u^2(x)ds\Bigg\},$$

$$H(\lambda, n, \alpha) = \frac{1}{\lambda(\lambda+n-2)+\frac{1}{4}(4-n-\alpha)^2}, \quad \alpha \leq 4-n. \tag{2.3.3}$$

Proof. For $\alpha = 4-n$ the required inequality (2.3.3) coincides with (2.2.3). Now, let $\alpha < 4-n$. We shall show that $u(0) = 0$. In fact, from the representation $u(0) = u(x) - \big(u(x) - u(0)\big)$ by the Cauchy inequality we have $\frac{1}{2}|u(0)|^2 \leq |u(x)|^2 + |u(x) - u(0)|^2$. Putting $v(x) = u(x) - u(0)$ we obtain

$$\frac{1}{2}|u(0)|^2 \int\limits_{G_0^d} r^{\alpha-4}dx \leq \int\limits_{G_0^d} r^{\alpha-4}u^2(x)dx + \int\limits_{G_0^d} r^{\alpha-4}|v|^2dx < \infty. \tag{2.3.4}$$

(The first integral from the right is finite due to $u \in \overset{\circ}{\mathbf{W}}{}^1_{\alpha-2}(G)$, and the second integral is also finite, by virtue of Theorem 2.6.) Since $\int\limits_{G_0^d} r^{\alpha-4}dx = \operatorname{meas}\Omega \int\limits_0^d r^{\alpha+n-5}dr = \infty$, by $\alpha+n-4 < 0$, the assumption $u(0) \neq 0$ contradicts (2.3.4). Thus $u(0) = 0$. Therefore we can use Theorem 2.6:

$$\int\limits_{G_0^d} ar^{\alpha-4}u^2dx \leq \frac{4}{|4-n-\alpha|^2}\int\limits_{G_0^d} ar^{\alpha-2}u_r^2dx. \tag{2.3.5}$$

Multiplying $(W)_2$ by $r^{n-5+\alpha}$ and integrating over $r \in (0, d)$ we obtain

$$\int\limits_{G_0^d} ar^{\alpha-4}u^2dx \leq \frac{1}{\lambda(\lambda+n-2)}\Bigg\{\int\limits_{G_0^d} ar^{\alpha-2}\frac{1}{r^2}|\nabla_\omega u|^2dx + \int\limits_{\Sigma_0^d} r^{\alpha-3}\sigma(\omega)u^2(x)ds$$

$$+ \int\limits_{\Gamma_0^d} r^{\alpha-3}\gamma(\omega)u^2(x)ds\Bigg\}. \tag{2.3.6}$$

Adding the inequalities (2.3.6), (2.3.5) and using the formula $|\nabla u|^2 = \left(\frac{\partial u}{\partial r}\right)^2 + \frac{1}{r^2}|\nabla_\omega u|^2$, we get the desired inequality (2.3.3). $\square$

Lemma 2.9. *Let $u \in C^0(\overline{G_\varepsilon^d}) \cap W^1(G_\varepsilon^d)$, $u(\varepsilon) = 0$ and $\sigma(\omega)$, $\gamma(\omega)$ be positive bounded piecewise smooth functions. Then for any $\varepsilon > 0$, it holds that*

$$\int\limits_{G_\varepsilon^d} a r^{\alpha-4} u^2 dx \leq H(\lambda, n, \alpha) \left\{ \int\limits_{G_\varepsilon^d} a r^{\alpha-2} |\nabla u|^2 dx + \int\limits_{\Sigma_\varepsilon^d} r^{\alpha-3} \sigma(\omega) u^2(x) ds + \int\limits_{\Gamma_\varepsilon^d} r^{\alpha-3} \gamma(\omega) u^2(x) ds \right\}, \quad (2.3.7)$$

where $H(\lambda, n, \alpha)$ is determined by (2.3.3).

Proof. We have inequality $(W)_2$. Multiplying it by $r^{n-5+\alpha}$ and integrating over $r \in (\varepsilon, d)$ we obtain (2.3.7) for $\alpha = 4 - n$. If $\alpha < 4 - n$, we consider the inequality (2.3.2) and integrate it over Ω; then we have

$$\frac{1}{4}(4 - n - \alpha)^2 \int\limits_{G_\varepsilon^d} r^{\alpha-4} u^2 dx \leq \int\limits_{G_\varepsilon^d} r^{\alpha-2} u_r^2 dx.$$

Adding this inequality to the above one for $\alpha = 4 - n$ and using the formula $|\nabla u|^2 = \left(\frac{\partial u}{\partial r}\right)^2 + \frac{1}{r^2} |\nabla_\omega u|^2$, the desired result follows. $\square$

Lemma 2.10. *Let $u \in \mathbf{C}^0(\overline{G}) \cap \mathbf{W}^1(G)$, $u(0) = 0$ and $\sigma(\omega)$, $\gamma(\omega)$ be positive bounded piecewise smooth functions. Then, for any $\varepsilon > 0$, it holds that*

$$\int\limits_{G_0^d} a r_\varepsilon^{\alpha-4} u^2 dx \leq H(\lambda, n, \alpha) \left\{ \int\limits_{G_0^d} a r_\varepsilon^{\alpha-2} |\nabla u|^2 dx + \int\limits_{\Sigma_0^d} r_\varepsilon^{\alpha-3} \sigma(\omega) u^2(x) ds + \int\limits_{\Gamma_0^d} r_\varepsilon^{\alpha-3} \gamma(\omega) u^2(x) ds \right\}, \quad (2.3.8)$$

$$H(\lambda, n, \alpha) = \frac{1}{\lambda(\lambda + n - 2) + \frac{1}{4}(4 - n - \alpha)^2}, \quad \alpha \leq 4 - n.$$

Proof. We perform the change of variables $y_i = x_i - \varepsilon l_i$, $i = 1, \dots, n$ and use the inequality (2.3.7); thus we obtain

$$\int\limits_{G_0^d} a r_\varepsilon^{\alpha-4} u^2(x)dx = \int\limits_{G_\varepsilon^{d_\varepsilon}} a|y|^{\alpha-4}u^2(y+\varepsilon l)dy$$

$$\leq H(\lambda,n,\alpha)\Bigg\{ \int\limits_{G_\varepsilon^{d_\varepsilon}} a|y|^{\alpha-2}|\nabla_y u(y+\varepsilon l)|^2 dy + \int\limits_{\Sigma_\varepsilon^{d_\varepsilon}} |y|^{\alpha-3}\sigma(y+\varepsilon l)u^2(y+\varepsilon l)ds$$

$$+ \int\limits_{\Gamma_\varepsilon^{d_\varepsilon}} |y|^{\alpha-3}\gamma(y+\varepsilon l)u^2(y+\varepsilon l)ds \Bigg\}$$

$$= H(\lambda,n,\alpha)\Bigg\{ \int\limits_{G_0^d} a r_\varepsilon^{\alpha-2}|\nabla u|^2 dx + \int\limits_{\Sigma_0^d} r_\varepsilon^{\alpha-3}\sigma(x)u^2(x)ds + \int\limits_{\Gamma_0^d} r_\varepsilon^{\alpha-3}\gamma(x)u^2(x)ds \Bigg\}.$$

□

2.4 Auxiliary integro-differential inequalities

Lemma 2.11. *Let G_0^d be the conical domain, $v(\varrho,\cdot) \in \mathbf{W}^{1,m}(\Omega)$ for almost all $\varrho \in (0,d)$ and*

$$V(\varrho) = \int\limits_{G_0^\varrho} a|\nabla v|^m dx + \int\limits_{\Sigma_0^\varrho} \frac{\sigma(\omega)}{r^{m-1}}|v|^m ds + \int\limits_{\Gamma_0^\varrho} \frac{\gamma(\omega)}{r^{m-1}}|v|^m ds < \infty. \tag{2.4.1}$$

Let $\vartheta(m)$ be the smallest positive eigenvalue of the problem $(NEVP)$ and $\gamma(\omega)$ be a positive bounded piecewise smooth function on $\partial\Omega$, $\sigma(\omega)$ be a positive continuous function on σ_0. Then for almost all $\varrho \in (0,d)$,

$$\int\limits_{\Omega_\varrho} av\frac{\partial v}{\partial r}|\nabla v|^{m-2} d\Omega_\varrho \leq \Xi(m)\cdot \frac{\varrho}{m\vartheta^{\frac{1}{m}}}V'(\varrho), \tag{2.4.2}$$

where

$$\Xi(m) = \begin{cases} \left(\frac{m}{2}\right)^{\frac{3}{2}}, & m \geq 2, \\ (m-1)^{\frac{m-1}{m}} \cdot 2^{\frac{2-m}{2}}, & 1 < m \leq 2. \end{cases} \tag{2.4.3}$$

Proof. Writing the function $V(\varrho)$ in spherical coordinates

$$V(\varrho) = \int\limits_0^\varrho r^{n-1}\Bigg(\int\limits_\Omega a|\nabla v(r,\omega)|^m d\Omega \Bigg) dr + \int\limits_0^\varrho r^{n-m-1}\Bigg(\int\limits_{\partial\Omega} \gamma(\omega)|v(r,\omega)|^m d\sigma \Bigg) dr$$

$$+ \int\limits_0^\varrho r^{n-m-1}\Bigg(\int\limits_{\sigma_0} \sigma(\omega)|v(r,\omega)|^m d\sigma \Bigg) dr$$

and differentiating it with respect to ϱ we obtain

$$V'(\varrho) = \varrho^{n-1} \int_{\Omega} a|\nabla v(\varrho,\omega)|^m d\Omega + \varrho^{n-m-1}\left(\int_{\partial\Omega} \gamma(\omega)|v(\varrho,\omega)|^m d\sigma + \int_{\sigma_0} \sigma(\omega)|v(\varrho,\omega)|^m d\sigma \right). \quad (2.4.4)$$

$\boxed{m \geq 2.}$

Using the Cauchy inequality and next the Young inequality with $p = \frac{m}{2}$, $p' = \frac{m}{m-2}$ we have

$$\int_{\Omega_\varrho} av\frac{\partial v}{\partial r}|\nabla v|^{m-2} d\Omega_\varrho = \varrho^{n-1} \int_{\Omega} av\frac{\partial v}{\partial r}|\nabla v|^{m-2}\Big|_{r=\varrho} d\Omega$$

$$= \varrho^n \int_{\Omega} a\left(\frac{v}{\varrho}\right)\cdot\left(\frac{\partial v}{\partial r}\right)\cdot|\nabla v|^{m-2}\Big|_{r=\varrho} d\Omega$$

$$\leq \varrho^n \int_{\Omega} a\left\{\frac{\varepsilon}{2}\left|\frac{v}{\varrho}\right|^2 + \frac{1}{2\varepsilon}\left|\frac{\partial v}{\partial r}\right|^2\right\}\cdot|\nabla v|^{m-2}\Big|_{r=\varrho} d\Omega$$

$$\leq \varrho^n \int_{\Omega} a\left\{\frac{\varepsilon\delta}{m}\left|\frac{v}{\varrho}\right|^m + \frac{m-2}{m}\cdot\frac{\varepsilon}{2}\delta^{-\frac{2}{m-2}}|\nabla v|^m + \frac{1}{2\varepsilon}\left|\frac{\partial v}{\partial r}\right|^2\cdot|\nabla v|^{m-2}\right\}\Big|_{r=\varrho} d\Omega,$$

for all ε, $\delta > 0$. Applying now the Friedrichs-Wirtinger type inequality $(W)_m$ we obtain

$$\int_{\Omega_\varrho} av\frac{\partial v}{\partial r}|\nabla v|^{m-2} d\Omega_\varrho \leq \frac{\varepsilon\delta\varrho^n}{m\vartheta(m)}\left\{\int_{\sigma_0} \sigma(\omega)\left|\frac{v(\varrho,\omega)}{\varrho}\right|^m d\sigma + \int_{\partial\Omega} \gamma(\omega)\left|\frac{v(\varrho,\omega)}{\varrho}\right|^m d\sigma\right\}$$
$$+ \varrho^n \int_{\Omega} a\left\{\frac{\varepsilon\delta}{m\vartheta(m)}\left|\frac{\nabla_\omega v}{\varrho}\right|^m + (m-2)\frac{\varepsilon}{2m}\delta^{-\frac{2}{m-2}}|\nabla v|^m\right.$$
$$\left. + \frac{1}{2\varepsilon}\left|\frac{\partial v}{\partial r}\right|^2\cdot|\nabla v|^{m-2}\right\}\Big|_{r=\varrho} d\Omega, \ \forall\varepsilon,\ \delta > 0. \quad (2.4.5)$$

Because $|\nabla v|^2 = v_r^2 + \frac{1}{r^2}|\nabla_\omega v|^2$ and $\left|\frac{\nabla_\omega v}{\varrho}\right| \leq |\nabla v|$, we get:

$$\frac{\varepsilon\delta}{m\vartheta(m)}\left|\frac{\nabla_\omega v}{\varrho}\right|^m + \frac{1}{2\varepsilon}\left|\frac{\partial v}{\partial r}\right|^2\cdot|\nabla v|^{m-2}$$
$$\leq \left\{\frac{\varepsilon\delta}{m\vartheta(m)}\left|\frac{\nabla_\omega v}{\varrho}\right|^2 + \frac{1}{2\varepsilon}\left|\frac{\partial v}{\partial r}\right|^2\right\}\cdot|\nabla v|^{m-2} = \frac{\varepsilon\delta}{m\vartheta(m)}|\nabla v|^m,$$

if we choose $\varepsilon > 0$ from the equality

$$\frac{1}{2\varepsilon} = \frac{\varepsilon\delta}{m\vartheta(m)}. \tag{2.4.6}$$

Therefore from (2.4.5) it follows that

$$\int_{\Omega_\varrho} av\frac{\partial v}{\partial r}|\nabla v|^{m-2}d\Omega_\varrho \le \frac{\varepsilon}{m}\left(\frac{\delta}{\vartheta(m)} + \frac{m-2}{2}\delta^{-\frac{2}{m-2}}\right)\varrho^n\int_\Omega a|\nabla v(\varrho,\omega)|^m d\Omega$$
$$+\frac{\varepsilon\delta}{m\vartheta(m)}\varrho^{n-m}\left\{\int_{\sigma_0}\sigma(\omega)|v(\varrho,\omega)|^m d\sigma + \int_{\partial\Omega}\gamma(\omega)|v(\varrho,\omega)|^m d\sigma\right\},\ \forall\delta > 0. \tag{2.4.7}$$

Let us put $\delta = \vartheta^{\frac{m-2}{m}}(m)$. Then, by (2.4.6), $\varepsilon = \sqrt{\frac{m}{2}}\vartheta^{\frac{1}{m}}(m)$. Thus from (2.4.7) and (2.4.4) we derive the required (2.4.2).

$$\boxed{1 < m \le 2.}$$

Because of $|\nabla v| \ge |v_r|$ and $1 < m < 2$, we have $|\nabla v|^{m-2} \le |v_r|^{m-2}$. Therefore using the Young inequality with $p = m$, $p' = \frac{m}{m-1}$:

$$\int_{\Omega_\varrho} av\frac{\partial v}{\partial r}|\nabla v|^{m-2}d\Omega_\varrho = \varrho^{n-1}\int_\Omega av\frac{\partial v}{\partial r}|\nabla v|^{m-2}\Big|_{r=\varrho} d\Omega \le \varrho^n\int_\Omega a\left|\frac{v}{\varrho}\right|\cdot|v_r|^{m-1}\Big|_{r=\varrho} d\Omega$$
$$\le \varrho^n\int_\Omega a\left\{\frac{\varepsilon}{m}\left|\frac{v}{\varrho}\right|^m + \frac{m-1}{m}\varepsilon^{-\frac{1}{m-1}}\left|\frac{\partial v}{\partial r}\right|^m\right\}\Big|_{r=\varrho} d\Omega,\ \forall\varepsilon > 0.$$

Next, applying the Friedrichs-Wirtinger type inequality $(W)_m$ we obtain

$$\int_{\Omega_\varrho} av\frac{\partial v}{\partial r}|\nabla v|^{m-2}d\Omega_\varrho \le \frac{\varepsilon\varrho^n}{m\vartheta(m)}\left\{\int_{\sigma_0}\sigma(\omega)\left|\frac{v(\varrho,\omega)}{\varrho}\right|^m d\sigma + \int_{\partial\Omega}\gamma(\omega)\left|\frac{v(\varrho,\omega)}{\varrho}\right|^m d\sigma\right\}$$
$$+\frac{1}{m}\varrho^n\int_\Omega a\left\{\frac{\varepsilon}{\vartheta(m)}\left|\frac{\nabla_\omega v}{\varrho}\right|^m + (m-1)\varepsilon^{-\frac{1}{m-1}}\left|\frac{\partial v}{\partial r}\right|^m\right\}\Big|_{r=\varrho} d\Omega,\quad \forall\varepsilon > 0.$$

Now we choose

$$\varepsilon = \langle(m-1)\vartheta(m)\rangle^{\frac{m-1}{m}} \implies (m-1)\varepsilon^{-\frac{1}{m-1}} = \frac{\varepsilon}{\vartheta(m)} \tag{2.4.8}$$

and therefore the above inequality gives

$$\int_{\Omega_\varrho} av\frac{\partial v}{\partial r}|\nabla v|^{m-2}d\Omega_\varrho \le \frac{\varepsilon\varrho^n}{m\vartheta(m)}\Bigg\{\int_{\sigma_0}\sigma(\omega)\left|\frac{v(\varrho,\omega)}{\varrho}\right|^m d\sigma + \int_{\partial\Omega}\gamma(\omega)\left|\frac{v(\varrho,\omega)}{\varrho}\right|^m d\sigma$$
$$+\int_\Omega a\left(\left|\frac{\nabla_\omega v}{\varrho}\right|^m + \left|\frac{\partial v}{\partial r}\right|^m\right)\Bigg|_{r=\varrho} d\Omega\Bigg\}. \tag{2.4.9}$$

But, because of $|\nabla v|^2 = v_r^2 + \frac{1}{r^2}|\nabla_\omega v|^2$ and the Jensen inequality (1.5.5), we get

$$\left|\frac{\nabla_\omega v}{\varrho}\right|^m + \left|\frac{\partial v}{\partial r}\right|^m = \left(\left|\frac{\nabla_\omega v}{\varrho}\right|^2\right)^{\frac{m}{2}} + (v_r^2)^{\frac{m}{2}} \le 2^{\frac{2-m}{2}}|\nabla v|^m, \quad 1 < m \le 2.$$

Hence and from (2.4.9) it follows that

$$\int_{\Omega_\varrho} av\frac{\partial v}{\partial r}|\nabla v|^{m-2}d\Omega_\varrho \le \frac{2^{\frac{2-m}{2}}\varepsilon}{m\vartheta(m)}\varrho^n\Bigg(\int_{\sigma_0}\sigma(\omega)\left|\frac{v(\varrho,\omega)}{\varrho}\right|^m d\sigma + \int_{\partial\Omega}\gamma(\omega)\left|\frac{v(\varrho,\omega)}{\varrho}\right|^m d\sigma$$
$$+\int_\Omega a|\nabla v|^m\Big|_{r=\varrho} d\Omega\Bigg).$$

Substituting here ε from (2.4.8) and recalling the definition (2.4.3)-(2.4.4) we get the desired inequality (2.4.2). □

For the case $m = 2$ we establish the best possible result.

Lemma 2.12. *Let G_0^d be the conical domain and $\nabla u(\varrho,\cdot) \in \mathbf{L}_2(\Omega)$ a.e. $\varrho \in (0,d)$ and λ be defined by (2.2.1). Assume that for a.e. $\varrho \in (0,d)$,*

$$U(\rho) = \int_{G_o^\varrho} ar^{2-n}|\nabla u|^2 dx + \int_{\Sigma_0^\varrho} r^{1-n}\sigma(\omega)u^2(x)ds + \int_{\Gamma_0^\varrho} r^{1-n}\gamma(\omega)u^2(x)ds < \infty. \tag{2.4.10}$$

Then

$$\int_\Omega a\left(\varrho u\frac{\partial u}{\partial r} + \frac{n-2}{2}u^2\right)\Bigg|_{r=\varrho} d\Omega \le \frac{\varrho}{2\lambda}U'(\varrho). \tag{2.4.11}$$

Proof. Writing $U(\varrho)$ in spherical coordinates,

$$U(\varrho) = \int_0^{\varrho} r^{2-n}\left(\int_{\Omega} a|\nabla u|^2 d\Omega\right) r^{n-1} dr$$
$$+ \int_0^{\varrho} r^{1-n}\left(\int_{\sigma_0} \sigma(\omega)|u|^2 d\sigma + \int_{\partial\Omega} \gamma(\omega)|u|^2 d\sigma\right) r^{n-2} dr$$
$$= \int_0^{\varrho} r \int_{\Omega} a\left(u_r^2 + \frac{1}{r^2}|\nabla_\omega u|^2\right) d\Omega dr + \int_0^{\varrho} \frac{1}{r}\left(\int_{\sigma_0} \sigma(\omega)|u|^2 d\sigma + \int_{\partial\Omega} \gamma(\omega)|u|^2 d\sigma\right) dr$$

and differentiating with respect to ϱ, we obtain

$$U'(\varrho) = \int_{\Omega} a\left(\varrho\left(\frac{\partial u}{\partial r}\right)^2 + \frac{1}{\varrho}|\nabla_\omega u|^2\right)\Bigg|_{r=\varrho} d\Omega$$
$$+ \frac{1}{\varrho}\left(\int_{\sigma_0} \sigma(\omega)u^2(\varrho,\omega)d\sigma + \int_{\partial\Omega} \gamma(\omega)u^2(\varrho,\omega)d\sigma\right). \tag{2.4.12}$$

Moreover, by the Cauchy inequality, we have for all $\varepsilon > 0$: $\rho u\frac{\partial u}{\partial r} \leq \frac{\varepsilon}{2}u^2 + \frac{1}{2\varepsilon}\rho^2\left(\frac{\partial u}{\partial r}\right)^2$. Then

$$\int_{\Omega} a\left(\varrho u\frac{\partial u}{\partial r} + \frac{n-2}{2}u^2\right)\Bigg|_{r=\varrho} d\Omega \leq \frac{\varepsilon + n - 2}{2}\int_{\Omega} au^2 d\Omega + \frac{\varrho^2}{2\varepsilon}\int_{\Omega} a\left(\frac{\partial u}{\partial r}\right)^2 d\Omega.$$

Thus choosing $\varepsilon = \lambda$ we obtain, by the Friedrichs-Wirtinger inequality $(W)_2$,

$$\int_{\Omega} a\left(\varrho u\frac{\partial u}{\partial r} + \frac{n-2}{2}u^2\right)\Bigg|_{r=\varrho} d\Omega$$
$$\leq \frac{\varepsilon + n - 2}{2\lambda(\lambda + n - 2)}\int_{\Omega} a|\nabla_\omega u|^2 d\Omega + \frac{\varrho^2}{2\varepsilon}\int_{\Omega} a\left(\frac{\partial u}{\partial r}\right)^2 d\Omega$$
$$+ \frac{\varepsilon + n - 2}{2\lambda(\lambda + n - 2)}\left(\int_{\sigma_0} \sigma(\omega)u^2(\varrho,\omega)d\sigma + \int_{\partial\Omega} \gamma(\omega)u^2(\varrho,\omega)d\sigma\right) = \frac{\varrho}{2\lambda}U'(\varrho). \quad \square$$

Chapter 3

Best possible estimates of solutions to the transmission problem for linear elliptic divergence second-order equations in a conical domain

3.1 Introduction

Let $G \subset \mathbb{R}^n$, $n \geq 2$ be a bounded domain with the boundary ∂G that is a smooth surface everywhere except at the origin $\mathcal{O} \in \partial G$ and near the point $\mathcal{O}$ it is a conical surface with the vertex at $\mathcal{O}$. We assume that $G = G_+ \cup G_- \cup \Sigma_0$ is divided into two subdomains G_+ and G_- by a $\Sigma_0 = G \cap \{x_n = 0\}$, where $\mathcal{O} \in \overline{\Sigma_0}$ (see Figure 1). We consider the elliptic transmission problem

$$\begin{cases} \mathcal{L}[u] \equiv \frac{\partial}{\partial x_i}\left(a^{ij}(x)u_{x_j}\right) + a^i(x)u_{x_i} + a(x)u = f(x), & x \in G \setminus \Sigma_0; \\ [u]_{\overline{\Sigma_0}} = 0, \quad \mathcal{S}[u] \equiv \left[\frac{\partial u}{\partial \nu}\right]_{\Sigma_0} + \frac{1}{|x|}\sigma\left(\frac{x}{|x|}\right)u(x) = h(x), & x \in \Sigma_0; \\ \mathcal{B}[u] \equiv \frac{\partial u}{\partial \nu} + \frac{1}{|x|}\gamma\left(\frac{x}{|x|}\right)u = g(x), & x \in \partial G \setminus \{\Sigma_0 \cup \mathcal{O}\} \end{cases} \tag{L}$$

(summation over repeated indices from 1 to n is understood); here:

- $\frac{\partial}{\partial \nu} = a^{ij}(x)n_i\frac{\partial}{\partial x_j}$,
- $\left[\frac{\partial u}{\partial \nu}\right]_{\Sigma_0}$ denotes the saltus of the co-normal derivative of the function $u(x)$ on

M. Borsuk, *Transmission Problems for Elliptic Second-Order Equations in Non-Smooth Domains*, Frontiers in Mathematics, DOI 10.1007/978-3-0346-0477-2_4, © Springer Basel AG 2010

crossing Σ_0, i.e.,

$$\left[\frac{\partial u}{\partial \nu}\right]_{\Sigma_0} = a_+^{ij}(x)\frac{\partial u_+}{\partial x_j}n_i\Big|_{\Sigma_0} - a_-^{ij}(x)\frac{\partial u_-}{\partial x_j}n_i\Big|_{\Sigma_0}.$$

In this chapter we obtain **the best possible estimates** of the weak solutions of the problem (L) near the conical boundary point. Analogous results were established earlier in [14] for the Dirichlet and Robin problems in the case of a conical domain without interfaces.

Definition 3.1. A function $u(x)$ is called a *weak* solution of the problem (L) provided that $u(x) \in \mathbf{C}^0(\overline{G}) \cap \overset{\circ}{\mathbf{W}}{}^1_0(G)$ and satisfies the integral identity

$$\int_G \{a^{ij}(x)u_{x_j}\eta_{x_i} - a^i(x)u_{x_i}\eta(x) - a(x)u\eta(x)\}\,dx + \int_{\Sigma_0} \frac{1}{r}\sigma(\omega)u(x)\eta(x)ds$$
$$+ \int_{\partial G} \frac{1}{r}\gamma(\omega)u(x)\eta(x)ds = \int_{\partial G} g(x)\eta(x)ds + \int_{\Sigma_0} h(x)\eta(x)ds - \int_G f(x)\eta(x)dx \quad (II)$$

for all functions $\eta(x) \in \mathbf{C}^0(\overline{G}) \cap \overset{\circ}{\mathbf{W}}{}^1_0(G)$.

Without loss of generality, we assume that there exists $d > 0$ such that G_0^d is *a rotational cone* with the vertex at $\mathcal{O}$ and the aperture ω_0, thus

$$\Gamma_0^d = \left\{(r,\omega)\Big| x_1^2 = \cot^2\frac{\omega_0}{2}\sum_{i=2}^n x_i^2;\ r \in (0,d),\ \omega_1 = \frac{\omega_0}{2},\ \omega_0 \in (0, 2\pi)\right\}. \quad (3.1.1)$$

Lemma 3.2. *Let $u(x)$ be a weak solution of (L). For any function $\eta(x) \in \mathbf{C}^0(\overline{G}) \cap \overset{\circ}{\mathbf{W}}{}^1_0(G)$ the equality*

$$\int_{G_0^\varrho} \Big\{a^{ij}(x)u_{x_j}\eta_{x_i} + \big(f(x) - a^i(x)u_{x_i} - a(x)u\big)\eta(x)\Big\}dx$$
$$= \int_{\Omega_\varrho} a^{ij}(x)u_{x_j}\eta(x)\cos(r,x_i)d\Omega_\varrho + \int_{\Gamma_0^\varrho} \left(g(x) - \frac{1}{r}\gamma(\omega)u(x)\right)\eta(x)ds$$
$$+ \int_{\Sigma_0^\varrho} \left(h(x) - \frac{1}{r}\sigma(\omega)u(x)\right)\eta(x)ds \qquad (II)_{loc}$$

holds for a.e. $\varrho \in (0,d)$.

Proof. Let $\chi_\varrho(x)$ be the characteristic function of the set G_0^ϱ. We consider the integral identity (II) replacing in it $\eta(x)$ by $\eta(x)\chi_\varrho(x)$. As a result we obtain

$$\int\limits_{G_0^\varrho}\Big\{a^{ij}(x)u_{x_j}\eta_{x_i}+\big(f(x)-a^i(x)u_{x_i}-a(x)u\big)\eta(x)\Big\}dx$$

$$=-\int\limits_{G_0^\varrho}a^{ij}(x)u_{x_j}\eta(x)\chi_{x_i}dx+\int\limits_{\Gamma_0^\varrho}\left(g(x)-\frac{1}{r}\gamma(\omega)u(x)\right)\eta(x)ds$$

$$+\int\limits_{\Sigma_0^\varrho}\left(h(x)-\frac{1}{r}\sigma(\omega)u(x)\right)\eta(x)ds.$$

Because of formula (7′) Subsection 3 §1 Chapt. 3 [28],

$$\chi_{x_i}=-\frac{x_i}{r}\delta(\varrho-r),$$

where $\delta(\varrho-r)$ is the Dirac distribution lumped on the sphere $r=\varrho$, we get (see Example 4 Subsection 3 §1 Chapt. 3 [28])

$$-\int\limits_{G_0^\varrho}a^{ij}(x)u_{x_j}\eta(x)\chi_{x_i}dx=\int\limits_{G_0^\varrho}a^{ij}(x)u_{x_j}\eta(x)\frac{x_i}{r}\delta(\varrho-r)dx$$

$$=\int\limits_{\Omega_\varrho}a^{ij}(x)u_{x_j}\eta(x)\cos(r,x_i)d\Omega_\varrho.$$

Hence it follows the required statement. □

Assumptions.

(a) *the condition of the uniform ellipticity:*

$$\nu_\pm\xi^2\le a_\pm^{ij}(x)\xi_i\xi_j\le\mu_\pm\xi^2,\quad\forall x\in\overline{G_\pm},\quad\forall\xi\in\mathbb{R}^n;$$
$$\nu_\pm,\mu_\pm=const>0,\ and\ a^{ij}(0)=a\delta_i^j,$$

where δ_i^j *is the Kronecker symbol and* $a=\begin{cases}a_+, & x\in\overline{G_+},\\ a_-, & x\in\overline{G_-},\end{cases}$ $a_\pm$ *are positive constants; we write:*

$$a_*=\min\{a_+,a_-\}>0,\quad a^*=\max\{a_+,a_-\}>0;$$
$$\nu_*=\min\{\nu_-,\nu_+\};\qquad \mu^*=\max(\mu_-,\mu_+);$$

(b) $a^{ij}(x)\in\mathbf{C}^0(\overline{G})$, $a^i(x)\in\mathbf{L}_p(G)$, $a(x),f(x)\in\mathbf{L}_{p/2}(G)\cap\mathbf{L}_2(G)$; $n<p\le 2n$, *for which the inequalities*

$$\left(\sum_{i,j=1}^n|a_\pm^{ij}(x)-a_\pm^{ij}(y)|^2\right)^{\frac12}\le a_\pm\mathcal{A}(|x-y|);$$

$$|x|\Big(\sum_{i=1}^{n}|a^i_{\pm}(x)|^2\Big)^{\frac{1}{2}}+|x|^2|a_{\pm}(x)|\le a_{\pm}\mathcal{A}(|x|)$$

hold for $x,y\in\overline{G}$, *where* $\mathcal{A}(r)$ *is a monotonically increasing, non-negative function,* **continuous at 0**, $\mathcal{A}(0)=0$;

(c) $a(x)\le 0$ *in* G; $\sigma(\omega)\ge\nu_0>0$ *on* σ_0; $\gamma(\omega)\ge\nu_0>0$ *on* ∂G;

(d) *there exist numbers* $f_1\ge 0$, $g_1\ge 0$, $h_1\ge 0$, $s>1$ *such that*

$$|f(x)|\le f_1|x|^{s-2},\quad |g(x)|\le g_1|x|^{s-1},\quad |h(x)|\le h_1|x|^{s-1};$$

(e) $\gamma(\omega)$ *is a positive bounded piecewise smooth function on* $\partial\Omega$, $\sigma(\omega)$ *is a positive continuous function on* σ_0.

Our main result is the following theorem.

Theorem 3.3. *Let* u *be a weak solution of the problem (L) and* λ *be defined by (2.2.1). Let assumptions (a)–(e) be satisfied with* $\mathcal{A}(r)$ *Dini-continuous at zero. Then there exist* $d\in(0,1)$ *and positive constants* C_0, C_1, C_2 *depending only on* $n,\nu_*,\mu^*,p,\lambda,\Big\|\sum_{i=1}^{n}|a^i(x)|^2\Big\|_{\mathbf{L}_{p/2}(G)}$, ω_0, s, *meas* G, *diam* G *and on the quantity* $\int_0^1\frac{\mathcal{A}(r)}{r}dr$ *such that the inequality*

$$|u(x)|\le C_0\Big(\|u\|_{2,G}+f_1+\frac{1}{\sqrt{\nu_0}}g_1+\frac{1}{\sqrt{\nu_0}}h_1\Big)\cdot\begin{cases}|x|^{\lambda}, & \text{if } s>\lambda,\\ |x|^{\lambda}\ln\Big(\frac{1}{|x|}\Big), & \text{if } s=\lambda,\\ |x|^{s}, & \text{if } s<\lambda\end{cases}\tag{3.1.2}$$

holds for all $x\in G_0^d$. *If, in addition,*

$$a^{ij}(x)\in\mathbf{C}^1(G),\ \sigma(\omega)\in C^1(\sigma_0),\ \gamma(\omega)\in\mathbf{C}^1(\partial G),\ f(x)\in\mathbf{V}^0_{p,2p-n}(G),$$
$$h(x)\in V^{1-1/p}_{p,2p-n}(\sigma_0),\ g(x)\in\mathbf{V}^{1-1/p}_{p,2p-n}(\partial G);\ p>n$$

and there is the number

$$\tau_s=:\sup_{\varrho>0}\varrho^{-s}\Big(\|h\|_{V^{1-\frac{1}{p}}_{p,2p-n}(\Sigma^{\varrho}_{\varrho/2})}+\|g\|_{\mathbf{V}^{1-\frac{1}{p}}_{p,2p-n}(\Gamma^{\varrho}_{\varrho/2})}\Big),\tag{3.1.3}$$

then for all $x\in G_0^d$ *the inequality*

$$|\nabla u(x)|\le C_1\Big(\|u\|_{2,G}+f_1+\frac{1}{\sqrt{\nu_0}}g_1+\frac{1}{\sqrt{\nu_0}}h_1+\tau_s\Big)\cdot\begin{cases}|x|^{\lambda-1}, & \text{if } s>\lambda,\\ |x|^{\lambda-1}\ln\Big(\frac{1}{|x|}\Big), & \text{if } s=\lambda,\\ |x|^{s-1}, & \text{if } s<\lambda\end{cases}\tag{3.1.4}$$

holds. Furthermore, if $u \in \mathbf{V}^2_{p,2p-n}(G)$, *then*

$$\|u\|_{\mathbf{V}^2_{p,2p-n}(G_0^\varrho)} \le C_2\Big(\|u\|_{2,G} + f_1 + \frac{1}{\sqrt{\nu_0}}g_1 + \frac{1}{\sqrt{\nu_0}}h_1 + \tau_s\Big)\cdot \begin{cases} \varrho^\lambda, & \text{if } s > \lambda, \\ \varrho^\lambda \ln\big(\frac{1}{\varrho}\big), & \text{if } s = \lambda, \\ \varrho^s, & \text{if } s < \lambda. \end{cases} \quad (3.1.5)$$

3.2 Local estimate at the boundary

We derive here a result asserting the local boundedness (near the conical point) of the weak solution of problem (L).

Theorem 3.4. *Let* $u(x)$ *be a weak solution of the problem (L) and assumptions (a)–(c) be satisfied. If, in addition,* $h(x) \in L_\infty(\Sigma_0)$, $g(x) \in L_\infty(\partial G)$, *then the inequality*

$$\sup_{G_0^{\varkappa\varrho}} |u(x)| \le C\left\{\varrho^{-n/t}\|u\|_{t,G_0^\varrho} + \varrho^{2(1-n/p)}\|f\|_{p/2,G_0^\varrho} + \varrho\left(\|g\|_{\infty,\Gamma_0^\varrho} + \|h\|_{\infty,\Sigma_0^\varrho}\right)\right\} \quad (3.2.1)$$

holds for any $t > 0$, $\varkappa \in (0,1)$ *and* $\varrho \in (0,d)$, *where* $C > 0$ *is a constant depending only on* $n, \nu_*, \mu^*, t, p, \varkappa, \left\|\sum_{i=1}^n |a^i(x)|^2\right\|_{\mathbf{L}_{p/2}(G)}$.

Proof. We apply the Moser iteration method. First we assume that $t \ge 2$. We consider the integral identity (II) and make the coordinate transformation $x = \varrho x'$. Let G' be the image of G, $\partial G'$ be the image of ∂G, Σ_0' be the image of Σ_0, and we have $dx = \varrho^n dx'$, $ds = \varrho^{n-1}ds'$. In addition, we denote

$$v(x') = u(\varrho x'),\ \mathcal{F}(x') = \varrho^2 f(\varrho x'),\ \mathcal{G}(x') = \varrho g(\varrho x'),\ \mathcal{H}(x') = \varrho h(\varrho x'). \quad (3.2.2)$$

Then (II) means

$$\begin{aligned} &\int_{G'} \left\{a^{ij}(\varrho x')v_{x_j'}\eta_{x_i'} - \varrho a^i(\varrho x')v_{x_i'}\eta(x') - \varrho^2 a(\varrho x')v(x')\eta(x')\right\}dx' \\ &+ \int_{\Sigma_0'} \frac{1}{|x'|}\sigma(\omega)v(x')\eta(x')ds' + \int_{\partial G'} \frac{1}{|x'|}\gamma(\omega)v(x')\eta(x')ds' \\ &= \int_{\partial G'} \mathcal{G}(x')\eta(x')ds' + \int_{\Sigma_0'} \mathcal{H}(x')\eta(x')ds' - \int_{G'} \mathcal{F}(x')\eta(x')dx' \end{aligned} \quad (II)'$$

for all $\eta(x') \in \mathbf{C}^0(\overline{G'}) \cap \overset{\circ}{\mathbf{W}}{}^1_0(G')$. Now we define the quantity k by

$$k = k(\varrho) = \nu_*^{-1}\left(\|\mathcal{F}\|_{p/2,G_0^1} + \|\mathcal{G}\|_{\infty,\Gamma_0^1} + \|\mathcal{H}\|_{\infty,\Sigma_0^1}\right) \quad (3.2.3)$$

and set

$$\overline{v}(x') = |v(x')| + k. \tag{3.2.4}$$

Now we observe that

$$|\mathcal{F}|\overline{v} = \frac{1}{k}|\mathcal{F}| \cdot k\overline{v} = \frac{1}{k}|\mathcal{F}|(\overline{v} - |v|) \cdot \overline{v} = \frac{1}{k}|\mathcal{F}| \cdot \overline{v}^2 - \frac{1}{k}|\mathcal{F}| \cdot |v|\overline{v} \leq \frac{1}{k}|\mathcal{F}| \cdot \overline{v}^2; \tag{3.2.5}$$

$$|\mathcal{H}|\overline{v} \leq \frac{1}{k}|\mathcal{H}| \cdot \overline{v}^2; \quad |\mathcal{G}|\overline{v} \leq \frac{1}{k}|\mathcal{G}| \cdot \overline{v}^2$$

in the same way. As the test function in the integral identity $(II)'$ we choose $\eta(x') = \zeta^2(|x'|)v\overline{v}^{t-2}(x')$, where $\zeta(|x'|) \in \mathbf{C}_0^\infty([0,1])$ is a non-negative function to be further specified. By the chain and the product rules, η is a valid test function in $(II)'$ and also

$$\eta_{x'_i} = \overline{v}^{t-2}v_{x'_i}\zeta^2(|x'|) + (t-2)\overline{v}^{t-3}|v|v_{x'_i}\zeta^2(|x'|) + 2\zeta\zeta_{x'_i}v\overline{v}^{t-2}(x'),$$

so that, by substitution into $(II)'$ taking into account that $a(\varrho x') \leq 0$ in G', $v \leq |v| \leq \overline{v}$ and $t \geq 2$, we obtain

$$\begin{aligned}
&\int\limits_{G_0^1} a^{ij}(\varrho x')v_{x'_i}v_{x'_j}\overline{v}^{t-2}\zeta^2(|x'|)dx' + \int\limits_{\Sigma_0^1} \frac{1}{|x'|}\sigma(\omega)v^2\overline{v}^{t-2}(x')\zeta^2(|x'|)ds' \\
&+ \int\limits_{\Gamma_0^1} \frac{1}{|x'|}\gamma(\omega)v^2\overline{v}^{t-2}(x')\zeta^2(|x'|)ds' \leq \varrho \int\limits_{G_0^1} |a^i(\varrho x')v_{x'_i}|\overline{v}^{t-1}(x')\zeta^2(|x'|)dx' \\
&+ 2\int\limits_{G_0^1} |a^{ij}(\varrho x')\zeta_{x'_i}v_{x'_j}|\overline{v}^{t-1}\zeta(|x'|)dx' + \int\limits_{\Gamma_0^1} |\mathcal{G}(x')|\overline{v}^{t-1}(x')\zeta^2(|x'|)ds' \\
&+ \int\limits_{\Sigma_0^1} |\mathcal{H}(x')|\overline{v}^{t-1}(x')\zeta^2(|x'|)ds' + \int\limits_{G_0^1} |\mathcal{F}(x')|\overline{v}^{t-1}(x')\zeta^2(|x'|)dx'.
\end{aligned}$$

By the ellipticity condition, assumption (c) and taking into account (3.2.5), it follows that

$$\begin{aligned}
&\int\limits_{G_0^1} \nu|\nabla' v|^2 \cdot \overline{v}^{t-2}\zeta^2(|x'|)dx' \leq \int\limits_{G_0^1} \Bigg(2\mu|\nabla' v| \cdot |\nabla'\zeta| \cdot \overline{v}^{t-1}\zeta(|x'|) + \frac{1}{k}|\mathcal{F}(x')| \cdot \overline{v}^t\zeta^2(|x'|) \\
&+ \varrho\left(\sum_{i=1}^n |a^i(x)|^2\right)^{\frac{1}{2}} |\nabla' v| \cdot \overline{v}^{t-1}\zeta^2(|x'|)\Bigg)dx' + \frac{1}{k}\|\mathcal{G}\|_{\infty,\Gamma_0^1} \cdot \int\limits_{\Gamma_0^1} \overline{v}^t(x')\zeta^2(|x'|)ds' \\
&+ \frac{1}{k}\|\mathcal{H}\|_{\infty,\Sigma_0^1} \cdot \int\limits_{\Sigma_0^1} \overline{v}^t(x')\zeta^2(|x'|)ds'.
\end{aligned} \tag{3.2.6}$$

We estimate every term by the Cauchy inequality for any $\varepsilon > 0$:

$$2\mu|\nabla' v| \cdot |\nabla'\zeta| \cdot \overline{v}^{t-1}\zeta \le 2\left(\sqrt{\nu}|\nabla' v| \cdot \overline{v}^{\frac{t}{2}-1}\zeta\right) \cdot \left(\frac{\mu^*}{\sqrt{\nu_*}}\overline{v}^{\frac{t}{2}}|\nabla'\zeta|\right)$$

$$\le \varepsilon\nu|\nabla' v|^2 \cdot \overline{v}^{t-2}\zeta^2 + \frac{(\mu^*)^2}{\varepsilon\nu_*}\overline{v}^t|\nabla'\zeta|^2;$$

$$\varrho\left(\sum_{i=1}^n |a^i(x)|^2\right)^{\frac{1}{2}} \cdot |\nabla' v| \cdot \overline{v}^{t-1}\zeta^2$$

$$\le \left(\sqrt{\nu}|\nabla' v| \cdot \overline{v}^{\frac{t}{2}-1}\right) \cdot \left(\frac{1}{\sqrt{\nu_*}}\overline{v}^{\frac{t}{2}} \cdot \varrho\left(\sum_{i=1}^n |a^i(x)|^2\right)^{\frac{1}{2}}\right)\zeta^2$$

$$\le \frac{\varepsilon}{2}\nu|\nabla' v|^2 \cdot \overline{v}^{t-2}\zeta^2 + \frac{1}{2\varepsilon\nu_*}\varrho^2\sum_{i=1}^n |a^i(x)|^2 \cdot \overline{v}^t\zeta^2.$$

For the estimating integrals over the boundaries we apply the inequality (1.5.12). Then from (3.2.6) it follows that

$$\int_{G_0^1} \nu|\nabla' v|^2 \cdot \overline{v}^{t-2}\zeta^2(|x'|)dx' \le \frac{3}{2}\varepsilon\int_{G_0^1} \nu|\nabla' v|^2 \cdot \overline{v}^{t-2}\zeta^2(|x'|)dx' \tag{3.2.7}$$

$$+ \int_{G_0^1}\left\{\frac{(\mu^*)^2}{\varepsilon\nu_*}\overline{v}^t \cdot |\nabla'\zeta|^2 + \left(\frac{1}{2\varepsilon\nu_*}\varrho^2\sum_{i=1}^n |a^i(x)|^2 + \frac{1}{k}|\mathcal{F}(x')|\right) \cdot \overline{v}^t\zeta^2(|x'|)\right\}dx'$$

$$+ \frac{1}{k}\left(\|\mathcal{G}\|_{\infty,\Gamma_0^1} + \|\mathcal{H}\|_{\infty,\Sigma_0^1}\right) \cdot \int_{G_0^1}\left(\delta|\nabla'(\zeta\overline{v}^{t/2})|^2 + \frac{1}{\delta}c_0\overline{v}^t\zeta^2\right)dx', \quad \forall\varepsilon,\delta > 0.$$

From relations

$$|\nabla'(\zeta\overline{v}^{t/2})|^2 \le 2\left(\zeta^2|\nabla'(\overline{v}^{t/2})|^2 + \overline{v}^t|\nabla'\zeta|^2\right), \quad |\nabla'(\overline{v}^{t/2})|^2 = \frac{t^2}{4}\overline{v}^{t-2}|\nabla' v|^2 \tag{3.2.8}$$

follows the inequality

$$|\nabla'(\zeta\overline{v}^{t/2})|^2 \le \frac{t^2}{2}\overline{v}^{t-2}|\nabla' v|^2\zeta^2 + 2\overline{v}^t|\nabla'\zeta|^2. \tag{3.2.9}$$

Now, from (3.2.7) and (3.2.9), by choosing $\varepsilon = \frac{1}{3}$ we find that

$$\frac{1}{2}\int_{G_0^1} \nu|\nabla' v|^2 \cdot \overline{v}^{t-2}\zeta^2(|x'|)dx'$$

$$\le \frac{\delta t^2}{2\nu_*} \cdot \frac{\|\mathcal{G}\|_{\infty,\Gamma_0^1} + \|\mathcal{H}\|_{\infty,\Sigma_0^1}}{k} \cdot \int_{G_0^1} \nu|\nabla' v|^2 \cdot \overline{v}^{t-2}\zeta^2(|x'|)dx' \tag{3.2.10}$$

$$+\int\limits_{G_0^1}\left(\frac{3\varrho^2}{2\nu_*}\sum_{i=1}^n|a^i(x)|^2+\frac{|\mathcal{F}(x')|+c_0\delta^{-1}\left(\|\mathcal{G}\|_{\infty,\Gamma_0^1}+\|\mathcal{H}\|_{\infty,\Sigma_0^1}\right)}{k}\right)\cdot\overline{v}^t\zeta^2(|x'|)dx'$$

$$+\int\limits_{G_0^1}\left\{\frac{3(\mu^*)^2}{\nu_*}+\frac{2\delta}{k}\left(\|\mathcal{G}\|_{\infty,\Gamma_0^1}+\|\mathcal{H}\|_{\infty,\Sigma_0^1}\right)\right\}\overline{v}^t\cdot|\nabla'\zeta|^2dx',\quad\forall\delta\in(0,1].$$

We choose now $\delta=\frac{1}{2t^2}$; by the definition of the number k in (3.2.3), the last inequality (3.2.10) can be rewritten in the following way:

$$\int\limits_{G_0^1}\nu|\nabla'v|^2\cdot\overline{v}^{t-2}\zeta^2(|x'|)dx'$$

$$\le 8c_0\nu_*t^2\cdot\int\limits_{G_0^1}\zeta^2(|x'|)\overline{v}^t(x')dx'+\left(\frac{12(\mu^*)^2}{\nu_*}+\frac{4\nu_*}{t^2}\right)\cdot\int\limits_{G_0^1}|\nabla'\zeta|^2\overline{v}^t(x')dx'$$

$$+4\int\limits_{G_0^1}\left(\frac{3\varrho^2}{2\nu_*}\sum_{i=1}^n|a^i(x)|^2+\frac{|\mathcal{F}(x')|}{k}\right)\cdot\overline{v}^t\zeta^2(|x'|)dx'.$$

But, by (3.2.8), the last inequality means

$$\int\limits_{G_0^1}\nu|\nabla'(\overline{v}^{t/2})|^2\zeta^2(|x'|)dx'$$

$$\le 2c_0\nu_*t^4\cdot\int\limits_{G_0^1}\zeta^2(|x'|)\overline{v}^t(x')dx'+\left(\frac{3(\mu^*)^2t^2}{\nu_*}+\nu_*\right)\cdot\int\limits_{G_0^1}|\nabla'\zeta|^2\overline{v}^t(x')dx'$$

$$+t^2\int\limits_{G_0^1}\left(\frac{3\varrho^2}{2\nu_*}\sum_{i=1}^n|a^i(x)|^2+\frac{|\mathcal{F}(x')|}{k}\right)\cdot\overline{v}^t\zeta^2(|x'|)dx'.$$

Since $t\ge 2$, we can rewrite the above inequality as

$$\int\limits_{G_0^1}\nu|\nabla'(\overline{v}^{t/2})|^2\zeta^2(|x'|)dx'\le C_1t^4\int\limits_{G_0^1}\nu\left(|\nabla'\zeta|^2+\zeta^2(|x'|)\right)\overline{v}^tdx'$$

$$+C_2t^2\int\limits_{G_0^1}\nu\left(\varrho^2\sum_{i=1}^n|a^i(x)|^2+\frac{|\mathcal{F}(x')|}{k}\right)\cdot\overline{v}^t\zeta^2(|x'|)dx',\quad(3.2.11)$$

where constants C_1,C_2 depend only on c_0,ν_*,μ^* and are independent of t. Setting

$$w(x')=\sqrt{\nu}\cdot\overline{v}^{t/2}(x')\quad(3.2.12)$$

from (3.2.11) we obtain

$$\int_{G_0^1} |\nabla' w|^2 \zeta^2(|x'|)dx' \le C_1 t^4 \int_{G_0^1} \left(|\nabla'\zeta|^2 + \zeta^2(|x'|)\right) w^2(x')dx' \\ + C_2 t^2 \int_{G_0^1} \left(\varrho^2 \sum_{i=1}^n |a^i(x)|^2 + \frac{|\mathcal{F}(x')|}{k} \right) \cdot w^2(x')\zeta^2(|x'|)dx'. \quad (3.2.13)$$

The desired iteration process can now be developed from (3.2.13). By the Sobolev Imbedding Theorem 1.15, we have

$$\|\zeta w\|^2_{\frac{2\widetilde{n}}{\widetilde{n}-2}, G_0^1} \le C^* \cdot \int_{G_0^1} \left(\left(|\nabla'\zeta|^2 + \zeta^2\right) w^2(x') + \zeta^2(|x'|)|\nabla' w|^2 \right) dx', \quad (3.2.14)$$

where $\widetilde{n} = n$ for $n > 2$, $\widetilde{2} \in (2, p)$ and C^* depends only on n. Using the Hölder inequality for integrals

$$\int_{G_0^1} |F(x')| \cdot w^2(x')\zeta^2(x')dx' \le \|F\|_{p/2, G_0^1} \cdot \|w\zeta\|^2_{\frac{2p}{p-2}, G_0^1}, \quad p > 2, \quad (3.2.15)$$

we get from (3.2.13)–(3.2.15):

$$\|\zeta w\|^2_{\frac{2\widetilde{n}}{\widetilde{n}-2}, G_0^1} \le C_3 t^4 \int_{G_0^1} \left(|\nabla'\zeta|^2 + \zeta^2(|x'|)\right) w^2(x')dx' \\ + C_4 t^2 \left\| \varrho^2 \sum_{i=1}^n |a^i(x)|^2 + \frac{|\mathcal{F}(x')|}{k} \right\|_{p/2, G_0^1} \cdot \|w\zeta\|^2_{\frac{2p}{p-2}, G_0^1}, \quad p > n. \quad (3.2.16)$$

By the interpolation inequality for L_p-norms

$$\|\zeta w\|_{\frac{2p}{p-2}, G_0^1} \le \varepsilon \|\zeta w\|_{\frac{2\widetilde{n}}{\widetilde{n}-2}, G_0^1} + \varepsilon^{\frac{\widetilde{n}}{\widetilde{n}-p}} \|\zeta w\|_{2, G_0^1}, \quad p > n, \ \forall \varepsilon > 0, \quad (3.2.17)$$

from (3.2.16)–(3.2.17) it follows that

$$\|\zeta w\|_{\frac{2\widetilde{n}}{\widetilde{n}-2}, G_0^1} \\ \le t\sqrt{C_4} \left\| \varrho^2 \sum_{i=1}^n |a^i(x)|^2 + \frac{|\mathcal{F}(x')|}{k} \right\|^{1/2}_{p/2, G_0^1} \times \left(\varepsilon \|w\zeta\|_{\frac{2\widetilde{n}}{\widetilde{n}-2}, G_0^1} + \varepsilon^{\frac{\widetilde{n}}{\widetilde{n}-p}} \|\zeta w\|_{2, G_0^1} \right) \\ + t^2 \sqrt{C_3} \cdot \|(\zeta + |\nabla'\zeta|)w\|_{2, G_0^1}, \quad p > n, \ \forall \varepsilon > 0.$$

Choosing $\varepsilon = \frac{1}{2t\sqrt{C_4}} \left\| \varrho^2 \sum_{i=1}^{n} |a^i(x)|^2 + \frac{|\mathcal{F}(x')|}{k} \right\|_{p/2,G_0^1}^{-\frac{1}{2}}$, we obtain

$$\|\zeta w\|_{\frac{2\tilde{n}}{\tilde{n}-2},G_0^1} \le Ct^{\frac{p}{p-\tilde{n}}} \|(\zeta + |\nabla'\zeta|)w\|_{2,G_0^1}, \quad n < p \le 2n, \tag{3.2.18}$$

where C depends only on $c_0, n, \nu_*, \mu^*, p, \left\| \sum_{i=1}^{n} |a^i(x)|^2 \right\|_{p/2,G}$ and is independent of t. Recalling the definition of w by (3.2.12), we finally establish from (3.2.18) the inequality

$$\|\zeta \cdot \overline{v}^{t/2}\|_{\frac{2\tilde{n}}{\tilde{n}-2},G_0^1} \le Ct^{\frac{p}{p-\tilde{n}}} \|(\zeta + |\nabla'\zeta|) \cdot \overline{v}^{t/2}\|_{2,G_0^1}, \quad n < p \le 2n. \tag{3.2.19}$$

This inequality can now be iterated to yield the desired estimate.

For all $\varkappa \in (0,1)$ we define sets $G'_{(j)} \equiv G_0^{\varkappa+(1-\varkappa)2^{-j}}$, $j = 0, 1, 2, \ldots$. It is easy to verify that $G_0^{\varkappa} \equiv G'_{(\infty)} \subset \cdots \subset G'_{(j+1)} \subset G'_{(j)} \subset \cdots \subset G'_{(0)} \equiv G_0^1$. Now we consider the sequence of cut-off functions $\zeta_j(x') \in \mathbf{C}^\infty(G'_{(j)})$ such that

$$\begin{aligned}
&0 \le \zeta_j(x') \le 1 \text{ in } G'_{(j)} \text{ and } \zeta_j(x') \equiv 1 \text{ in } G'_{(j+1)},\\
&\zeta_j(x') \equiv 0 \quad \text{for} \quad |x'| > \varkappa + 2^{-j}(1-\varkappa);\\
&|\nabla'\zeta_j| \le \tfrac{2^{j+1}}{1-\varkappa} \quad \text{for} \quad \varkappa + 2^{-j-1}(1-\varkappa) < |x'| < \varkappa + 2^{-j}(1-\varkappa).
\end{aligned}$$

We define also the number sequence $t_j = t\left(\frac{\tilde{n}}{\tilde{n}-2}\right)^j$, $j = 0, 1, 2, \ldots$. Now we rewrite the inequality (3.2.19) replacing $\zeta(|x'|)$ by $\zeta_j(x')$ and t by t_j; then taking t_j-th root, we obtain

$$\|\overline{v}\|_{t_{j+1},G'_{(j+1)}} \le \left(\frac{C}{1-\varkappa}\right)^{2/t_j} \cdot 4^{\frac{j}{t_j}} \cdot (t_j)^{\frac{2p}{p-\tilde{n}} \cdot \frac{1}{t_j}} \|\overline{v}\|_{t_j,G'_{(j)}}.$$

After iteration, we find that

$$\|\overline{v}\|_{t_{j+1},G'_{(j+1)}} \le \left\{ \frac{Ct^{\frac{p}{p-\tilde{n}}}}{1-\varkappa} \cdot \left(\frac{\tilde{n}}{\tilde{n}-2}\right)^{\frac{\tilde{n}}{p-\tilde{n}}} \right\}^{2\sum_{j=0}^{\infty} \frac{1}{t_j}} \cdot 4^{\sum_{j=0}^{\infty} \frac{j}{t_j}} \cdot \|\overline{v}\|_{t,G_0^1}. \tag{3.2.20}$$

Notice that the series $\sum_{j=0}^{\infty} \frac{j}{t_j}$ is convergent by the d'Alembert ratio test, but the series $\sum_{j=0}^{\infty} \frac{1}{t_j} = \frac{\tilde{n}}{2t}$ as a geometric series. Therefore from (3.2.20) we get $\|\overline{v}\|_{t_{j+1},G'_{(j+1)}} \le C\|\overline{v}\|_{t,G_0^1}$. Consequently, letting $j \to \infty$, we have $\sup_{x' \in G_0^{\varkappa}} \overline{v}(x') \le C\|\overline{v}\|_{t,G_0^1}$. Hence, because of the definitions of the function $\overline{v}(x')$ by (3.2.4) and of the number k by (3.2.3), we obtain

$$\sup_{x' \in G_0^{\varkappa}} |v(x')| \le C\Big(\|v\|_{t,G_0^1} + \|\mathcal{F}\|_{p/2,G_0^1} + \|\mathcal{G}\|_{\infty,\Gamma_0^1} + \|\mathcal{H}\|_{\infty,\Sigma_0^1}\Big).$$

Returning to the variables x, u, by (3.2.2), we obtain the required estimate (3.2.1) in the case $t \geq 2$.

Let now $0 < t < 2$. We consider (3.2.1) with $t = 2$:

$$\sup_{x\in G_0^{\varkappa\varrho}} |u(x)| \leq C\varrho^{-\frac{n}{2}} \|u\|_{2,G_0^\varrho} + K(\varrho), \tag{3.2.21}$$

where

$$K(\varrho) = C\left\{\varrho^{2(1-n/p)}\|f\|_{p/2,G_0^\varrho} + \varrho\left(\|g\|_{\infty,\Gamma_0^\varrho} + \|h\|_{\infty,\Sigma_0^\varrho}\right)\right\}.$$

Now, using the Young inequality with $q = \frac{2}{t}$ and $q' = \frac{2}{2-t}$ we can write

$$C\varrho^{-\frac{n}{2}}\|u\|_{2,G_0^\varrho} = C\varrho^{-\frac{n}{2}}\left(\int_{G_0^\varrho} u^t \cdot u^{2-t}\right)^{1/2}$$

$$\leq \left(\sup_{G_0^\varrho}|u(x)|\right)^{1-t/2} \cdot C\varrho^{-\frac{n}{2}}\|u\|_{t,G_0^\varrho}^{t/2} \leq \frac{1}{2}\sup_{G_0^\varrho}|u(x)| + C_1\varrho^{-\frac{n}{t}}\|u\|_{t,G_0^\varrho}. \tag{3.2.22}$$

Let us define the function $\psi(s) = \sup_{x\in G_0^s} |u(x)|$. Then from (3.2.21)–(3.2.22) it follows that

$$\psi(\varkappa\varrho) \leq \frac{1}{2}\psi(\varrho) + C_1\varrho^{-\frac{n}{t}}\|u\|_{t,G_0^\varrho} + K(\varrho), \quad \varkappa \in (0,1). \tag{3.2.23}$$

Further we apply Lemma 1.24. Letting $r = \varkappa\varrho$, $R = \varrho$, $\delta = \frac{1}{2}$, $\alpha = \frac{n}{t}$, $A = C_1\|u\|_{t,G_0^\varrho}$, $B = K(\varrho)$ from (3.2.23) we obtain the validity of required estimate (3.2.1) in the case $0 < t < 2$. Thus, the proof of Theorem 3.4 is complete. □

3.3 Global integral estimates

In this section we derive a global estimate for the Dirichlet integral.

Theorem 3.5. *Let $u(x)$ be a weak solution of problem (L) and assumptions (a)–(c) be satisfied. If, in addition, $h(x) \in L_2(\Sigma_0)$, $g(x) \in \mathbf{L}_2(\partial G)$, then the inequality*

$$\int_G \nu|\nabla u|^2 dx + \int_{\Sigma_0} \frac{\sigma(\omega)}{r} u^2(x) ds + \int_{\partial G} \frac{\gamma(\omega)}{r} u^2(x) ds$$

$$\leq C\left\{\int_G u^2(x)dx + \int_G f^2(x)dx + \frac{1}{\nu_0}\int_{\Sigma_0} h^2(x)ds + \frac{1}{\nu_0}\int_{\partial G} g^2(x)ds\right\} \tag{3.3.1}$$

holds, where constant $C > 0$ depends only on $p, n, \nu_*, \left\|\sum_{i=1}^n |a^i(x)|^2\right\|_{\mathbf{L}_{p/2}(G)}$ and $\operatorname{meas} G$, $\operatorname{diam} G$.

Proof. We put in *(II)* $\eta(x) = u(x)$ and apply the classical Hölder inequality. By assumptions (a), (c), we obtain

$$\int_G \nu|\nabla u|^2 dx + \int_{\Sigma_0} \frac{\sigma(\omega)}{r} u^2(x)ds + \int_{\partial G} \frac{\gamma(\omega)}{r} u^2(x)ds \tag{3.3.2}$$

$$\leq \int_G \sqrt{\sum_{i=1}^n |a^i(x)|^2}|u||\nabla u|dx + \int_{\Sigma_0} |u||h(x)|ds + \int_{\partial G} |u||g(x)|ds + \int_G |u||f(x)|dx.$$

Further, by assumptions (b), (c), the Cauchy inequality and the integral Hölder inequality, we have:

$$\int_G \sqrt{\sum_{i=1}^n |a^i(x)|^2}|u||\nabla u|dx \leq \frac{\varepsilon}{2}\int_G \nu|\nabla u|^2 dx + \frac{1}{2\varepsilon\nu_*}\int_G \sum_{i=1}^n |a^i(x)|^2|u|^2 dx$$

$$\leq \frac{\varepsilon}{2}\int_G \nu|\nabla u|^2 dx + \frac{1}{2\varepsilon\nu_*}\left(\int_G \left(\sum_{i=1}^n |a^i(x)|^2\right)^{\frac{p}{2}} dx\right)^{\frac{2}{p}} \cdot \left(\int_G |u|^{\frac{2p}{p-2}} dx\right)^{\frac{p-2}{p}},$$

$p > 2$, for all $\varepsilon > 0$. Now we apply the inequality (1.6.5), Theorem 1.16; hence it follows that

$$\int_G \sqrt{\sum_{i=1}^n |a^i(x)|^2}|u||\nabla u|dx \leq \frac{\varepsilon}{2}\int_G \nu|\nabla u|^2 dx + \frac{1}{2\varepsilon\nu_*^2}\left\|\sum_{i=1}^n |a^i(x)|^2\right\|_{\mathbf{L}_{p/2}(G)}$$

$$\times \int_G \left(\delta\nu|\nabla u|^2 + c(\delta, p, n, meas\, G)u^2\right) dx,\ \forall\varepsilon > 0,\ \forall\delta > 0,\ p > n. \tag{3.3.3}$$

If we choose $\delta = \frac{\varepsilon^2\nu_*^2}{\left\|\sum_{i=1}^n |a^i(x)|^2\right\|_{\mathbf{L}_{p/2}(G)}}$, then from (3.3.2)–(3.3.3) we get

$$(1-\varepsilon)\int_G \nu|\nabla u|^2 dx + \int_{\Sigma_0} \frac{\sigma(\omega)}{r} u^2(x)ds + \int_{\partial G} \frac{\gamma(\omega)}{r} u^2(x)ds$$

$$\leq c\left(\varepsilon, p, n, \nu_*, \left\|\sum_{i=1}^n |a^i(x)|^2\right\|_{\mathbf{L}_{p/2}(G)}, meas\, G\right) \cdot \int_G |u|^2 dx + \int_{\Sigma_0} |u||h(x)|ds$$

$$+ \int_{\partial G} |u||g(x)|ds + \int_G |u||f(x)|dx. \tag{3.3.4}$$

By the Cauchy inequality, by virtue of the assumption (c), we have

$$\int_{\Sigma_0} |u||h(x)|ds = \int_{\Sigma_0} \left(\sqrt{\frac{\sigma(\omega)}{r}}|u|\right)\left(\sqrt{\frac{r}{\sigma(\omega)}}|h(x)|\right)ds$$
$$\leq \frac{1}{2}\int_{\Sigma_0} \frac{\sigma(\omega)}{r}u^2(x)ds + \frac{diamG}{2\nu_0}\int_{\Sigma_0} h^2(x)ds;$$

$$\int_{\partial G} |u||g(x)|ds = \int_{\partial G} \left(\sqrt{\frac{\gamma(\omega)}{r}}|u|\right)\left(\sqrt{\frac{r}{\gamma(\omega)}}|g(x)|\right)ds$$
$$\leq \frac{1}{2}\int_{\partial G} \frac{\gamma(\omega)}{r}u^2(x)ds + \frac{diamG}{2\nu_0}\int_{\partial G} g^2(x)ds;$$

$$\int_G |u||f(x)|dx \leq \frac{1}{2}\int_G |u|^2dx + \frac{1}{2}\int_G |f|^2dx.$$

Hence and from (3.3.4) with $\varepsilon = \frac{1}{2}$ we derive the required inequality (3.3.1) □

Now we will derive a global estimate for the weighted Dirichlet integral.

Theorem 3.6. [1] *Let $u(x)$ be a weak solution of problem (L) and λ be as above in* (2.2.1). *Let assumptions (a)–(c) be satisfied with function $\mathcal{A}(r)$ that is continuous at zero. If, in addition,*

$$f(x) \in \overset{\circ}{\mathbf{W}}{}^0_\alpha(G), \int_{\Sigma_0} r^{\alpha-1}h^2(x)ds < \infty, \int_{\partial G} r^{\alpha-1}g^2(x)ds < \infty, \quad 4-n \leq \alpha \leq 2,$$

then $u(x) \in \overset{\circ}{\mathbf{W}}{}^1_{\alpha-2}(G)$ and

$$\int_G a\left(r^{\alpha-2}|\nabla u|^2 + r^{\alpha-4}u^2\right)dx + \int_{\Sigma_0} r^{\alpha-3}\sigma(\omega)u^2(x)ds + \int_{\partial G} r^{\alpha-3}\gamma(\omega)u^2(x)ds$$
$$\leq C\Big\{\int_G \left(u^2 + (1+r^\alpha)f^2(x)\right)dx + \int_{\Sigma_0} r^{\alpha-1}h^2(x)ds + \int_{\partial G} r^{\alpha-1}g^2(x)ds\Big\}, \quad (3.3.5)$$

where the constant $C > 0$ depends only on p, n, ν_*, μ^*, a_*, ν_0, α, λ, $\left\|\sum_{i=1}^n |a^i(x)|^2\right\|_{\mathbf{L}_{p/2}(G)}$ and meas G.

Proof. We put in (II) $\eta(x) = r_\varepsilon^{\alpha-2}u(x)$, $\eta_{x_i} = r_\varepsilon^{\alpha-2}u_{x_i} + (\alpha-2)r_\varepsilon^{\alpha-3}\frac{x_i-\varepsilon l_i}{r_\varepsilon}u(x)$. As a result we have

---
[1] See also below Subsection 3.5.2

$$\int_G ar_\varepsilon^{\alpha-2}|\nabla u|^2dx + \int_{\Sigma_0} r^{-1}r_\varepsilon^{\alpha-2}\sigma(\omega)u^2(x)ds + \int_{\partial G} r^{-1}r_\varepsilon^{\alpha-2}\gamma(\omega)u^2(x)ds$$
$$= \frac{2-\alpha}{2}\int_G ar_\varepsilon^{\alpha-4}(x_i-\varepsilon l_i)(u^2)_{x_i}dx$$
$$+(2-\alpha)\int_G \big(a^{ij}(x)-a^{ij}(0)\big)r_\varepsilon^{\alpha-4}(x_i-\varepsilon l_i)u_{x_j}u(x)dx$$
$$-\int_G \big(a^{ij}(x)-a^{ij}(0)\big)r_\varepsilon^{\alpha-2}u_{x_i}u_{x_j}dx + \int_G \big(a^i(x)u_{x_i}+a(x)u-f(x)\big)r_\varepsilon^{\alpha-2}u(x)dx$$
$$+\int_{\Sigma_0} r_\varepsilon^{\alpha-2}u(x)h(x)ds + \int_{\partial G} r_\varepsilon^{\alpha-2}u(x)g(x)ds. \tag{3.3.6}$$

We transform the first integral on the right-hand side by integrating by parts:

$$\int_G ar_\varepsilon^{\alpha-4}(x_i-\varepsilon l_i)\frac{\partial u^2}{\partial x_i}dx$$
$$= \int_{G_+} a_+r_\varepsilon^{\alpha-4}(x_i-\varepsilon l_i)\frac{\partial u_+^2}{\partial x_i}dx + \int_{G_-} a_-r_\varepsilon^{\alpha-4}(x_i-\varepsilon l_i)\frac{\partial u_-^2}{\partial x_i}dx$$
$$= -\int_G au^2\frac{\partial}{\partial x_i}\Big(r_\varepsilon^{\alpha-4}(x_i-\varepsilon l_i)\Big)dx + \int_{\partial G_+} a_+u_+^2r_\varepsilon^{\alpha-4}(x_i-\varepsilon l_i)n_ids$$
$$+\int_{\partial G_-} a_-u_-^2r_\varepsilon^{\alpha-4}(x_i-\varepsilon l_i)n_ids$$
$$= -\int_G au^2\frac{\partial}{\partial x_i}\Big(r_\varepsilon^{\alpha-4}(x_i-\varepsilon l_i)\Big)dx + \int_{\partial G} au^2r_\varepsilon^{\alpha-4}(x_i-\varepsilon l_i)n_ids$$
$$+[a]_{\Sigma_0}\int_{\Sigma_0} u^2r_\varepsilon^{\alpha-4}(x_i-\varepsilon l_i)n_ids, \tag{3.3.7}$$

because of $[u]_{\Sigma_0}=0$. Now, we make elementary calculations:

1) $\frac{\partial}{\partial x_i}\big(r_\varepsilon^{\alpha-4}(x_i-\varepsilon l_i)\big) = nr_\varepsilon^{\alpha-4}+(\alpha-4)(x_i-\varepsilon l_i)r_\varepsilon^{\alpha-5}\frac{x_i-\varepsilon l_i}{r_\varepsilon} = (n+\alpha-4)r_\varepsilon^{\alpha-4}$;

2) because of $n_i\Big|_{\Sigma_0} = \cos(x_n,x_i) = \delta_i^n$, $(x_i-\varepsilon l_i)n_i\Big|_{\Sigma_0} = \delta_i^n(x_i-\varepsilon l_i)\Big|_{\Sigma_0}$ $= (x_n-\varepsilon l_n)\Big|_{\Sigma_0} = x_n\Big|_{\Sigma_0} = 0$, since $\Sigma_0=\{x_n=0\}\cap G$ and $l_n=0$;

3) from the representation $\partial G = \Gamma_0^d \cup \Gamma_d$ and by (1.2.9), $(x_i - \varepsilon l_i)n_i\Big|_{\Gamma_0^d} = -\varepsilon \sin\frac{\omega_0}{2}$, therefore

$$\int_{\partial G} au^2 r_\varepsilon^{\alpha-4}(x_i - \varepsilon l_i)n_i ds = -\varepsilon \sin\frac{\omega_0}{2}\int_{\Gamma_0^d} au^2 r_\varepsilon^{\alpha-4} ds + \int_{\Gamma_d} au^2 r_\varepsilon^{\alpha-4}(x_i - \varepsilon l_i)n_i ds.$$

Hence and from (3.3.7) it follows that

$$\frac{2-\alpha}{2}\int_G a r_\varepsilon^{\alpha-4}(x_i - \varepsilon l_i)\frac{\partial u^2}{\partial x_i}dx = \frac{(2-\alpha)(4-n-\alpha)}{2}\int_G a r_\varepsilon^{\alpha-4}u^2 dx$$
$$-\varepsilon\frac{2-\alpha}{2}\sin\frac{\omega_0}{2}\int_{\Gamma_0^d} au^2 r_\varepsilon^{\alpha-4} ds + \frac{2-\alpha}{2}\int_{\Gamma_d} au^2 r_\varepsilon^{\alpha-4}(x_i - \varepsilon l_i)n_i ds. \quad (3.3.8)$$

From (3.3.6), (3.3.8) with regard to $4-n \le \alpha \le 2$ we obtain the following equality:

$$\int_G a r_\varepsilon^{\alpha-2}|\nabla u|^2 dx + \varepsilon\frac{2-\alpha}{2}\sin\frac{\omega_0}{2}\int_{\Gamma_0^d} au^2 r_\varepsilon^{\alpha-4} ds + \int_{\Sigma_0}\frac{1}{r}r_\varepsilon^{\alpha-2}\sigma(\omega)u^2(x)ds$$
$$+\int_{\partial G}\frac{1}{r}r_\varepsilon^{\alpha-2}\gamma(\omega)u^2(x)ds \le \frac{2-\alpha}{2}\int_{\Gamma_d} au^2 r_\varepsilon^{\alpha-4}(x_i - \varepsilon l_i)n_i ds$$
$$+(2-\alpha)\int_G \big(a^{ij}(x) - a^{ij}(0)\big)u_{x_j} r_\varepsilon^{\alpha-4}(x_i - \varepsilon l_i)u(x)dx \quad (3.3.9)$$
$$-\int_G r_\varepsilon^{\alpha-2}\big(a^{ij}(x) - a^{ij}(0)\big)u_{x_i}u_{x_j}dx + \int_G \big(a^i(x)u_{x_i} + a(x)u - f(x)\big)r_\varepsilon^{\alpha-2}u(x)dx$$
$$+\int_{\Sigma_0} r_\varepsilon^{\alpha-2}u(x)h(x)ds + \int_{\partial G} r_\varepsilon^{\alpha-2}u(x)g(x)ds.$$

Next we estimate the integral over Γ_d. Because on Γ_d:

$$r_\varepsilon \ge hr \ge hd \quad \Rightarrow \quad (\alpha-3)\ln r_\varepsilon \le (\alpha-3)\ln(hd),$$

since $\alpha \le 2$, we have $r_\varepsilon^{\alpha-3}|_{\Gamma_d} \le (hd)^{\alpha-3}$ and therefore:

$$\frac{2-\alpha}{2}\int_{\Gamma_d} au^2 r_\varepsilon^{\alpha-4}(x_i - \varepsilon l_i)n_i ds \le \frac{2-\alpha}{2}\int_{\Gamma_d} a r_\varepsilon^{\alpha-3}u^2 ds \le \frac{2-\alpha}{2}(hd)^{\alpha-3}\int_{\Gamma_d} au^2 ds$$
$$\le c_\delta\int_{G_d} u^2 dx + \delta\int_{G_d}|\nabla u|^2 dx,\ \forall\delta > 0 \quad (3.3.10)$$

by (1.5.12). By the Cauchy inequality and $\gamma(\omega) \geq \nu_0 > 0$,

$$ug \leq \left(r^{\frac{1}{2}} \frac{1}{\sqrt{\gamma(\omega)}} |g|\right) \left(r^{-\frac{1}{2}} \sqrt{\gamma(\omega)} |u|\right) \leq \frac{\delta}{2} r^{-1} \gamma(\omega) u^2 + \frac{1}{2\delta\nu_0} r g^2(x), \ \forall \delta > 0;$$

taking into account property 1) of r_ε we obtain

$$\int\limits_{\partial G} r_\varepsilon^{\alpha-2} |u||g| ds \leq \frac{\delta}{2} \int\limits_{\partial G} r_\varepsilon^{\alpha-2} \frac{1}{r} \gamma(\omega) u^2 ds + \frac{1}{2\delta\nu_0} \int\limits_{\partial G} r^{\alpha-1} g^2(x) ds, \ \forall \delta > 0. \quad (3.3.11)$$

Similarly, because of $\sigma(\omega) \geq \nu_0 > 0$,

$$\int\limits_{\Sigma_0} r_\varepsilon^{\alpha-2} |u||h(x)| ds \leq \frac{\delta}{2} \int\limits_{\Sigma_0} r_\varepsilon^{\alpha-2} \frac{1}{r} \sigma(\omega) u^2 ds + \frac{1}{2\delta\nu_0} \int\limits_{\Sigma_0} r^{\alpha-1} h^2(x) ds, \ \forall \delta > 0 \quad (3.3.12)$$

and

$$\int\limits_G r_\varepsilon^{\alpha-2} u f(x) dx \leq \frac{\delta}{2} \int\limits_G a r^{-2} r_\varepsilon^{\alpha-2} u^2 dx + \frac{1}{2a_*\delta} \int\limits_G r^\alpha f^2(x) dx, \ \forall \delta > 0. \quad (3.3.13)$$

Further, we use the representation $G = G_0^d \cup G_d$. We estimate integrals over G_0^d. By assumption (b) and the Cauchy inequality, we obtain

$$\begin{aligned} &\int\limits_{G_0^d} \Big\{ \left(a^{ij}(x) - a^{ij}(0)\right) \left(r_\varepsilon^{\alpha-2} u_{x_i} u_{x_j} + r_\varepsilon^{\alpha-4} (x_i - \varepsilon l_i) u(x) u_{x_j}\right) \\ &\qquad + r_\varepsilon^{\alpha-2} a^i(x) u_{x_i} u(x) + r_\varepsilon^{\alpha-2} a(x) u^2(x) \Big\} dx \\ &\leq \mathcal{A}(d) \int\limits_{G_0^d} a \Big(r_\varepsilon^{\alpha-2} |\nabla u|^2 + r_\varepsilon^{\alpha-3} |\nabla u| \cdot |u(x)| + r^{-1} r_\varepsilon^{\alpha-2} |\nabla u| \cdot |u(x)| \\ &\qquad + r^{-2} r_\varepsilon^{\alpha-2} u^2(x) \Big) dx \\ &\leq 2\mathcal{A}(d) \int\limits_{G_0^d} a \left(r_\varepsilon^{\alpha-2} |\nabla u|^2 + r^{-2} r_\varepsilon^{\alpha-2} u^2 + r_\varepsilon^{\alpha-4} u^2\right) dx. \end{aligned} \quad (3.3.14)$$

Next, we estimate integrals over G_d. By assumptions (a), (c) and the Cauchy inequality and taking into account the inequality (3.3.3), we get

$$\begin{aligned} &\int\limits_{G_d} \Big\{ \left(a^{ij}(x) - a^{ij}(0)\right) \left(r_\varepsilon^{\alpha-2} u_{x_i} u_{x_j} + r_\varepsilon^{\alpha-4} (x_i - \varepsilon l_i) u(x) u_{x_j}\right) \\ &\qquad + r_\varepsilon^{\alpha-2} a^i(x) u_{x_i} u(x) + r_\varepsilon^{\alpha-2} a(x) u^2(x) \Big\} dx \end{aligned}$$

$$\leq \mu^* \int\limits_{G_d} \left(3r_\varepsilon^{\alpha-2}|\nabla u|^2 + r_\varepsilon^{\alpha-4}|u|^2\right) dx + \int\limits_{G_d} r_\varepsilon^{\alpha-2}|u||\nabla u|\sqrt{\sum_{i=1}^n |a^i(x)|^2 dx}$$

$$\leq C\left(p, n, \nu_*, \mu^*, \alpha, d, \left\|\sum_{i=1}^n |a^i(x)|^2\right\|_{\mathbf{L}_{p/2}(G)}\right) \cdot \int\limits_{G_d} \left(\nu|\nabla u|^2 + u^2\right) dx. \quad (3.3.15)$$

Thus, from (3.3.9)–(3.3.15) we derive:

$$\int\limits_G a r_\varepsilon^{\alpha-2}|\nabla u|^2 dx + \varepsilon\frac{2-\alpha}{2}\sin\frac{\omega_0}{2}\int\limits_{\Gamma_0^d} a u^2 r_\varepsilon^{\alpha-4} ds + \int\limits_{\Sigma_0} \frac{1}{r} r_\varepsilon^{\alpha-2}\sigma(\omega)u^2(x) ds$$

$$+ \int\limits_{\partial G} \frac{1}{r} r_\varepsilon^{\alpha-2}\gamma(\omega)u^2(x) ds$$

$$\leq 2\mathcal{A}(d)\int\limits_{G_0^d} a\left(r_\varepsilon^{\alpha-2}|\nabla u|^2 + r^{-2}r_\varepsilon^{\alpha-2}u^2 + r_\varepsilon^{\alpha-4}u^2\right) dx \quad (3.3.16)$$

$$+ C\left(p, n, \nu_*, \mu^*, \alpha, d, \left\|\sum_{i=1}^n |a^i(x)|^2\right\|_{\mathbf{L}_{p/2}(G)}\right) \cdot \int\limits_{G_d} \left(\nu|\nabla u|^2 + u^2\right) dx$$

$$+ \frac{1}{2\delta\nu_0}\int\limits_{\partial G} r^{\alpha-1}g^2(x) ds + \frac{1}{2\delta\nu_0}\int\limits_{\Sigma_0} r^{\alpha-1}h^2(x) ds + \frac{\delta}{2}\int\limits_G a r^{-2}r_\varepsilon^{\alpha-2}u^2 dx$$

$$+ \frac{1}{2a_*\delta}\int\limits_G r^\alpha f^2(x) dx + \frac{\delta}{2}\int\limits_{\Sigma_0} r_\varepsilon^{\alpha-2}\frac{1}{r}\sigma(\omega)u^2 ds + \frac{\delta}{2}\int\limits_{\partial G} r_\varepsilon^{\alpha-2}\frac{1}{r}\gamma(\omega)u^2 ds, \ \forall\delta > 0.$$

By the inequality $r_\varepsilon \geq hr$ (see §1.3), we have $r_\varepsilon^{\alpha-4} \leq h^{-2}r^{-2}r_\varepsilon^{\alpha-2}$. Hence, by Lemma 2.5, from (3.3.16) it follows that

$$\int\limits_G a r_\varepsilon^{\alpha-2}|\nabla u|^2 dx + \int\limits_{\Sigma_0} \frac{1}{r} r_\varepsilon^{\alpha-2}\sigma(\omega)u^2(x) ds + \int\limits_{\partial G} \frac{1}{r} r_\varepsilon^{\alpha-2}\gamma(\omega)u^2(x) ds$$

$$\leq c(\lambda, \omega_0)\left(\delta + \mathcal{A}(d)\right)$$

$$\times \left\{\int\limits_G a r_\varepsilon^{\alpha-2}|\nabla u|^2 dx + \int\limits_{\Sigma_0} r^{-1}r_\varepsilon^{\alpha-2}\sigma(\omega)u^2(x) ds + \int\limits_{\partial G} r^{-1}r_\varepsilon^{\alpha-2}\gamma(\omega)u^2 ds\right\}$$

$$+ C\left(p, n, \nu_*, \mu^*, \alpha, d, \left\|\sum_{i=1}^n |a^i(x)|^2\right\|_{\mathbf{L}_{p/2}(G)}\right) \cdot \int\limits_G \left(\nu|\nabla u|^2 + u^2\right) dx \quad (3.3.17)$$

$$+ \frac{1}{2\delta\nu_0}\int\limits_{\partial G} r^{\alpha-1}g^2(x) ds + \frac{1}{2\delta\nu_0}\int\limits_{\Sigma_0} r^{\alpha-1}h^2(x) ds + \frac{1}{2a_*\delta}\int\limits_G r^\alpha f^2(x) dx, \ \forall\delta > 0.$$

Choosing $\delta = \frac{1}{4c(\lambda,\omega_0)}$ and next $d > 0$ such that, by the continuity of $\mathcal{A}(r)$ at zero, $c(\lambda, \omega_0)\mathcal{A}(d) \le \frac{1}{4}$ we derive

$$\int\limits_G a r_\varepsilon^{\alpha-2}|\nabla u|^2 dx + \int\limits_{\Sigma_0} \frac{1}{r} r_\varepsilon^{\alpha-2}\sigma(\omega)u^2(x)ds + \int\limits_{\partial G} \frac{1}{r} r_\varepsilon^{\alpha-2}\gamma(\omega)u^2(x)ds$$
$$\le C\left(p, n, \nu_*, \mu^*, a_*, \alpha, \lambda, \left\|\sum_{i=1}^n |a^i(x)|^2\right\|_{\mathbf{L}_{p/2}(G)}\right)\left\{\int\limits_G \left(\nu|\nabla u|^2 + u^2 + r^\alpha f^2(x)\right) dx\right.$$
$$\left. + \frac{1}{\nu_0}\int\limits_{\partial G} r^{\alpha-1} g^2(x)ds + \frac{1}{\nu_0}\int\limits_{\Sigma_0} r^{\alpha-1}h^2(x)ds\right\}, \ \forall \varepsilon > 0. \tag{3.3.18}$$

Now, we observe that the right-hand side of (3.3.18) does not depend on ε. Therefore we can perform the passage to the limit as $\varepsilon \to +0$ by the Fatou Theorem. Then we get

$$\int\limits_G a r^{\alpha-2}|\nabla u|^2 dx + \int\limits_{\Sigma_0} r^{\alpha-3}\sigma(\omega)u^2(x)ds + \int\limits_{\partial G} r^{\alpha-3}\gamma(\omega)u^2(x)ds$$
$$\le C\left(p, n, \nu_*, \mu^*, a_*, \alpha, \lambda, \left\|\sum_{i=1}^n |a^i(x)|^2\right\|_{\mathbf{L}_{p/2}(G)}\right)\left\{\int\limits_G \left(\nu|\nabla u|^2 + u^2 + r^\alpha f^2(x)\right) dx\right.$$
$$\left. + \frac{1}{\nu_0}\int\limits_{\partial G} r^{\alpha-1} g^2(x)ds + \frac{1}{\nu_0}\int\limits_{\Sigma_0} r^{\alpha-1}h^2(x)ds\right\}. \tag{3.3.19}$$

Applying Theorem 3.5 and Corollary 2.4 (see inequality (2.2.3)), from (3.3.19) follows the required estimate (3.3.5). □

3.4 Local integral weighted estimates

Theorem 3.7. *Let $u(x)$ be a weak solution of problem (L) and assumptions (a)–(d) be satisfied with $\mathcal{A}(r)$ which is Dini-continuous at zero. Let λ be as above in (2.2.1). Then $u(x) \in \overset{\circ}{\mathbf{W}}{}^1_{2-n}(G)$ and there exist $d \in (0, \frac{1}{e})$ and a constant $C > 0$ depending only on $n, s, \lambda, a_*, \omega_0$ and on $\int\limits_0^1 \frac{\mathcal{A}(r)}{r}dr$ such that the inequality*

$$\int\limits_{G_0^\varrho} a\left(r^{2-n}|\nabla u|^2 + r^{-n}u^2(x)\right) dx + \int\limits_{\Sigma_0^\varrho} r^{1-n}\sigma(\omega)u^2(x)ds + \int\limits_{\Gamma_0^\varrho} r^{1-n}\gamma(\omega)u^2(x)ds$$

$$\leq C\Big(\|u\|^2_{2,G} + f_1^2 + \frac{1}{\nu_0}g_1^2 + \frac{1}{\nu_0}h_1^2\Big) \cdot \begin{cases} \varrho^{2\lambda}, & \textit{if } s > \lambda, \\ \varrho^{2\lambda}\ln^2\big(\frac{1}{\varrho}\big), & \textit{if } s = \lambda, \\ \varrho^{2s}, & \textit{if } s < \lambda \end{cases} \quad (3.4.1)$$

holds for almost all $\varrho \in (0, d)$.

Proof. By Theorem 3.6 $u(x) \in \overset{\circ}{\mathbf{W}}{}^1_{2-n}(G)$, so it is enough to prove the estimate (3.4.1). Putting $\eta(x) = r^{2-n}u(x)$ in $(II)_{loc}$, according to the definition (2.4.10) we have

$$U(\varrho) = \varrho\int_{\Omega} au(x)\frac{\partial u}{\partial r}\Big|_{r=\varrho} d\Omega + \int_{\Omega_\varrho} r^{2-n}u(x)\left(a^{ij}(x) - a^{ij}(0)\right)u_{x_j}\cos(r, x_i)d\Omega_\varrho$$

$$+ \int_{\Gamma_0^\varrho} r^{2-n}u(x)g(x)ds + \int_{\Sigma_0^\varrho} r^{2-n}u(x)h(x)ds + \int_{G_0^\varrho}\Big\{-r^{2-n}\left(a^{ij}(x) - a^{ij}(0)\right)u_{x_i}u_{x_j}$$

$$+ (n-2)r^{-n}u(x)a^{ij}(x)x_iu_{x_j} + r^{2-n}u(x)a^i(x)u_{x_i} + r^{2-n}a(x)u^2(x)$$

$$- r^{2-n}u(x)f(x)\Big\}dx. \quad (3.4.2)$$

Now, we transform some integrals from the right-hand side:

$$(n-2)\int_{G_0^\varrho} r^{-n}u(x)a^{ij}(x)x_iu_{x_j}dx$$

$$= (n-2)\int_{G_0^\varrho} r^{-n}x_i\left\langle \frac{a}{2}\cdot\frac{\partial u^2}{\partial x_i} + u(x)\left(a^{ij}(x) - a^{ij}(0)\right)u_{x_j}\right\rangle dx; \quad (3.4.3)$$

by the Gauss-Ostrogradskiy divergence theorem,

$$\int_{G_0^\varrho} ar^{-n}x_i\frac{\partial u^2}{\partial x_i}dx = -\int_{G_0^\varrho} au^2(x)\left(nr^{-n} - nr^{-n}\right)dx + \varrho^{-n}\int_{\Omega_\varrho} au^2(x)x_in_id\Omega_\varrho$$

$$+ [a]_{\Sigma_0}\int_{\Sigma_0^\varrho} r^{-n}u^2(x)x_in_ids + \int_{\Gamma_0^\varrho} ar^{-n}u^2(x)x_in_ids. \quad (3.4.4)$$

Hence, since by Lemma 2.1

$$n_i\Big|_{\Gamma_0^\varrho} = 0 \quad \text{and} \quad n_i\Big|_{\Omega_\varrho} = \varrho; \quad x_in_i\Big|_{\Sigma_0} = x_i\cos(x_n, x_i)\Big|_{\Sigma_0} = x_n\Big|_{\Sigma_0} = 0,$$

we have

$$\frac{n-2}{2}\int_{G_0^\varrho} ar^{-n}x_i\frac{\partial u^2}{\partial x_i}dx = \frac{n-2}{2}\int_{\Omega} au^2(x)d\Omega. \quad (3.4.5)$$

By $a(x) \le 0$ and Lemma 2.12, from (3.4.2)–(3.4.5) we derive

$$U(\varrho) \le \frac{\varrho}{2\lambda}U'(\varrho) + \varrho^{2-n}\int\limits_{\Omega_\varrho} u(x)\left(a^{ij}(x) - a^{ij}(0)\right)u_{x_j}\cos(r,x_i)d\Omega_\varrho$$

$$+\int\limits_{\Gamma_0^\varrho} r^{2-n}u(x)g(x)ds + \int\limits_{\Sigma_0^\varrho} r^{2-n}u(x)h(x)ds + \int\limits_{G_0^\varrho}\Big\{-r^{2-n}\left(a^{ij}(x) - a^{ij}(0)\right)u_{x_i}u_{x_j}$$

$$+(n-2)r^{-n}u(x)\left(a^{ij}(x) - a^{ij}(0)\right)x_iu_{x_j} + r^{2-n}u(x)a^i(x)u_{x_i} - r^{2-n}u(x)f(x)\Big\}dx.$$

Hence, by virtue of assumption (b), it follows that

$$U(\varrho) \le \frac{\varrho}{2\lambda}U'(\varrho) + \varrho\mathcal{A}(\varrho)\int\limits_{\Omega} a|u||\nabla u|d\Omega + \int\limits_{\Gamma_0^\varrho} r^{2-n}|u(x)||g(x)|ds$$

$$+\int\limits_{\Sigma_0^\varrho} r^{2-n}|u(x)||h(x)|ds + c_1(n)\mathcal{A}(\varrho)\int\limits_{G_0^\varrho} a\left(r^{2-n}|\nabla u|^2 + r^{1-n}|u||\nabla u|\right)dx$$

$$+\int\limits_{G_0^\varrho} r^{2-n}|u(x)||f(x)|dx. \tag{3.4.6}$$

Further, we shall derive an upper bound for each integral from the right-hand side. Applying the Cauchy and Friedrichs-Wirtinger inequalities (see $(W)_2$) according to (2.4.12), we obtain

$$\int\limits_{\Omega} a\varrho|u||\nabla u|d\Omega \le \frac{1}{2}\int\limits_{\Omega} a\left(\varrho^2|\nabla u|^2 + |u|^2\right)d\Omega \le c_2(\lambda)\varrho U'(\varrho); \tag{3.4.7}$$

$$\int\limits_{G_0^\varrho} ar^{1-n}|u||\nabla u|dx \le \int\limits_{G_0^\varrho} a\left(r^{2-n}|\nabla u|^2 + r^{-n}|u|^2\right)dx \le c_3(\lambda)U(\varrho) \tag{3.4.8}$$

by virtue of inequality (2.2.3); and for all $\delta > 0$,

$$\int\limits_{\Gamma_0^\varrho} r^{2-n}|u||g|ds = \int\limits_{\Gamma_0^\varrho}\left(r^{\frac{1-n}{2}}\sqrt{\gamma(\omega)}|u|\right)\left(r^{\frac{3-n}{2}}\frac{1}{\sqrt{\gamma(\omega)}}|g|\right)ds$$

$$\le \frac{\delta}{2}\int\limits_{\Gamma_0^\varrho} r^{1-n}\gamma(\omega)|u|^2ds + \frac{1}{2\delta\nu_0}\int\limits_{\Gamma_0^\varrho} r^{3-n}|g|^2ds; \tag{3.4.9}$$

$$\int\limits_{\Sigma_0^\varrho} r^{2-n}|u||h|ds = \int\limits_{\Sigma_0^\varrho}\left(r^{\frac{1-n}{2}}\sqrt{\sigma(\omega)}|u|\right)\left(r^{\frac{3-n}{2}}\frac{1}{\sqrt{\sigma(\omega)}}|g|\right)ds$$

$$\leq \frac{\delta}{2}\int\limits_{\Sigma_0^\varrho} r^{1-n}\sigma(\omega)|u|^2 ds + \frac{1}{2\delta\nu_0}\int\limits_{\Sigma_0^\varrho} r^{3-n}|h|^2 ds; \qquad (3.4.10)$$

$$\int\limits_{G_0^\varrho} r^{2-n}|u(x)||f(x)|dx \leq \frac{\delta}{2a_*}\int\limits_{G_0^\varrho} ar^{-n}|u|^2 dx + \frac{1}{2\delta}\int\limits_{G_0^\varrho} r^{4-n}|f|^2 dx$$

$$\leq \frac{\delta}{2a_*}c_4(\lambda)U(\varrho) + \frac{1}{2\delta}\int\limits_{G_0^\varrho} r^{4-n}|f|^2 dx \qquad (3.4.11)$$

by virtue of inequality (2.2.3). Thus, from (3.4.6)–(3.4.11) we get

$$[1 - c_5(n,\lambda,a_*)(\delta + \mathcal{A}(\varrho))]\, U(\varrho) \leq \frac{\varrho}{2\lambda}\left(1 + c_6(\lambda)\mathcal{A}(\varrho)\right) U'(\varrho) +$$

$$+ \frac{1}{2\delta}\left\{\int\limits_{G_0^\varrho} r^{4-n}|f|^2 dx + \frac{1}{\nu_0}\int\limits_{\Gamma_0^\varrho} r^{3-n}|g|^2 ds + \frac{1}{\nu_0}\int\limits_{\Sigma_0^\varrho} r^{3-n}|h|^2 ds\right\}, \ \forall\delta > 0. \quad (3.4.12)$$

However, by the condition (d),

$$\int\limits_{G_0^\varrho} r^{4-n}|f|^2 dx + \frac{1}{\nu_0}\int\limits_{\Gamma_0^\varrho} r^{3-n}|g|^2 ds + \frac{1}{\nu_0}\int\limits_{\Sigma_0^\varrho} r^{3-n}|h|^2 ds$$

$$\leq \frac{1}{2s}c_0\left(f_1^2 + \frac{1}{\nu_0}g_1^2 + \frac{1}{\nu_0}h_1^2\right)\cdot \varrho^{2s},$$

where c_0 depends only on $meas\,\Omega$, $meas\,\partial\Omega$, $meas\,\sigma_0$. Hence we derive the differential inequality (CP) from §2.4 with

$$\mathcal{P}(\varrho) = \frac{2\lambda}{\varrho}\cdot[1 - c_5(n,\lambda,a_*)(\delta + \mathcal{A}(\varrho))], \ \forall\delta > 0; \quad \mathcal{N}(\varrho) \equiv 0;$$

$$\mathcal{Q}(\varrho) = \frac{\lambda}{s}c_0\left(f_1^2 + \frac{1}{\nu_0}g_1^2 + \frac{1}{\nu_0}h_1^2\right)\cdot\delta^{-1}\varrho^{2s-1}, \ \forall\delta > 0; \qquad (3.4.13)$$

$$U_0 = C\Big\{\int\limits_G \left(u^2 + (1 + r^{4-n})f^2(x)\right) dx + \int\limits_{\Sigma_0} r^{3-n}h^2(x)ds + \int\limits_{\partial G} r^{3-n}g^2(x)ds\Big\},$$

due to (3.3.5) with $\alpha = 4 - n$.

1) $s > \lambda$. In this case we put $\delta = \varrho^\varepsilon$, for all $\varepsilon > 0$. Then

$$\mathcal{P}(\varrho) = \frac{2\lambda}{\varrho}\cdot[1 - c_5(n,\lambda,a_*)(\varrho^\varepsilon + \mathcal{A}(\varrho))];$$

$$\mathcal{Q}(\varrho) = \frac{\lambda}{s}c_0\left(f_1^2 + \frac{1}{\nu_0}g_1^2 + \frac{1}{\nu_0}h_1^2\right)\cdot\varrho^{2s-1-\varepsilon}.$$

Since $\mathcal{P}(\varrho) = \frac{2\lambda}{\varrho} - \frac{\mathcal{K}(\varrho)}{\varrho}$, where $\mathcal{K}(\varrho)$ satisfies the Dini condition at zero we obtain

$$-\int\limits_{\varrho}^{\tau} \mathcal{P}(s)ds = -2\lambda \ln\left(\frac{\tau}{\varrho}\right) + \int\limits_{\varrho}^{\tau} \frac{\mathcal{K}(s)}{s} ds \le \ln\left(\frac{\varrho}{\tau}\right)^{2\lambda} + \int\limits_{0}^{d} \frac{\mathcal{K}(r)}{r} dr \implies$$

$$\exp\left(-\int\limits_{\varrho}^{d} \mathcal{P}(\tau)d\tau\right) \le \left(\frac{\varrho}{d}\right)^{2\lambda} \exp\left(\int\limits_{0}^{d} \frac{\mathcal{K}(\tau)}{\tau} d\tau\right) = K_0 \left(\frac{\varrho}{d}\right)^{2\lambda};$$

$$\exp\left(-\int\limits_{\varrho}^{\tau} \mathcal{P}(\tau)d\tau\right) \le \left(\frac{\varrho}{\tau}\right)^{2\lambda} \exp\left(\int\limits_{0}^{d} \frac{\mathcal{K}(\tau)}{\tau} d\tau\right) = K_0 \left(\frac{\varrho}{\tau}\right)^{2\lambda}.$$

We have also:

$$\int\limits_{\varrho}^{d} \mathcal{Q}(\tau) \exp\Big(-\int\limits_{\varrho}^{\tau} \mathcal{P}(\sigma)d\sigma\Big)d\tau \le \frac{\lambda c_0 K_0}{s} \left(f_1^2 + \frac{1}{\nu_0} g_1^2 + \frac{1}{\nu_0} h_1^2\right) \varrho^{2\lambda} \int\limits_{\varrho}^{d} \tau^{2s-2\lambda-\varepsilon-1} d\tau$$
$$\le \frac{\lambda c_0 K_0}{s} \left(f_1^2 + \frac{1}{\nu_0} g_1^2 + \frac{1}{\nu_0} h_1^2\right) \cdot \frac{d^{s-\lambda}}{s-\lambda} \varrho^{2\lambda},$$

since $s > \lambda$ and we can choose $\varepsilon = s - \lambda$.

Now we apply Theorem 1.21: then from (1.7.1), by virtue of the inequalities deduced above and according to (2.2.3) with $\alpha = 4 - n$, we obtain the statement (3.4.1) for $s > \lambda$.

2) $s = \lambda$. In this case we can take in (3.4.13) any function $\delta(\varrho) > 0$ instead of $\delta > 0$. Then we obtain the problem (CP) with

$$\mathcal{P}(\varrho) = \frac{2\lambda(1-\delta(\varrho))}{\varrho} - c_5 \frac{\mathcal{A}(\varrho)}{\varrho}; \quad \mathcal{N}(\varrho) = 0;$$
$$\mathcal{Q}(\varrho) = c_0 \left(f_1^2 + \frac{1}{\nu_0} g_1^2 + \frac{1}{\nu_0} h_1^2\right) \cdot \delta^{-1}(\varrho)\varrho^{2\lambda-1}.$$

Now, we choose $\delta(\varrho) = \frac{1}{2\lambda \ln\left(\frac{ed}{\varrho}\right)}$, $0 < \varrho < d$, where e is the Euler number. Then we derive

$$-\int\limits_{\varrho}^{\tau} \mathcal{P}(\sigma)d\sigma \le \ln\Big(\frac{\varrho}{\tau}\Big)^{2\lambda} + \int\limits_{\varrho}^{\tau} \frac{d\sigma}{\sigma \ln\Big(\frac{ed}{\sigma}\Big)} + c_5 \int\limits_{0}^{d} \frac{\mathcal{A}(\tau)}{\tau} d\tau$$
$$= \ln\Big(\frac{\varrho}{\tau}\Big)^{2\lambda} + \ln\left(\frac{\ln\Big(\frac{ed}{\varrho}\Big)}{\ln\Big(\frac{ed}{\tau}\Big)}\right) + c_5 \int\limits_{0}^{d} \frac{\mathcal{A}(\tau)}{\tau} d\tau \implies$$

$$\exp\Big(-\int_\varrho^\tau \mathcal{P}(\sigma)d\sigma\Big) \le \Big(\frac{\varrho}{\tau}\Big)^{2\lambda} \cdot \frac{\ln\big(\frac{ed}{\varrho}\big)}{\ln\big(\frac{ed}{\tau}\big)} \exp\Big(c_5 \int_0^d \frac{\mathcal{A}(\tau)}{\tau} d\tau\Big),$$

$$\exp\Big(-\int_\varrho^d \mathcal{P}(\tau)d\tau\Big) \le \Big(\frac{\varrho}{d}\Big)^{2\lambda} \ln\Big(\frac{ed}{\varrho}\Big) \exp\Big(c_5 \int_0^d \frac{\mathcal{A}(\tau)}{\tau} d\tau\Big)$$

and therefore

$$\int_\varrho^d \mathcal{Q}(\tau) \exp\Big(-\int_\varrho^\tau \mathcal{P}(\sigma)d\sigma\Big) d\tau$$

$$\le c_6 \left(f_1^2 + \frac{1}{\nu_0} g_1^2 + \frac{1}{\nu_0} h_1^2\right) \varrho^{2\lambda} \ln\Big(\frac{ed}{\varrho}\Big) \cdot \int_\varrho^d \frac{d\tau}{\tau\delta(\tau)\ln\big(\frac{ed}{\tau}\big)}$$

$$\le 2\lambda c_6 \left(f_1^2 + \frac{1}{\nu_0} g_1^2 + \frac{1}{\nu_0} h_1^2\right) \cdot \varrho^{2\lambda} \ln^2\Big(\frac{ed}{\varrho}\Big).$$

Now we apply Theorem 1.21, and from (1.7.1), by virtue of the inequalities deduced above, we establish

$$U(\varrho) \le \widetilde{c}_6 (U_0 + f_1^2 + \frac{1}{\gamma_0} g_1^2 + \frac{1}{\sigma_0} h_1^2) \varrho^{2\lambda} \ln^2 \frac{1}{\varrho}, \quad 0 < \varrho < d < \frac{1}{e}.$$

Thus, the statement (3.4.1) for $s = \lambda$ is proved.

3) $0 < s < \lambda$. Analogously to case 1) according to (3.4.13) we derive

$$\exp\Big(-\int_\varrho^d \mathcal{P}(\tau)d\tau\Big) \le \Big(\frac{\varrho}{d}\Big)^{2\lambda(1-\delta)} \exp\Big(c_5 \int_0^d \frac{\mathcal{A}(\tau)}{\tau} d\tau\Big) = c_7 \Big(\frac{\varrho}{d}\Big)^{2\lambda(1-\delta)}.$$

Therefore, if we choose $\delta \in (0, \frac{\lambda - s}{\lambda})$, then

$$\int_\varrho^d \mathcal{Q}(\tau) \exp\Big(-\int_\varrho^\tau \mathcal{P}(\sigma)d\sigma\Big) d\tau$$

$$\le c_0 \left(f_1^2 + \frac{1}{\nu_0} g_1^2 + \frac{1}{\nu_0} h_1^2\right) \cdot \delta^{-1} \varrho^{2\lambda(1-\delta)} \int_\varrho^d \tau^{2s - 2\lambda(1-\delta) - 1} d\tau$$

$$\le c_8 \left(f_1^2 + \frac{1}{\nu_0} g_1^2 + \frac{1}{\nu_0} h_1^2\right) \cdot \varrho^{2s},$$

Now we apply Theorem 1.21, and then from (1.7.1), by virtue of the inequalities deduced above, we get

$$\begin{aligned} U(\varrho) &\le c_9\left(U_0\varrho^{2\lambda(1-\delta)} + \left(f_1^2 + \frac{1}{\nu_0}g_1^2 + \frac{1}{\nu_0}h_1^2\right)\cdot \varrho^{2s}\right) \\ &\le c_{10}(U_0 + f_1^2 + \frac{1}{\nu_0}g_1^2 + \frac{1}{\nu_0}h_1^2)\varrho^{2s}, \end{aligned}$$

since $\delta \in (0, \frac{\lambda - s}{\lambda})$. Thus, the statement (3.4.1) for $s < \lambda$ is proved. □

3.5 The power modulus of continuity at the conical point for weak solutions

3.5.1 Proof of Theorem 3.3

Let us define the function

$$\psi(\varrho) = \begin{cases} \varrho^\lambda, & \text{if } s > \lambda, \\ \varrho^\lambda \ln\left(\frac{1}{\varrho}\right), & \text{if } s = \lambda, \\ \varrho^s, & \text{if } s < \lambda \end{cases} \tag{3.5.1}$$

for $0 < \varrho < d$.

By Theorem 3.4 for the local bound of the weak solution modulus,

$$\sup_{G_0^{\varrho/2}} |u(x)| \le C\left\{\varrho^{-n/2}\|u\|_{2,G_0^\varrho} + \varrho^{2(1-n/p)}\|f\|_{p/2,G_0^\varrho} + \varrho\left(\|g\|_{\infty,\Gamma_0^\varrho} + \|h\|_{\infty,\Sigma_0^\varrho}\right)\right\} \tag{3.5.2}$$

where $C = C\left(n, \nu_*, \mu^*, p, \left\|\sum_{i=1}^n |a^i(x)|^2\right\|_{\mathbf{L}_{p/2}(G)}\right)$ and $p > n$. Next, by Theorem 3.7,

$$\begin{aligned} \varrho^{-n/2}\|u\|_{2,G_0^\varrho} &\le 2^{n/2}\left(\int_{G_0^\varrho} r^{-n}u^2(x)dx\right)^{1/2} \\ &\le C\left(\|u\|_{2,G} + f_1 + \frac{1}{\sqrt{\nu_0}}g_1 + \frac{1}{\sqrt{\nu_0}}h_1\right)\psi(\varrho). \end{aligned} \tag{3.5.3}$$

Further, by the assumption (d),

$$\varrho^{2(1-n/p)}\|f\|_{p/2,G_0^\varrho} + \varrho\left(\|g\|_{\infty,\Gamma_0^\varrho} + \|h\|_{\infty,\Sigma_0^\varrho}\right) \le c\left(f_1 + \frac{1}{\sqrt{\nu_0}}g_1 + \frac{1}{\sqrt{\nu_0}}h_1\right)\psi(\varrho). \tag{3.5.4}$$

Now, from (3.5.2)–(3.5.4) it follows that

$$\sup_{G^{\varrho/2}_{\varrho/4}} |u(x)| \le C\Big(\|u\|_{2,G} + f_1 + \frac{1}{\sqrt{\nu_0}} g_1 + \frac{1}{\sqrt{\nu_0}} h_1\Big)\psi(\varrho).$$

Putting $|x| = \frac{1}{3}\varrho$ we get finally the required estimate (3.1.2).

Now we consider two sets $G^{2\varrho}_{\varrho/4}$ and $G^{\varrho}_{\varrho/2} \subset G^{2\varrho}_{\varrho/4}$, $\varrho > 0$. We perform the change of variables $x = \varrho x'$ and $u(\varrho x') = \psi(\varrho)v(x')$. Then the function $v(x')$ satisfies the problem

$$\begin{cases} \frac{\partial}{\partial x'_i}\left(a^{ij}(\varrho x')v_{x'_j}\right) + \varrho a^i(\varrho x')v_{x'_i} + \varrho^2 a(\varrho x')v = \frac{\varrho^2}{\psi(\varrho)} f(\varrho x'), \ x \in G^2_{1/4}, \\ [v(x')]_{\Sigma^2_{1/4}} = 0, \ \left[\frac{\partial v}{\partial \nu'}\right]_{\Sigma^2_{1/4}} + \frac{1}{|x'|}\sigma(\omega)v(x') = \frac{\varrho}{\psi(\varrho)} h(\varrho x'), \ x \in \Sigma^2_{1/4}, \\ \frac{\partial v}{\partial \nu'} + \frac{1}{|x'|}\gamma(\omega)v(x') = \frac{\varrho}{\psi(\varrho)} g(\rho x'), \ x \in \Gamma^2_{1/4}. \end{cases} \qquad (L'')$$

By the Sobolev Embedding Theorem 1.19,

$$\sup_{x' \in G^1_{1/2}} |\nabla' v(x')| \le c\|v\|_{\mathbf{W}^{2,p}(G^1_{1/2})}, \quad p > n. \qquad (3.5.5)$$

On the strength of the local L^p *a priori* estimate [66, 67] for the solution of the equation of the (L'') inside the domains $\left(G^2_{1/4}\right)_{\pm}$ and near smooth portions of the boundaries $\Sigma^2_{1/4} \cup \Gamma^2_{1/4}$ we have

$$\begin{aligned} &\|v\|_{\mathbf{W}^{2,p}(G^1_{1/2})} \qquad (3.5.6)\\ &\le c\frac{\varrho}{\psi(\varrho)}\left\{\varrho\|f\|_{\mathbf{L}^p(G^2_{1/4})} + \|h\|_{W^{1-1/p,p}(\Sigma^2_{1/4})} + \|g\|_{\mathbf{W}^{1-1/p,p}(\Gamma^2_{1/4})}\right\} + c\|v\|_{\mathbf{L}^p(G^2_{1/4})}. \end{aligned}$$

Returning back to the variables x, from (3.5.5) and (3.5.6) it follows that

$$\begin{aligned} &\sup_{G^{\varrho}_{\varrho/2}} |\nabla u| \\ &\le c\varrho^{1-n/p}\Big\{\varrho^{-2}\|u\|_{\mathbf{L}^p(G^{2\varrho}_{\varrho/4})} + \|f\|_{p,G^{2\varrho}_{\varrho/4}} + \|g\|_{\mathbf{V}^{1-1/p}_{p,0}(\Gamma^{2\varrho}_{\varrho/4})} + \|h\|_{V^{1-1/p}_{p,0}(\Sigma^{2\varrho}_{\varrho/4})}\Big\} \end{aligned}$$

and

$$\begin{aligned} &\varrho^{2-n/p}\|u\|_{\mathbf{V}^2_{p,0}(G^{\varrho}_{\varrho/2})} \\ &\le c\varrho^{2-n/p}\Big\{\varrho^{-2}\|u\|_{\mathbf{L}^p(G^{2\varrho}_{\varrho/4})} + \|f\|_{p,G^{2\varrho}_{\varrho/4}} + \|g\|_{\mathbf{V}^{1-1/p}_{p,0}(\Gamma^{2\varrho}_{\varrho/4})} + \|h\|_{V^{1-1/p}_{p,0}(\Sigma^{2\varrho}_{\varrho/4})}\Big\} \end{aligned}$$

or

$$\sup_{G^{\varrho}_{\varrho/2}} |\nabla u|$$
$$\le c\varrho^{-1}\Big\{|u|_{0,G^{2\varrho}_{\varrho/4}} + \|f\|_{\mathbf{V}^0_{p,2p-n}(G^{2\varrho}_{\varrho/4})} + \|g\|_{\mathbf{V}^{1-1/p}_{p,2p-n}(\Gamma^{2\varrho}_{\varrho/4})} + \|h\|_{V^{1-1/p}_{p,2p-n}(\Sigma^{2\varrho}_{\varrho/4})}\Big\}$$

and

$$\|u\|_{\mathbf{V}^2_{p,2p-n}(G^{\varrho}_{\varrho/2})}$$
$$\le c\Big\{|u|_{0,G^{2\varrho}_{\varrho/4}} + \|f\|_{\mathbf{V}^0_{p,2p-n}(G^{2\varrho}_{\varrho/4})} + \|g\|_{\mathbf{V}^{1-1/p}_{p,2p-n}(\Gamma^{2\varrho}_{\varrho/4})} + \|h\|_{V^{1-1/p}_{p,2p-n}(\Sigma^{2\varrho}_{\varrho/4})}\Big\}.$$

Hence, by virtue of (3.1.2), (3.1.3) and assumption (d), desired results (3.1.4) and (3.1.5) follow. □

3.5.2 Remark to Theorem 3.7

Now we can state that Theorem 3.6 is true for $\alpha \in (4-n-2\lambda, 2]$, if a neighborhood of the conic point is convex. In fact, from estimate (3.1.2) we obtain $u(0) = 0$ and therefore we can apply Lemma 2.10 to the proof of Theorem 3.6 for $\alpha \in (4-n-2\lambda, 4-n)$. In this case the equality (3.3.8) can be rewritten, by virtue of (2.3.8), in the form of the inequality

$$\frac{2-\alpha}{2}\int_G a r_\varepsilon^{\alpha-4}(x_i - \varepsilon l_i)\frac{\partial u^2}{\partial x_i}dx \le \frac{2-\alpha}{2}\int_{\Gamma_d} a u^2 r_\varepsilon^{\alpha-4}(x_i-\varepsilon l_i)\cos(\overrightarrow{n}, x_i)ds$$
$$+\frac{(2-\alpha)(4-n-\alpha)}{2}H(\lambda,n,\alpha)\Big\{\int_{G_0^d} r_\varepsilon^{\alpha-2}|\nabla u|^2dx + \int_{\Sigma_0^d} r_\varepsilon^{\alpha-3}\sigma(\omega)u^2(x)ds$$
$$+\int_{\Gamma_0^d} r_\varepsilon^{\alpha-3}\gamma(\omega)u^2(x)ds\Big\},$$

where $H(\lambda, n, \alpha)$ is determined by (2.3.3). By virtue of the convexity of G_0^d and the first property of r_ε (see §1.3), we have $r_\varepsilon \ge r$. Therefore (3.3.17) takes the form

$$\left(1-\frac{(2-\alpha)(4-n-\alpha)}{2}H(\lambda,n,\alpha)\right)\Big\{\int_G a r_\varepsilon^{\alpha-2}|\nabla u|^2dx + \int_{\Sigma_0}\frac{1}{r}r_\varepsilon^{\alpha-2}\sigma(\omega)u^2(x)ds$$
$$+\int_{\partial G}\frac{1}{r}r_\varepsilon^{\alpha-2}\gamma(\omega)u^2(x)ds\Big\} \le c(\lambda,\omega_0)\left(\delta+\mathcal{A}(d)\right)\Big\{\int_G a r_\varepsilon^{\alpha-2}|\nabla u|^2dx$$
$$+\int_{\Sigma_0} r^{-1}r_\varepsilon^{\alpha-2}\sigma(\omega)u^2(x)ds + \int_{\partial G} r^{-1}r_\varepsilon^{\alpha-2}\gamma(\omega)u^2ds\Big\} \tag{3.5.7}$$

$$+ C\left(p, n, \nu_*, \mu^*, \alpha, d, \left\| \sum_{i=1}^{n} |a^i(x)|^2 \right\|_{\mathbf{L}_{p/2}(G)}\right) \int_G (\nu |\nabla u|^2 + u^2)\, dx$$

$$+ \frac{1}{2\delta\nu_0} \int_{\partial G} r^{\alpha-1} g^2(x) ds + \frac{1}{2\delta\nu_0} \int_{\Sigma_0} r^{\alpha-1} h^2(x) ds + \frac{1}{2a_*\delta} \int_G r^\alpha f^2(x) dx,\ \forall \delta > 0.$$

But, by (2.3.3) and $4-n-2\lambda < \alpha < 4-n$, we verify that $1-\frac{(2-\alpha)(4-n-\alpha)}{2}H(\lambda, n, \alpha) > 0$. Therefore from (3.5.7) we conclude the validity of (3.3.18) and hence we obtain the required statement.

3.6 Appendix

We consider in detail the two-dimensional transmission problem for the Laplace operator in an angular domain and investigate the corresponding eigenvalue problem. Let domain $G \subset \mathbb{R}^2$ lie inside the corner $G_0 = \{(r, \omega)\,|r > 0; -\frac{\omega_0}{2} < \omega < \frac{\omega_0}{2}\}$, $\omega_0 \in]0, 2\pi[$; $\mathcal{O} \in \partial G$ and in some neighborhood of $\mathcal{O}$ boundary ∂G coincide with sides of the corner: $\omega = -\frac{\omega_0}{2}$ and $\omega = \frac{\omega_0}{2}$. Let us write

$$\Gamma_\pm = \{(r, \omega) \mid r > 0;\ \omega = \pm\frac{\omega_0}{2}\}, \quad \Sigma_0 = \{(r, \omega) \mid r > 0;\ \omega = 0\}$$

and put $\sigma(\omega)\Big|_{\Sigma_0} = \sigma(0) = \sigma = const \geq 0$, $\gamma(\omega)\Big|_{\omega=\pm\frac{\omega_0}{2}} = \gamma_\pm = const > 0$. We consider the following problem:

$$\begin{cases} a_\pm \triangle u_\pm = f_\pm(x), & x \in G_\pm; \\ [u]_{\Sigma_0} = 0, \quad [a\frac{\partial u}{\partial \vec{n}}]_{\Sigma_0} + \frac{\sigma}{r} u(x) = h(x), & x \in \Sigma_0; \\ \alpha_\pm a_\pm \frac{\partial u_\pm}{\partial \vec{n}} + \frac{1}{r}\gamma_\pm u_\pm(x) = g_\pm(x), & x \in \Gamma_\pm \setminus \mathcal{O}, \end{cases} \tag{3.6.1}$$

where $\alpha_\pm \in \{0; 1\}$. It is well known that the homogeneous problem ($f(x) = h(x) = g(x) = 0$) has a solution of the form $u(r, \omega) = r^\lambda \psi(\omega)$, where λ^2 is an eigenvalue and $\psi(\omega)$ is an associated regular eigenfunction of the problem

$$\begin{cases} \psi''_+ + \lambda^2 \psi_+(\omega) = 0, & \omega \in \left(0, \frac{\omega_0}{2}\right); \\ \psi''_- + \lambda^2 \psi_-(\omega) = 0, & \omega \in \left(-\frac{\omega_0}{2}, 0\right); \\ \psi_+(0) = \psi_-(0); \quad a_+\psi'_+(0) - a_-\psi'_-(0) = \sigma\psi(0); & \\ \pm\alpha_\pm a_\pm \psi'(\pm\frac{\omega_0}{2}) + \gamma_\pm \psi(\pm\frac{\omega_0}{2}) = 0. & \end{cases} \tag{3.6.2}$$

1) case $\lambda = 0$. In this case we have $\psi_\pm(\omega) = A_\pm \cdot \omega + B_\pm$. From boundary conditions $B_+ = B_- = B$ and for the finding of A_+, A_-, B we have the system

$$\begin{cases} a_+A_+ - a_-A_- - \sigma B &= 0, \\ \left(\alpha_+ a_+ + \frac{\omega_0}{2}\gamma_+\right) A_+ + \gamma_+ B &= 0, \\ -\left(\alpha_- a_- + \frac{\omega_0}{2}\gamma_-\right) A_- + \gamma_- B &= 0. \end{cases}$$

Since $A_+^2 + A_-^2 + B^2 \neq 0$, the system determinant must be equal to zero; this means the equality

$$\sigma\left(\alpha_+ a_+ + \frac{\omega_0}{2}\gamma_+\right)\left(\alpha_- a_- + \frac{\omega_0}{2}\gamma_-\right) + a_+\gamma_+\left(\alpha_- a_- + \frac{\omega_0}{2}\gamma_-\right) + a_-\gamma_-\left(\alpha_+ a_+ + \frac{\omega_0}{2}\gamma_+\right) = 0 \quad (3.6.3)$$

holds. Thus, if equality (3.6.3) is satisfied, then $\lambda = 0$ and we have the corresponding eigenfunction

$$\psi(\omega) = \begin{cases} a_-\gamma_-\left\{\left(\omega - \frac{\omega_0}{2}\right)\gamma_+ - \alpha_+ a_+\right\}, & \omega \in \left[0, \frac{\omega_0}{2}\right], \\ a_+\gamma_+\left\{\left(\omega + \frac{\omega_0}{2}\right)\gamma_- - \alpha_- a_-\right\}, & \omega \in \left[-\frac{\omega_0}{2}, 0\right], \end{cases} \qquad \text{if } \sigma = 0;$$

$$\psi(\omega) = \begin{cases} -\gamma_+\left(\alpha_- a_- + \frac{\omega_0}{2}\gamma_-\right)\left(\omega + \frac{a_+}{\sigma}\right) - \frac{a_-\gamma_-}{\sigma}\left(\alpha_+ a_+ + \frac{\omega_0}{2}\gamma_+\right), & \omega \in \left[0, \frac{\omega_0}{2}\right], \\ \gamma_-\left(\alpha_+ a_+ + \frac{\omega_0}{2}\gamma_+\right)\left(\omega - \frac{a_-}{\sigma}\right) - \frac{a_+\gamma_+}{\sigma}\left(\alpha_- a_- + \frac{\omega_0}{2}\gamma_-\right), & \omega \in \left[-\frac{\omega_0}{2}, 0\right], \end{cases}$$

if $\sigma \neq 0$.

2) case $\lambda \neq 0$. Solving the problem equations we have $\psi_\pm(\omega) = A_\pm \cos(\lambda\omega) + B_\pm \sin(\lambda\omega)$. From boundary conditions we obtain $A_+ = A_- = A$ and for the finding of A, B_+, B_- we have the system

$$\begin{cases} \sigma A - \lambda a_+ B_+ + \lambda a_- B_- &= 0, \\ \left(\gamma_+ \cos\frac{\lambda\omega_0}{2} - \lambda\alpha_+ a_+ \sin\frac{\lambda\omega_0}{2}\right) A + \left(\gamma_+ \sin\frac{\lambda\omega_0}{2} + \lambda\alpha_+ a_+ \cos\frac{\lambda\omega_0}{2}\right) B_+ &= 0, \\ \left(\gamma_- \cos\frac{\lambda\omega_0}{2} - \lambda\alpha_- a_- \sin\frac{\lambda\omega_0}{2}\right) A - \left(\gamma_- \sin\frac{\lambda\omega_0}{2} + \lambda\alpha_- a_- \cos\frac{\lambda\omega_0}{2}\right) B_- &= 0. \end{cases}$$

Since $A^2 + B_+^2 + B_-^2 \neq 0$, the system determinant must be equal to zero; this means that λ is determined from the transcendence equation

$$\begin{aligned} &\sigma(\lambda^2\alpha_+\alpha_- a_+ a_- + \gamma_+\gamma_-) + \lambda^2(a_+ - a_-)(\alpha_- a_-\gamma_+ - \alpha_+ a_+\gamma_-) \\ &+ \lambda[\sigma(\alpha_- a_-\gamma_+ + \alpha_+ a_+\gamma_-) + (a_+ + a_-)(\gamma_+\gamma_- - \lambda^2\alpha_+\alpha_- a_+ a_-)]\sin(\lambda\omega_0) \\ &+ [\sigma(\lambda^2\alpha_+\alpha_- a_+ a_- - \gamma_+\gamma_-) + \lambda^2(a_+ + a_-)(\alpha_- a_-\gamma_+ + \alpha_+ a_+\gamma_-)]\cos(\lambda\omega_0) = 0. \end{aligned} \quad (3.6.4)$$

Now we investigate some special cases of boundary conditions.

The Dirichlet problem: $\alpha_\pm = 0,\ \gamma_\pm = 1$.

Equation (3.6.4) takes the form $\sigma(1 - \cos(\lambda\omega_0)) + \lambda(a_+ + a_-)\sin(\lambda\omega_0) = 0$. Hence we derive $\lambda = \begin{cases} \frac{\pi}{\omega_0}, & \text{if } \sigma = 0, \\ \lambda^*, & \text{if } \sigma > 0, \end{cases}$ where λ^* is the least positive root of the transcendence equation $\tan\frac{\lambda\omega_0}{2} = -\frac{a_+ + a_-}{\sigma}\lambda$, and the corresponding eigenfunction

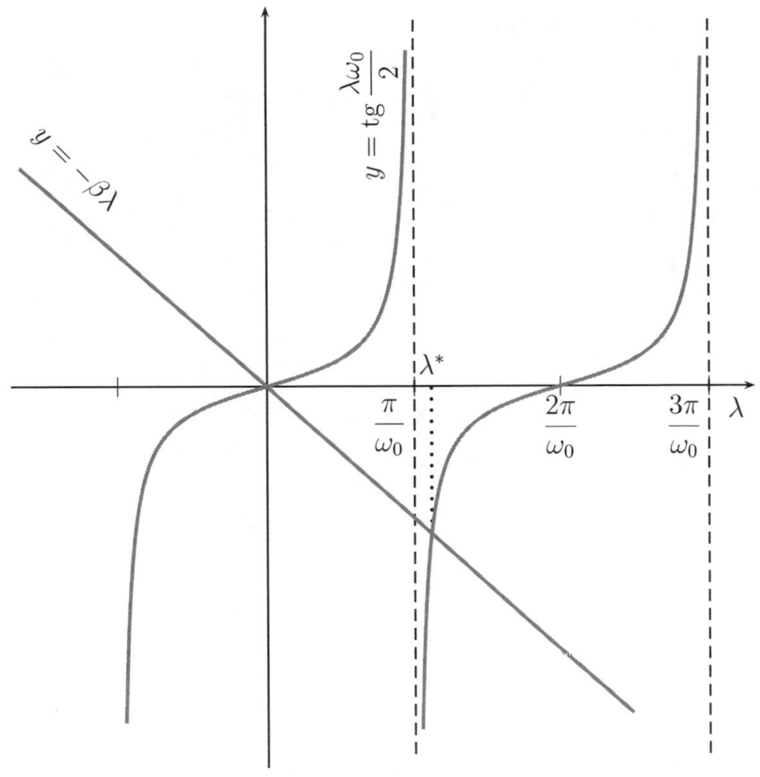

Figure 2

$\psi(\omega) = \begin{cases} \sin\lambda\left(\frac{\omega_0}{2} - \omega\right), & \omega \in \left[0, \frac{\omega_0}{2}\right], \\ \sin\lambda\left(\frac{\omega_0}{2} + \omega\right), & \omega \in \left[-\frac{\omega_0}{2}, 0\right]. \end{cases}$ By the graphical method (see Figure 2), we observe that $\frac{\pi}{\omega_0} < \lambda^* < \frac{2\pi}{\omega_0}$.

The Neumann problem: $\gamma\pm = 0,\ \alpha_\pm = 1$.

Equation (3.6.4) takes the form $\sigma(1+\cos(\lambda\omega_0))-\lambda(a_+ + a_-)\sin(\lambda\omega_0) = 0$. Hence we derive $\lambda = \begin{cases} \frac{\pi}{\omega_0}, & \text{if } \sigma = 0, \\ \lambda^*, & \text{if } \sigma > 0, \end{cases}$ where λ^* is the least positive root of the transcendence equation $\tan\frac{\lambda\omega_0}{2} = \frac{\sigma}{a_+ + a_-}\cdot\frac{1}{\lambda}$. For $\lambda = \frac{\pi}{\omega_0}$ we find the corresponding eigenfunction

$$\psi(\omega) = \begin{cases} a_- \sin\frac{\pi\omega}{\omega_0}, & \omega \in \left[0, \frac{\omega_0}{2}\right], \\ a_+ \sin\frac{\pi\omega}{\omega_0}, & \omega \in \left[-\frac{\omega_0}{2}, 0\right]. \end{cases}$$

For $\lambda = \lambda^*$ we find the corresponding eigenfunction

$$\psi(\omega) = \begin{cases} \cos\lambda^*\left(\omega - \frac{\omega_0}{2}\right), & \omega \in \left[0, \frac{\omega_0}{2}\right], \\ \cos\lambda^*\left(\omega + \frac{\omega_0}{2}\right), & \omega \in \left[-\frac{\omega_0}{2}, 0\right]. \end{cases}$$

By the graphical method (see Figure 3), we observe that $0 < \lambda^* < \frac{\pi}{\omega_0}$.

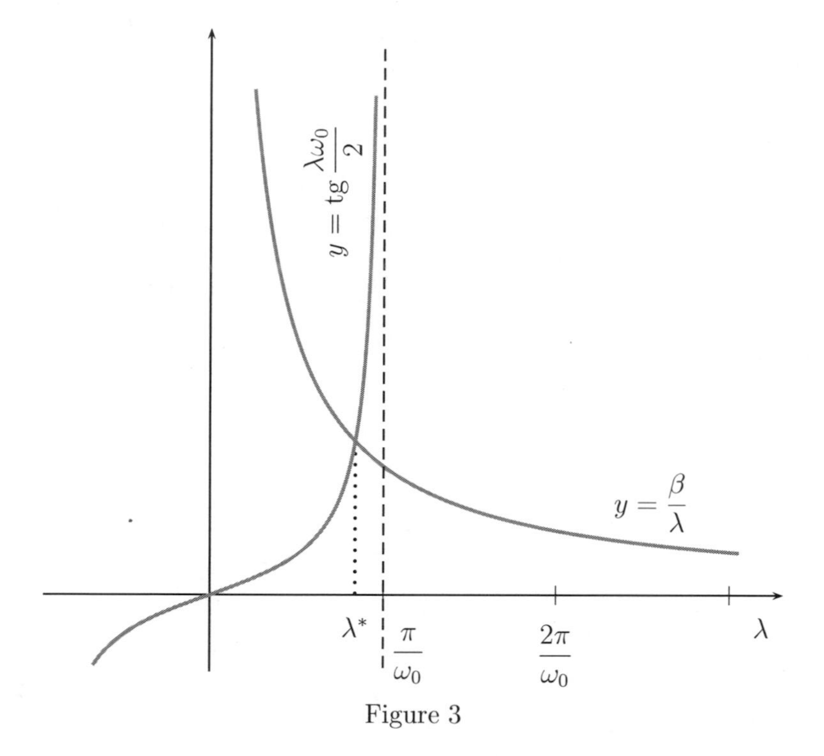

Figure 3

Mixed problem: $\alpha_+ = 1,\ \alpha_- = 0;\ \gamma_+ = 0,\ \gamma_- = 1.$

Equation (3.6.4) takes the form

$$\sigma\sin(\lambda\omega_0) + \lambda(a_+ + a_-)\cos(\lambda\omega_0) = \lambda(a_+ - a_-). \tag{3.6.5}$$

If $\sigma = 0$, we find $\lambda = \frac{2}{\omega_0}\arctan\sqrt{\frac{a_-}{a_+}} < \frac{\pi}{\omega_0}$ and the corresponding eigenfunction

$$\psi(\omega) = \begin{cases} \cos(\lambda\omega) + \sqrt{\frac{a_-}{a_+}}\cdot\sin(\lambda\omega), & \omega \in \left[0, \frac{\omega_0}{2}\right], \\ \cos(\lambda\omega) + \sqrt{\frac{a_+}{a_-}}\cdot\sin(\lambda\omega), & \omega \in \left[-\frac{\omega_0}{2}, 0\right]. \end{cases}$$

If $\sigma > 0$, then $\lambda = \lambda^*$, where λ^* is the least positive root of the transcendence equation (3.6.5), and the corresponding eigenfunction

$$\psi(\omega) = \begin{cases} \sin \frac{\lambda\omega_0}{2} \cos \lambda \left(\omega - \frac{\omega_0}{2}\right), & \omega \in \left[0, \frac{\omega_0}{2}\right], \\ \cos \frac{\lambda\omega_0}{2} \sin \lambda \left(\omega + \frac{\omega_0}{2}\right), & \omega \in \left[-\frac{\omega_0}{2}, 0\right]. \end{cases}$$

Rewriting equation (3.6.5) in the form $\tan \frac{\lambda\omega_0}{2} = \frac{2\lambda a_-}{\sqrt{4\lambda^2 a_+ a_- + \sigma^2} - \sigma}$, by the graphical method (see Figure 4), we observe that $\frac{2}{\omega_0} \arctan \sqrt{\frac{a_-}{a_+}} < \lambda^* < \frac{\pi}{\omega_0}$.

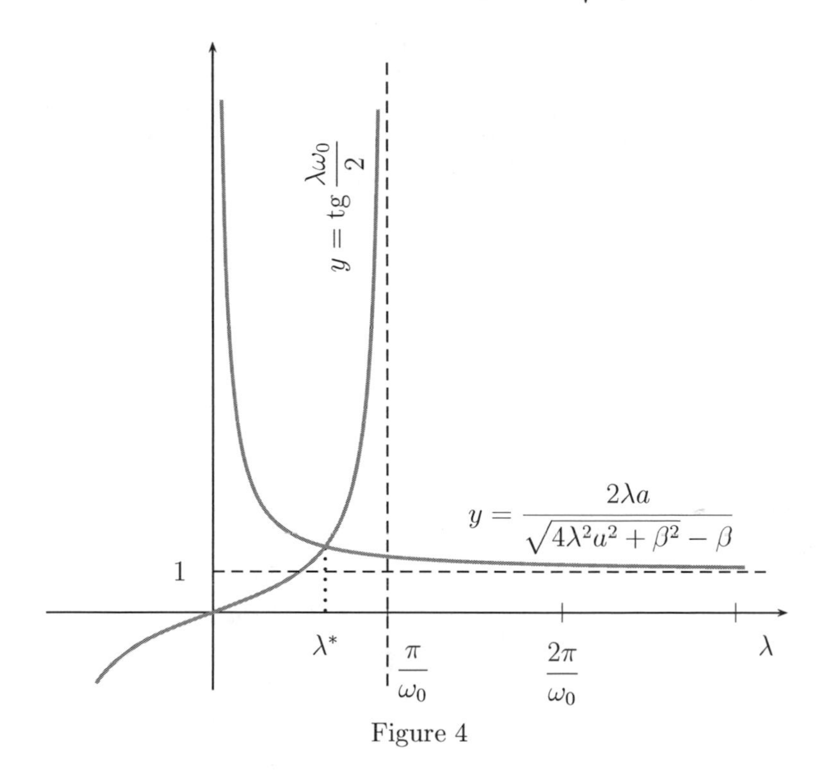

Figure 4

The Robin problem: $\alpha_\pm = 1$; $\gamma_\pm \neq 0$.

Direct calculations of the above system will give:

1) $\dfrac{\gamma_+}{\gamma_-} = \dfrac{a_+}{a_-}$. In this case we get either $\psi_\pm(\omega) = a_\mp \sin(\lambda^*\omega)$, where λ^* is the least positive root of the transcendence equation $\tan\left(\lambda\frac{\omega_0}{2}\right) = -\lambda\frac{a_+}{\gamma_+}$ and from the graphical solution (see Figure 2) we obtain $\frac{\pi}{\omega_0} < \lambda^* < \frac{2\pi}{\omega_0}$, or $\psi_\pm(\omega) = \cos(\lambda^*\omega) \pm \frac{\sigma}{a_+ + a_-} \cdot \frac{\sin(\lambda^*\omega)}{\lambda^*}$, where λ^* is the least positive root of the transcendence equation

$$\tan\left(\lambda\frac{\omega_0}{2}\right) = \lambda\frac{a_+\sigma + \gamma_+(a_+ + a_-)}{a_+(a_+ + a_-)\lambda^2 - \gamma_+\sigma}.$$

In particular, for $\sigma = 0$ we have: $\psi_\pm(\omega) = \cos(\lambda\omega)$, $\tan\left(\lambda\frac{\omega_0}{2}\right) = \frac{\gamma_+}{a_+\lambda}$, $0 < \lambda^* < \frac{\pi}{\omega_0}$ (see Figure 3).

2) $\frac{\gamma_+}{\gamma_-} \neq \frac{a_+}{a_-}$. In this case we get $A \neq 0 \implies \psi(0) \neq 0$ and

$$\psi_+(\omega) = \frac{\lambda^* a_+ \cos\lambda^*\left(\omega - \frac{\omega_0}{2}\right) - \gamma_+ \sin\lambda^*\left(\omega - \frac{\omega_0}{2}\right)}{\lambda^* a_+ \cos\left(\frac{\lambda^*\omega_0}{2}\right) + \gamma_+ \sin\left(\frac{\lambda^*\omega_0}{2}\right)}, \quad \omega \in \left(0, \frac{\omega_0}{2}\right);$$

$$\psi_-(\omega) = \frac{\lambda^* a_- \cos\lambda^*\left(\omega + \frac{\omega_0}{2}\right) + \gamma_- \sin\lambda^*\left(\omega + \frac{\omega_0}{2}\right)}{\lambda^* a_- \cos\left(\frac{\lambda^*\omega_0}{2}\right) + \gamma_- \sin\left(\frac{\lambda^*\omega_0}{2}\right)}, \quad \omega \in \left(-\frac{\omega_0}{2}, 0\right),$$

where λ^* is the least positive root of the transcendence equation

$$a_+\frac{\lambda a_+ \tan\left(\frac{\lambda\omega_0}{2}\right) - \gamma_+}{\lambda a_+ + \gamma_+ \tan\left(\frac{\lambda\omega_0}{2}\right)} + a_-\frac{\lambda a_- \tan\left(\frac{\lambda\omega_0}{2}\right) - \gamma_-}{\lambda a_- + \gamma_- \tan\left(\frac{\lambda\omega_0}{2}\right)} = \frac{\sigma}{\lambda}.$$

In particular, in the case of the problem without the interface ($a_+ = a_- = 1$, $\sigma = 0$) we obtain the least eigenvalue as the least positive root of the transcendence equation $\tan(\lambda\omega_0) = \frac{\lambda(\gamma_+ + \gamma_-)}{\lambda^2 - \gamma_+\gamma_-}$ and the corresponding eigenfunction $\psi(\omega) = \lambda\cos\left[\lambda\left(\omega - \frac{\omega_0}{2}\right)\right] - \gamma_+ \sin\left[\lambda\left(\omega - \frac{\omega_0}{2}\right)\right]$ (see §10.1.7 [14]). By the graphical method (see Figure 5), we observe that $\frac{\pi}{2\omega_0} < \lambda^* < \frac{\pi}{\omega_0}$.

3.7 Examples

Let us present some examples which demonstrate that the assumptions on coefficients of operator $\mathcal{L}$ are essential for the validity of Theorem 6.3.

Let the domain $G \subset \mathbb{R}^2$ be as in §3.6.

Example 3.1. Let us consider the function

$$u(r,\omega) = r^\lambda \left(\ln\frac{1}{r}\right)^{(\lambda-1)/(\lambda+1)} \cdot \sin(\lambda\omega) \cdot \begin{cases} a_-, & \omega \in \left[0, \frac{\omega_0}{2}\right], \\ a_+, & \omega \in \left[-\frac{\omega_0}{2}, 0\right], \end{cases} \quad a_\pm > 0,\ \lambda = \frac{\pi}{\omega_0}.$$

By direct calculations (see also the investigation of the Neumann problem in Appendix, §3.6), we verify that it satisfies the transmission problem

$$\begin{cases} \frac{\partial}{\partial x_i}\left(a^{ij}(x)u_{x_j}\right) + a^i(x)u_{x_i} = 0, & x \in G \setminus \Sigma_0; \\ [u]_{\Sigma_0} = 0, \quad \left[\frac{\partial u}{\partial \nu}\right]_{\Sigma_0} = 0, & x \in \Sigma_0; \\ \frac{\partial u}{\partial \nu} = 0, & x \in \partial G \setminus \{\Sigma_0 \cup \mathcal{O}\}, \end{cases}$$

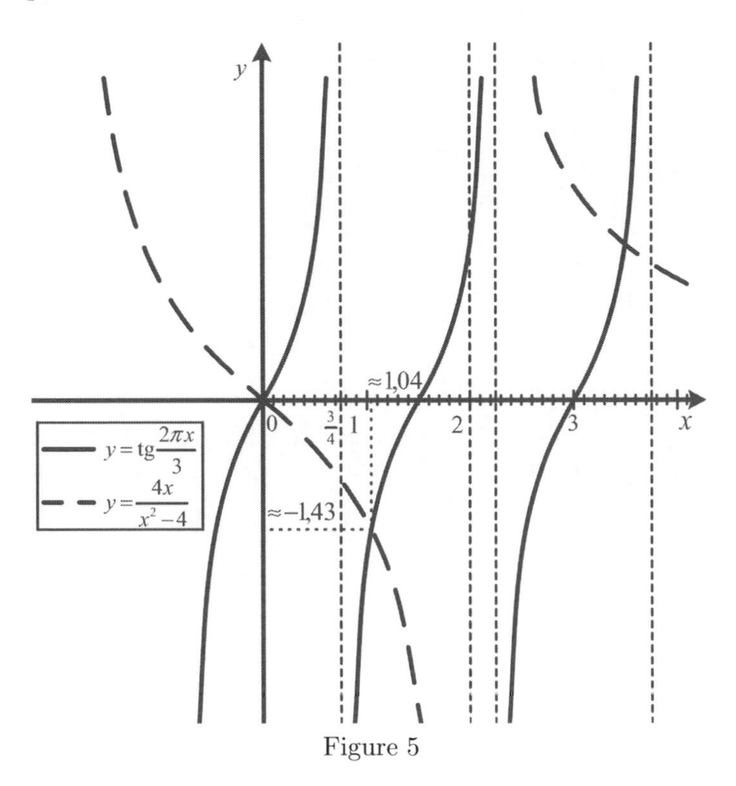

Figure 5

where

$$a^{11}(x) = a - \frac{2a}{\lambda+1} \cdot \frac{x_2^2}{r^2 \ln(1/r)}, \quad a^{12}(x) = a^{21}(x) = \frac{2a}{\lambda+1} \cdot \frac{x_1 x_2}{r^2 \ln(1/r)},$$

$$a^{22}(x) = a - \frac{2a}{\lambda+1} \cdot \frac{x_1^2}{r^2 \ln(1/r)}, \quad a^{ij}(0) = a\delta_i^j, \; i,j = 1,2;$$

$$a^1(x) = -\frac{1}{r}\mathcal{A}(r)\cos\omega, \quad a^2(x) = -\frac{1}{r}\mathcal{A}(r)\sin\omega,$$

$$\mathcal{A}(r) = \frac{2a}{(\lambda+1)\ln(1/r)}, \quad \Longrightarrow \int_0^d \frac{\mathcal{A}(r)}{r} dr = +\infty.$$

Clearly, the equation is uniformly elliptic in G_0^d for $0 < d < e^{-2}$ with the ellipticity constants $\nu = a - \frac{2a}{\ln(1/d)}$ and $\mu = a$. Thus, we observe that leading coefficients of the equation are continuous but not Dini-continuous at zero. From the explicit form of solution u we have

$$|u(x)| \le c|x|^{\lambda-\varepsilon}, \quad \|u\|_{\mathbf{V}^2_{p,2p-n}(G_0^\varrho)} \le c\varrho^{\lambda-\varepsilon} \tag{3.7.1}$$

for all $\varepsilon > 0$. This example shows that it is not possible to replace $\lambda - \varepsilon$ in (3.7.1) by λ without additional assumptions regarding the continuity modulus of leading coefficients of the equation at zero.

Example 3.2. Let $(\lambda, \psi(\omega))$ be a solution of eigenvalue problem (3.6.2) (see Appendix, §3.6). Then function $u(x) = r^{\lambda} \ln(\frac{1}{r})\psi(\omega)$ is a solution of the transmission problem

$$\begin{cases} \triangle u = -2\lambda r^{\lambda-2}\psi(\omega), & x \in G \setminus \Sigma_0; \\ [u]_{\Sigma_0} = 0, \quad \left[a\frac{\partial u}{\partial \vec{n}}\right]_{\Sigma_0} + \frac{\sigma}{r}u(x) = 0, & x \in \Sigma_0; \\ \alpha a\frac{\partial u}{\partial \vec{n}} + \frac{1}{r}\gamma u_{\pm}(x) = 0, & x \in \partial G \setminus \{\Sigma_0 \cup \mathcal{O}\}, \end{cases} \tag{3.7.2}$$

where $a > 0,\ \sigma > 0,\ \gamma > 0,\ \alpha \in \{0, 1\}$.

All assumptions of Theorem 3.3 are fulfilled with $s = \lambda$. This example shows the precision of assumption (d) and estimate (3.1.2) for $s = \lambda$.

Chapter 4

Transmission problem for the Laplace operator with N different media

4.1 Introduction

In this chapter we investigate the behavior of weak solutions to the transmission problem (LN) for the Laplace operator with N different media in a neighborhood of the boundary conical point:

$$\begin{cases} a\triangle u - pu(x) = f(x), & x \in G \setminus \bigcup\limits_{k=1}^{N-1} \Sigma_k; \\ [u]_{\overline{\Sigma_k}} = 0, \quad \mathcal{S}_k[u] \equiv \left[a\frac{\partial u}{\partial n_k}\right]_{\Sigma_k} + \frac{1}{|x|}\beta_k(\omega)u(x) = h_k(x), & x \in \Sigma_k, \\ & k = 1, \dots, N-1; \\ \mathcal{B}[u] \equiv \alpha(x) \cdot a\frac{\partial u}{\partial \vec{n}} + \frac{1}{|x|}\gamma(\omega)u(x) = g(x), & x \in \partial G \setminus \{\mathcal{O}\}, \end{cases} \tag{LN}$$

where $a > 0$, $p \geq 0$; $\alpha(x) = \begin{cases} 0, & \text{if } x \in \mathcal{D}, \\ 1, & \text{if } x \notin \mathcal{D}. \end{cases}$

A principal new feature of this chapter is the derivation of sharp estimates for the (LN) solutions in n-dimensional ($n \geq 2$) conic domains with N different media ($N \geq 2$). We demonstrate some examples.

Let $G \subset \mathbb{R}^n$, $n \geq 2$ be a bounded domain with boundary ∂G that is a smooth surface everywhere except at the origin $\mathcal{O} \in \partial G$ and near the point $\mathcal{O}$ it is a convex conical surface with vertex at $\mathcal{O}$ and the opening ω_0. We assume that

M. Borsuk, *Transmission Problems for Elliptic Second-Order Equations in Non-Smooth Domains*, Frontiers in Mathematics, DOI 10.1007/978-3-0346-0477-2_5, © Springer Basel AG 2010

$G = \bigcup_{i=1}^{N} G_i$ is divided into N sub-domains G_i, $i = 1, \dots, N$ by $(N-1)$ hyperplanes Σ_k, $k = 1, \dots, N-1$, where $\mathcal{O}$ belongs to every $\overline{\Sigma_k}$ and $G_i \cap G_j = \emptyset$, $i \neq j$. Let ϕ_i be openings at the vertex $\mathcal{O}$ in domains G_i. Let us define the value $\theta_k = \phi_1 + \phi_2 + \dots + \phi_k$, thus $\omega_0 = \theta_N$. In addition to Chapter 1 we introduce the following notation:

- Ω_i: a domain on the unit sphere S^{n-1} with boundary $\partial\Omega_i$ obtained by the intersection of the domain G_i with the sphere S^{n-1} $(i = 1, \dots, N)$; thus $\Omega = \bigcup_{i=1}^{N} \Omega_i$;
- $\Sigma = \sum_{k=1}^{N-1} \Sigma_k$, where $\Sigma_k = G \cap \{\omega_1 = \frac{\omega_0}{2} - \theta_k\}$, $k = 1, \dots, N-1$;
- $\sigma = \sum_{k=1}^{N-1} \sigma_k$, where $\sigma_k = \Sigma_k \cap \Omega$;
- $G_a^b = \{(r, \omega) \mid 0 \le a < r < b; \omega \in \Omega\} \cap G$: a layer in $\mathbb{R}^n$;
- $\Gamma_a^b = \{(r, \phi) \mid 0 \le a < r < b; \phi \in \partial\Omega\} \cap \partial G$: the lateral surface of layer G_a^b;
- $G_d = G \setminus G_0^d$, $\Gamma_d = \partial G \setminus \Gamma_0^d$, $d > 0$;
- $(G_i)_a^b = \{(r, \omega) \mid 0 \le a < r < b; \omega \in \Omega\} \cap G_i$ $i = 1, \dots, N$;
- $\Sigma_a^b = G_a^b \cap \Sigma$; $\Sigma_d = \Sigma \setminus \Sigma_0^d$, $d > 0$;
- $(\Sigma_k)_a^b = G_a^b \cap \Sigma_k$, $k = 1, \dots, N-1$;
- $\mathcal{D} \subseteq \partial G$: the part of the boundary ∂G where we consider the Dirichlet boundary condition.
- $u(x) = u_i(x)$, $x \in G_i$; $f(x) = f_i(x)$, $x \in G_i$; $a\Big|_{\overline{G_i}} = a_i$, etc.
- $[u]_{\Sigma_k}$ denotes the saltus of the function $u(x)$ on crossing Σ_k, i.e.,

$$[u]_{\Sigma_k} = u_k(\overline{x})\Big|_{\Sigma_k} - u_{k+1}(\overline{x})\Big|_{\Sigma_k}, \quad u_k(\overline{x})\Big|_{\Sigma_k} = \lim_{G_k \ni x \to \overline{x} \in \Sigma_k} u(x),$$
$$u_{k+1}(\overline{x})\Big|_{\Sigma_k} = \lim_{G_{k+1} \ni x \to \overline{x} \in \Sigma_k} u(x);$$

- $\left[a \frac{\partial u}{\partial \vec{n}_k}\right]_{\Sigma_k}$ denotes the saltus of the co-normal derivative of the function $u(x)$ on crossing Σ_k, i.e.,

$$\left[a \frac{\partial u}{\partial \vec{n}_k}\right]_{\Sigma_k} = a_k \frac{\partial u_k}{\partial \vec{n}_k}\Big|_{\Sigma_k} - a_{k+1} \frac{\partial u_{k+1}}{\partial \vec{n}_k}\Big|_{\Sigma_k},$$

where $\vec{n}_k$ denotes the outward unit vector with respect to G_k normal to Σ_k.

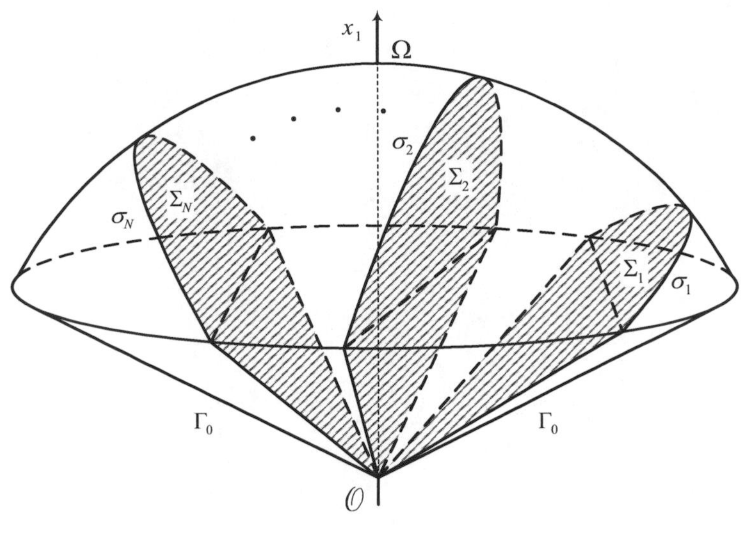

Figure 6

Without loss of generality we assume that there exists $d > 0$ such that G_0^d is *a convex rotational cone* with the vertex at $\mathcal{O}$ and the aperture ω_0, thus

$$\Gamma_0^d = \left\{ (r,\omega) \middle| x_1^2 = \cot^2 \frac{\omega_0}{2} \sum_{i=2}^{n} x_i^2;\ r \in (0,d),\ \omega_1 = \frac{\omega_0}{2},\ \omega_0 \in (0,\pi) \right\}. \tag{4.1.1}$$

We use the standard function spaces:

- $C^k(\overline{G_i})$ with the norm $|u_i|_{k,G_i}$,
- Lebesgue space $L_p(G_i)$, $p \geq 1$ with the norm $\|u_i\|_{p,G_i}$,
- the Sobolev space $W^{k,p}(G_i)$ with the norm $\|u_i\|_{k,p;G_i}$,

and introduce their direct sums

- $\mathbf{C}^k(\overline{G}) = C^k(\overline{G_1}) \dotplus \cdots \dotplus C^k(\overline{G_N})$ with the norm $|u|_{k,G} = \sum\limits_{i=1}^{N} |u_i|_{k,G_i}$;
- $\mathbf{L}_p(G) = L_p(G_1) \dotplus \cdots \dotplus L_p(G_N)$ with the norm $\|u\|_{\mathbf{L}_p(G)} = \sum\limits_{i=1}^{N} \left(\int\limits_{G_i} |u_i|^q dx \right)^{\frac{1}{p}}$;
- $\mathbf{W}^{k,p}(G) = W^{k,p}(G_1) \dotplus \cdots \dotplus W^{k,p}(G_N)$ with the norm

$$\|u\|_{k,p;G} = \sum_{i=1}^{N} \left(\int\limits_{G_i} \sum_{|\beta|=0}^{k} |D^\beta u_i|^p dx \right)^{\frac{1}{p}}.$$

We define the weighted Sobolev spaces:

- $\mathbf{V}^k_{p,\alpha}(G) = V^k_{p,\alpha}(G_1) \dotplus \cdots \dotplus V^k_{p,\alpha}(G_N)$ for integer $k \geq 0$ and real α, where $V^k_{p,\alpha}(G_i)$ denotes the space of all distributions $u \in \mathcal{D}'(G_i)$ satisfying for $i = 1, \dots, N$: $r^{\frac{\alpha}{p}+|\beta|-k}|D^\beta u_i| \in L_p(G_i)$; $\mathbf{V}^k_{p,\alpha}(G)$ is a Banach space for the norm

$$\|u\|_{\mathbf{V}^k_{p,\alpha}(G)} = \sum_{i=1}^{N} \Big(\int_{G_i} \sum_{|\beta|=0}^{k} r^{\alpha+p(|\beta|-k)} |D^\beta u_i|^p \, dx \Big)^{\frac{1}{p}}.$$

- $\mathbf{V}^{k-\frac{1}{p}}_{p,\alpha}(\partial G)$ is the space of functions φ, given on ∂G, with the norm

$$\|\varphi\|_{\mathbf{V}^{k-\frac{1}{p}}_{p,\alpha}(\partial G)} = \inf \|\Phi\|_{\mathbf{V}^k_{p,\alpha}(G)},$$

where the infimum is taken over all functions Φ such that $\Phi\Big|_{\partial G} = \varphi$ in the sense of traces.

We write

$$\mathbf{W}^k(G) \equiv \mathbf{W}^{k,2}(G), \quad \overset{\circ}{\mathbf{W}}{}^k_\alpha(G) \equiv \mathbf{V}^k_{2,\alpha}(G).$$

Definition 4.1. A function $u(x)$ is called *a weak* solution of problem (LN) provided that $u(x) \in \mathbf{C}^0(\overline{G}) \cap \overset{\circ}{\mathbf{W}}{}^1_0(G)$ and satisfies the integral identity

$$\int_G a u_{x_j} \eta_{x_j} dx + \int_\Sigma \frac{1}{r}\beta(\omega) u(x)\eta(x) ds + \int_{\partial G} \alpha(x)\frac{1}{r}\gamma(\omega)u(x)\eta(x)ds$$
$$= \int_{\partial G} \alpha(x) g(x)\eta(x) ds + \int_\Sigma h(x)\eta(x)ds - \int_G (pu(x) + f(x))\,\eta(x)dx \qquad (II)$$

for all functions $\eta(x) \in \mathbf{C}^0(\overline{G}) \cap \overset{\circ}{\mathbf{W}}{}^1_0(G)$.

The summation over repeated indices from 1 to n is understood; here:

$$\int_G f(x)dx = \sum_{i=1}^{N} \int_{G_i} f_i(x)dx, \quad \int_\Sigma h(x)ds = \sum_{k=1}^{N-1} \int_{\Sigma_k} h_k(x)ds, \text{ etc.}$$

Remark 4. In the Dirichlet boundary condition case ($\alpha(x) \equiv 0$) we assume, without loss of generality, that $g\Big|_{\partial G \cap \mathcal{D}} = 0 \implies u\Big|_{\partial G \cap \mathcal{D}} = 0$.

We assume that $M_0 = \max_{x \in \overline{G}} |u(x)|$ is known.

Lemma 4.2. *Let $u(x)$ be a weak solution of (LN). For any function $\eta(x) \in \mathbf{C}^0(\overline{G}) \cap \overset{\circ}{\mathbf{W}}{}^1_0(G)$ the equality*

$$\int\limits_{G_0^\varrho} \Big\{ a u_{x_i} \eta_{x_i} + (f(x) + p u(x))\, \eta(x) \Big\} dx = \int\limits_{\Omega_\varrho} a \frac{\partial u}{\partial r} \eta(x) d\Omega_\varrho$$

$$+ \int\limits_{\Gamma_0^\varrho} \alpha(x) \left(g(x) - \frac{1}{r} \gamma(\omega) u(x) \right) \eta(x) ds + \int\limits_{\Sigma_0^\varrho} \left(h(x) - \frac{1}{r} \beta(\omega) u(x) \right) \eta(x) ds \quad (II)_{loc}$$

holds for a.e. $\varrho \in (0, d)$.

Proof. The proof is analogous to the proof of Lemma 3.2 Chapter 3. □

Let us define the numbers

$$\begin{cases} a_* = \min\{a_1, \dots, a_N\} > 0, \ a^* = \max\{a_1, \dots, a_N\} > 0; \\ p^* = \max\{p_1, \dots, p_N\} \geq 0; \\ [a]_{\Sigma_k} = a_k - a_{k+1}, \ k = 1, \dots, N-1; \\ a_0 = \max\limits_{1 \leq k \leq N-1} \left| [a]_{\Sigma_k} \right|; \quad \widetilde{a} = \max(a^*, a_0). \end{cases} \tag{4.1.2}$$

Assumptions.

(a) $f(x) \in \mathbf{L}_{p/2}(G) \cap \mathbf{L}_2(G)$; $p > n$;

(b) $\gamma(\omega) \geq \gamma_0 > \widetilde{a} \tan \frac{\omega_0}{2}$ *on* ∂G; $\beta_k(\omega) \geq \beta_0 > \widetilde{a} \tan \frac{\omega_0}{2}$ *on* Σ_k, $k = 1, \dots, N-1$;

(c) *there exist numbers* $f_0 \geq 0$, $g_0 \geq 0$, $h_0 \geq 0$, $s > 1$, *such that*

$$|f(x)| \leq f_0 |x|^{s-2}, \ |g(x)| \leq g_0 |x|^{s-1}, \ |h_k(x)| \leq h_0 |x|^{s-1}, \ k = 1, \dots, N-1.$$

We formulate the main result as the following theorem. Let

$$\lambda = \frac{2 - n + \sqrt{(n-2)^2 + 4\vartheta}}{2}, \tag{4.1.3}$$

where ϑ is the smallest positive eigenvalue of the problem $(EVPN)$ (see Subsection 4.2.1).

Theorem 4.3. *Let u be a weak solution of problem (LN) and assumptions* (a)–(c) *be satisfied. Then there are $d \in (0,1)$ and $C_0 > 0$ depending only on n, a_*, a^*, λ, $\omega_0, f_0, h_0, g_0, \beta_0, \gamma_0, s, M_0$, meas G, diam G such that the inequality*

$$|u(x)| \leq C_0 \left(\|u\|_{2,G} + f_0 + \frac{1}{\sqrt{\gamma_0}} g_0 + \frac{1}{\sqrt{\beta_0}} h_0 \right) \cdot \begin{cases} |x|^\lambda, & \text{if } s > \lambda, \\ |x|^\lambda \ln\left(\frac{1}{|x|}\right), & \text{if } s = \lambda, \\ |x|^s, & \text{if } s < \lambda \end{cases} \tag{4.1.4}$$

holds for all $x \in G_0^d$. *If, in addition,*

$$\beta(\omega) \in \mathbf{C}^1(\Sigma),\ \gamma(\omega) \in \mathbf{C}^1(\partial G),$$
$$f(x) \in \mathbf{V}^0_{p,2p-n}(G),$$
$$h(x) \in V^{1-1/p}_{p,2p-n}(\Sigma),\ g(x) \in \mathbf{V}^{1-1/p}_{p,2p-n}(\partial G);\ p > n$$

and there is a number

$$\tau_s =: \sup_{\varrho>0} \varrho^{-s}\Big(\|h\|_{\mathbf{V}^{1-\frac{1}{p}}_{p,2p-n}(\Sigma^{\varrho}_{\varrho/2})} + \|g\|_{V^{1-\frac{1}{p}}_{p,2p-n}(\Gamma^{\varrho}_{\varrho/2})}\Big), \tag{4.1.5}$$

then

$$|\nabla u(x)| \le C_1\Big(\|u\|_{2,G} + f_0 + \frac{1}{\sqrt{\gamma_0}} g_0 + \frac{1}{\sqrt{\beta_0}} h_0 + \tau_s\Big) \cdot \begin{cases} |x|^{\lambda-1}, & \text{if } s > \lambda, \\ |x|^{\lambda-1} \ln\Big(\frac{1}{|x|}\Big), & \text{if } s = \lambda, \\ |x|^{s-1}, & \text{if } s < \lambda \end{cases} \tag{4.1.6}$$

for all $x \in G_0^d$. *Furthermore, the following is true:*

- $u \in \mathbf{V}^2_{p,2p-n}(G),\ p > n$ *and*

$$\|u\|_{\mathbf{V}^2_{p,2p-n}(G_0^{\varrho})} \le C_2\Big(\|u\|_{2,G} + f_0 + \frac{1}{\sqrt{\gamma_0}} g_0 + \frac{1}{\sqrt{\beta_0}} h_0 + \tau_s\Big) \cdot \begin{cases} \varrho^{\lambda}, & \text{if } s > \lambda, \\ \varrho^{\lambda} \ln\Big(\frac{1}{\varrho}\Big), & \text{if } s = \lambda, \\ \varrho^{s}, & \text{if } s < \lambda; \end{cases} \tag{4.1.7}$$

- *if*

$$f(x) \in \overset{\circ}{\mathbf{W}}{}^0_{\alpha}(G),\ \int_{\Sigma} r^{\alpha-1} h^2(x) ds < \infty,\ \int_{\partial G} r^{\alpha-1} g^2(x) ds < \infty,$$

where

$$4 - n - 2\lambda < \alpha \le 2, \tag{4.1.8}$$

then $u(x) \in \overset{\circ}{\mathbf{W}}{}^1_{\alpha-2}(G)$ *and*

$$\int_G a\left(r^{\alpha-2}|\nabla u|^2 + r^{\alpha-4}u^2\right) dx + \int_{\Sigma} r^{\alpha-3}\beta(\omega) u^2(x) ds + \int_{\partial G} \alpha(x) r^{\alpha-3}\gamma(\omega) u^2(x) ds \tag{4.1.9}$$
$$\le C\Big\{\int_G \left(u^2 + (1+r^{\alpha}) f^2(x)\right) dx + \int_{\Sigma} r^{\alpha-1} h^2(x) ds + \int_{\partial G} \alpha(x) r^{\alpha-1} g^2(x) ds\Big\},$$

where the constant $C > 0$ *depends only on* $n, a_*, a^*, \alpha, \lambda$ *and the domain* G.

4.2 Auxiliary statements and inequalities

In addition to Lemma 1.1 by means of direct calculation we obtain

Lemma 4.4.

$$x_i \cos(\vec{n}_k, x_i)|_{\Sigma_k} = 0, \quad k = 1, \dots, N-1 \tag{4.2.1}$$

(it is the equation of Σ_k*.)*

We will need some statements and inequalities.

4.2.1 The eigenvalue problem

Let $\Omega \subset S^{n-1}$ with smooth boundary $\partial\Omega$ be the intersection of the cone $\mathcal{C}$ with the unit sphere S^{n-1}. Let $\vec{\nu}$ be the exterior normal to $\partial\mathcal{C}$ at points of $\partial\Omega$ and $\vec{\tau}_k$ be the exterior with respect to Ω_k normal to Σ_k (lying in the plane tangent to Ω_k), $k = 1, \dots, N-1$. Let $\gamma(\omega)$, $\omega \in \partial\Omega$ be a positive bounded piecewise smooth function, $\beta_k(\omega)$ be a positive continuous function on Σ_k, $k = 1, \dots, N-1$. We consider the eigenvalue problem for the Laplace-Beltrami operator $\triangle_\omega$ on the unit sphere

$$\begin{cases} a_i\left(\triangle_\omega \psi_i + \vartheta\psi_i\right) = 0, \quad \omega \in \Omega_i, \ a_i \text{ are positive constants; } i = 1, \dots, N, \\ [\psi]_{\sigma_k} = 0, \quad \left[a\frac{\partial\psi}{\partial\vec{\tau_k}}\right]_{\sigma_k} + \beta_k(\omega)\psi\Big|_{\sigma_k} = 0, \quad k = 1, \dots, N-1, \\ \alpha(\omega)a\frac{\partial\psi}{\partial\vec{\nu}} + \gamma(\omega)\psi\Big|_{\partial\Omega} = 0, \end{cases} \tag{EVPN}$$

which consists in the determination of all values ϑ (eigenvalues) for which $(EVPN)$ has a non-zero weak solution (eigenfunction).

From the variational principle we obtain **the Friedrichs-Wirtinger type inequality**(see the proof for $N = 2$ in §2.2):

Theorem 4.5. *Let* ϑ *be the smallest positive eigenvalue of the problem* $(EVPN)$ *and* $\psi \in \mathbf{W}^1(\Omega)$*. Let* $\gamma(\omega)$ *be a positive bounded piecewise smooth function on* $\partial\Omega$*,* $\beta_k(\omega)$ *be a positive continuous function on* σ_k*,* $k = 1, \dots, N-1$*. Then*

$$\vartheta \int\limits_\Omega a\psi^2(\omega)d\Omega \le \int\limits_\Omega a|\nabla_\omega\psi|^2 d\Omega + \sum_{k=1}^{N-1} \int\limits_{\sigma_k} \beta_k(\omega)\psi^2(\omega)d\sigma + \int\limits_{\partial\Omega} \alpha(\omega)\gamma(\omega)\psi^2(\omega)d\sigma. \tag{4.2.2}$$

By (4.1.3), the Friedrichs-Wirtinger inequality can be written in the form

$$\lambda(\lambda+n-2) \int\limits_\Omega a\psi^2(\omega)d\Omega \le \int\limits_\Omega a|\nabla_\omega\psi|^2 d\Omega + \int\limits_\sigma \beta(\omega)\psi^2(\omega)d\sigma + \int\limits_{\partial\Omega} \alpha(\omega)\gamma(\omega)\psi^2(\omega)d\sigma, \tag{4.2.3}$$

for all $\psi(\omega) \in \mathbf{W}^1(\Omega)$.

Similarly as in §§2.2–2.3 we obtain

Corollary 4.6.

$$\int_{G_0^d} ar^{\alpha-4}u^2dx \le \frac{1}{\lambda(\lambda+n-2)}\Big\{\int_{G_0^d} ar^{\alpha-2}|\nabla u|^2dx + \int_{\Sigma_0^d} r^{\alpha-3}\beta(\omega)u^2(x)ds + \int_{\Gamma_0^d}\alpha(x)r^{\alpha-3}\gamma(\omega)u^2(x)ds\Big\}, \quad \forall\alpha, \tag{4.2.4}$$

for all $u \in \overset{\circ}{\mathbf{W}}{}^1_{\alpha-2}(G_0^d)$.

Lemma 4.7. *Let* $v \in \mathbf{C}^0(\overline{G}) \cap \mathbf{W}^1(G)$, $v(0) = 0$ *and* $\beta(\omega) > 0$, $\gamma(\omega) > 0$. *Then for any* $\varepsilon > 0$,

$$\int_{G_0^d} ar_\varepsilon^{\alpha-4}v^2dx \le H(\lambda,n,\alpha)\Big\{\int_{G_0^d} ar_\varepsilon^{\alpha-2}|\nabla v|^2dx + \int_{\Sigma_0^d} r_\varepsilon^{\alpha-3}\beta(\omega)v^2(x)ds + \int_{\Gamma_0^d}\alpha(x)r_\varepsilon^{\alpha-3}\gamma(\omega)v^2(x)ds\Big\}, \tag{4.2.5}$$

where $H(\lambda, n, \alpha)$ *is determined by* (2.3.3) *and* $\alpha \le 4 - n$.

Lemma 4.8. *Let* $v \in \mathbf{C}^0(\overline{G}) \cap \mathbf{W}^1(G)$ *and* $\beta(\omega) > 0$, $\gamma(\omega) > 0$. *Then for any* $\varepsilon > 0$,

$$\int_{G_0^d} ar_\varepsilon^{\alpha-2}r^{-2}v^2dx \le \frac{1}{\lambda(\lambda+n-2)}\Big\{\int_{G_0^d} ar_\varepsilon^{\alpha-2}|\nabla v|^2dx + \int_{\Sigma_0^d} r^{-1}r_\varepsilon^{\alpha-2}\beta(\omega)v^2(x)ds + \int_{\Gamma_0^d}\alpha(x)r^{-1}r_\varepsilon^{\alpha-2}\gamma(\omega)v^2ds\Big\}. \tag{4.2.6}$$

4.2.2 The comparison principle

We consider the second-order linear degenerate operator Q of the form

$$Q(u,\eta) \equiv \int_{G_0^d}\Big\{au_{x_i}\eta_{x_i} + (f(x)+pu(x))\,\eta(x)\Big\}dx - \int_{\Omega_d} a\frac{\partial u}{\partial r}\eta(x)d\Omega_d - \int_{\Gamma_0^d}\alpha(x)\left(g(x) - \frac{1}{r}\gamma(\omega)u(x)\right)\eta(x)ds - \int_{\Sigma_0^d}\left(h(x) - \frac{1}{r}\beta(\omega)u(x)\right)\eta(x)ds \tag{4.2.7}$$

for $u(x) \in \mathbf{C}^0(\overline{G_0^d}) \cap \overset{\circ}{\mathbf{W}}{}^1_0(G_0^d)$ and for all non-negative η belonging to $\mathbf{C}^0(\overline{G_0^d}) \cap \overset{\circ}{\mathbf{W}}{}^1_0(G_0^d)$.

Proposition 4.9. *Let $\beta(\omega)$, $\gamma(\omega)$ be positive piecewise functions, $0 < a_* \le a \le a^*$, $p \ge 0$ and $d \lll 1$. Let functions $u, w \in \mathbf{C}^0(\overline{G_0^d}) \cap \overset{\circ}{\mathbf{W}}{}_0^1(G_0^d)$ satisfy the inequality*

$$Q(u,\eta) \le Q(w,\eta) \tag{4.2.8}$$

for all non-negative $\eta \in \mathbf{C}^0(\overline{G_0^d}) \cap \overset{\circ}{\mathbf{W}}{}_0^1(G_0^d)$ and also the inequality

$$u(x) \le w(x), \quad x \in \Omega_d \cup \left(\Gamma_0^d \cap \mathcal{D}\right) \tag{4.2.9}$$

hold in the weak sense. Then $u(x) \le w(x)$ in G_0^d.

Proof. Let us define $z(x) = u(x) - w(x)$. Then we have

$$0 \ge Q(u,\eta) - Q(w,\eta) = \int\limits_{G_0^d} \Big\{ a z_{x_i}\eta_{x_i} + pz(x)\eta(x) \Big\} dx$$

$$+ \int\limits_{\Gamma_0^d} \alpha(x)\frac{1}{r}\gamma(\omega)z(x)\eta(x)ds + \int\limits_{\Sigma_0^d} \frac{1}{r}\beta(\omega)z(x)\eta(x)ds - \int\limits_{\Omega_d} a\frac{\partial z}{\partial r}\eta(x)d\Omega_d \tag{4.2.10}$$

for all non-negative $\eta \in \mathbf{C}^0(\overline{G_0^d}) \cap \overset{\circ}{\mathbf{W}}{}_0^1(G_0^d)$. We define the sets

$$(G_0^d)^+ := \{x \in G_0^d \mid u(x) > w(x)\} \subset G_0^d,$$
$$(\Sigma_0^d)^+ := \{x \in \Sigma_0^d \mid u(x) > w(x)\} \subset \Sigma_0^d,$$
$$(\Gamma_0^d)^+ := \{x \in \Gamma_0^d \mid u(x) > w(x)\} \subset \Gamma_0^d$$

and assume that $(G_0^d)^+ \ne \emptyset$. As the test function in the integral inequality (4.2.10), we choose $\eta = \max\{(u-w), 0\}$. Then it follows from (4.2.9) and (4.2.10) that

$$\int\limits_{(G_0^d)^+} \left(a|\nabla z|^2 + pz^2(x)\right) dx + \int\limits_{(\Gamma_0^d)^+} \alpha(x)\frac{1}{r}\gamma(\omega)z^2(x)ds + \int\limits_{(\Sigma_0^d)^+} \frac{1}{r}\beta(\omega)z^2(x)ds \le 0.$$

This implies that $z(x) = 0$ almost everywhere in $(G_0^d)^+$. Thus, we have finished with a contradiction to our assumption about the set $(G_0^d)^+ \ne \emptyset$. By this fact, Proposition 4.9 is proved. □

4.3 The barrier function. The preliminary estimate of the solution modulus

Let us define the linear operators

$$\mathcal{L}_i \equiv a_i \triangle, \ i = 1, \dots, N$$

and let us use numbers (4.1.2).

Lemma 4.10 (Existence of the barrier function). *Let us fix numbers $\beta_0 > \widetilde{a}\tan\frac{\omega_0}{2}$, $\gamma_0 > \widetilde{a}\tan\frac{\omega_0}{2}$, $\delta > 0$, $g_0 \geq 0$, $h_0 \geq 0$ and let $\gamma(\omega) \geq \gamma_0$ on ∂G and $\beta_k(\omega) \geq \beta_0$ on Σ_k, $k = 1,\dots,N-1$. Then there exist $m > 0$, depending only on ω_0, a number $\varkappa_0 \in (0, \frac{\delta_0}{\widetilde{a}}\cot\frac{\omega_0}{2} - 1)$ (where $\delta_0 = \min(\beta_0,\gamma_0)$), numbers $B > 0$, $d \in (0,1)$ and a function $w(x) \in \mathbf{C}^1(\overline{G_0}) \cap \mathbf{C}^2(G_0)$ that depend only on ω_0, the ellipticity constants a_*, a^* of the operators $\mathcal{L}_i$ and the quantities $\gamma_0, \beta_0, \delta_0, g_0, h_0, \omega_0$ such that for any $\varkappa \in (0; \min(\delta, \varkappa_0))$ the following inequalities hold:*

$$\mathcal{L}_i[w(x)] \leq -a_* m^2 |x|^{\varkappa-1};\quad x \in G_0^d \cap G_i;\ i = 1,\dots,N; \tag{4.3.1}$$

$$\mathcal{B}[w(x)] \geq g_0 |x|^{\delta};\quad x \in \Gamma_0^d; \tag{4.3.2}$$

$$\mathcal{S}_k[w(x)] \geq h_0 |x|^{\delta};\quad x \in \Sigma_k^d;\ k = 1,\dots,N-1; \tag{4.3.3}$$

$$0 \leq w(x) \leq c_0(\varkappa_0, B, \omega_0)|x|^{\varkappa+1};\quad x \in \overline{G_0^d}; \tag{4.3.4}$$

$$|\nabla w(x)| \leq c_1(\varkappa_0, B, \omega_0)|x|^{\varkappa};\quad x \in \overline{G_0^d}, \tag{4.3.5}$$

where operators $\mathcal{B}$, $\mathcal{S}_k$ were defined in problem (LN).

Proof. Let $(x, y, x') \in \mathbb{R}^n$, where $x = x_1$, $y = x_2$, $x' = (x_3,\dots,x_n)$. In $\{x_1 \geq 0\}$ we consider the cone K with the vertex in $\mathcal{O}$ such that $K \supset G_0^d$ (we recall that $G_0^d \subset \{x_1 \geq 0\}$). Let ∂K be the lateral surface of K and let $\partial K \cap yOx = \Gamma_\pm$ be $x = \pm my$, where $m = \cot\frac{\omega_0}{2}$, $0 < \omega_0 < \pi$ such that in the interior of K the inequality $x > m|y|$ holds. We shall consider the function:

$$w(x; y, x') \equiv x^{\varkappa-1}(x^2 - m^2y^2) + Bx^{\varkappa+1} \quad \text{with some } \varkappa \in (0;1),\ B > 0. \tag{4.3.6}$$

Let us calculate the operator $\mathcal{L}_i$ on the function (4.3.6). For $t = \frac{y}{x}$, $|t| < \frac{1}{m}$ we obtain:

$$\begin{aligned} \mathcal{L}_i w &= -m^2 x^{\varkappa-1} \varsigma_i(\varkappa), \quad \text{where} \\ \varsigma_i(\varkappa) &= a_i\left\{(\varkappa^2 - 3\varkappa + 2)t^2 + 2 - m^{-2}(1+B)(\varkappa^2 + \varkappa)\right\}. \end{aligned}$$

Since $\varsigma_i(0) = 2a_i(t^2+1) \geq 2a_i \geq 2a_*$ and since $\varsigma_i(\varkappa)$ are the square functions, there exists a number $\varkappa_0 > 0$ depending only on a_*, a^*, m such that $\varsigma_i(\varkappa) \geq a_*$ for $\varkappa \in [0; \varkappa_0]$. Therefore we obtain (4.3.1).

Now, let us write

$$\Gamma_\pm : x = \pm my,\ m = \cot\frac{\omega_0}{2},\ 0 < \omega_0 < \pi. \tag{4.3.7}$$

Then we have

$$\text{on } \Gamma_+ : \begin{cases} x &= r\cos\frac{\omega_0}{2}, \\ y &= r\sin\frac{\omega_0}{2} \end{cases} \quad \begin{cases} \angle(\vec{n}, x) = \frac{\pi}{2} + \frac{\omega_0}{2}, \\ \angle(\vec{n}, y) = \frac{\omega_0}{2}, \end{cases} \tag{4.3.8}$$

$$\text{on } \Gamma_- : \begin{cases} x &= r\cos\frac{\omega_0}{2}, \\ y &= -r\sin\frac{\omega_0}{2} \end{cases} \quad \begin{cases} \angle(\vec{n}, x) = \frac{\pi}{2} + \frac{\omega_0}{2}, \\ \angle(\vec{n}, y) = \pi + \frac{\omega_0}{2}, \end{cases} \tag{4.3.9}$$

$$\sin\frac{\omega_0}{2}=\frac{1}{\sqrt{1+m^2}},\qquad \cos\frac{\omega_0}{2}=\frac{m}{\sqrt{1+m^2}}. \tag{4.3.10}$$

Therefore, we obtain

$$w_x=(1+\varkappa)x^{\varkappa}(1+B)-(\varkappa-1)m^2y^2x^{\varkappa-2}\Rightarrow w_x\big|_{\Gamma_\pm}=[2+B(1+\varkappa)]x^{\varkappa},$$
$$w_y=-2m^2yx^{\varkappa-1}\Rightarrow w_y\big|_{\Gamma_\pm}=\mp 2mx^{\varkappa}. \tag{4.3.11}$$

Because of $\frac{\partial w}{\partial\vec{n}}\Big|_{\Gamma_\pm}=w_x\cos\angle(\vec{n},x)\Big|_{\Gamma_\pm}+w_y\cos\angle(\vec{n},y)\Big|_{\Gamma_\pm}$ and (4.3.8)–(4.3.11), we get: $\frac{\partial w}{\partial\vec{n}}\Big|_{\Gamma_\pm}=-r^{\varkappa}\frac{m^{\varkappa}}{(1+m^2)^{\frac{\varkappa+1}{2}}}\left\{2(1+m^2)+B(1+\varkappa)\right\}$. Hence, by (4.3.7) and $\gamma(\omega)\geq\gamma_0>\widetilde{a}\tan\frac{\omega_0}{2}$ on ∂G, it follows that

$$\begin{aligned}\mathcal{B}[w]\Big|_{\Gamma_\pm^d}&=\frac{m^{\varkappa}}{(1+m^2)^{\frac{\varkappa+1}{2}}}r^{\varkappa}\left\{Bm\gamma(\omega)-\alpha(x)a\left(B(1+\varkappa)+2(1+m^2)\right)\right\}\\&\geq\frac{m^{\varkappa}}{(1+m^2)^{\frac{\varkappa+1}{2}}}r^{\varkappa}\left\{Bm\gamma_0-Ba^*(1+\varkappa)-2a^*(1+m^2)\right\}.\end{aligned}$$

Since $m>\frac{a^*}{\gamma_0}$ and $\varkappa_0<m\frac{\gamma_0}{a^*}-1$, for $\varkappa\leq\varkappa_0$ we obtain

$$\mathcal{B}[w]\Big|_{\Gamma_\pm^d}\geq\frac{m^{\varkappa_0}r^{\varkappa}}{(1+m^2)^{\frac{\varkappa_0+1}{2}}}\left\{B(m\gamma_0-a^*-a^*\varkappa_0)-2a^*(1+m^2)\right\}\geq g_0r^{\delta}, \tag{4.3.12}$$

$0<r<d<1$, if we choose $\varkappa\leq\delta\implies r^{\varkappa}\geq r^{\delta}$. Hence it follows that

$$B\geq\left\{\frac{g_0(1+m^2)^{\frac{\varkappa_0+1}{2}}}{m^{\varkappa_0}}+2a^*(1+m^2)\right\}\cdot\frac{1}{m\gamma_0-a^*(1+\varkappa_0)}. \tag{4.3.13}$$

Thus, (4.3.2) is proved.

Now we prove (4.3.3). At first, we have on Σ_k:

$$x=r\cos\left(\frac{\omega_0}{2}-\theta_k\right),\ y=r\sin\left(\frac{\omega_0}{2}-\theta_k\right)\implies$$
$$y=m_kx,\ \text{where } m_k=\tan\left(\frac{\omega_0}{2}-\theta_k\right).$$

Therefore

$$\begin{aligned}w\Big|_{\Sigma_k}&=(B+1-m^2m_k^2)x^{1+\varkappa}=(B+1-m^2m_k^2)r^{1+\varkappa}\cos^{1+\varkappa}\left(\frac{\omega_0}{2}-\theta_k\right)\\&=\frac{B+1-m^2m_k^2}{(1+m_k^2)^{\frac{1+\varkappa}{2}}}r^{1+\varkappa}.\end{aligned}$$

Further,

$$\cos(\vec{n_k}, x) = \cos\left(\frac{\pi}{2} - \frac{\omega_0}{2} + \theta_k\right) = \sin\left(\frac{\omega_0}{2} - \theta_k\right) = \frac{m_k}{\sqrt{1+m_k^2}},$$
$$\cos(\vec{n_k}, y) = \cos\left(\pi - \frac{\omega_0}{2} + \theta_k\right) = -\cos\left(\frac{\omega_0}{2} - \theta_k\right) = -\frac{1}{\sqrt{1+m_k^2}}$$

on Σ_k, $k = 1, \dots, N-1$ and by virtue of

$$w_x\Big|_{\Sigma_k} = \frac{(1+\varkappa)(1+B) + (1-\varkappa)m^2 m_k^2}{(1+m_k^2)^{\frac{\varkappa}{2}}} \cdot r^\varkappa; \quad w_y\Big|_{\Sigma_k} = -\frac{2m^2 m_k^2}{(1+m_k^2)^{\frac{\varkappa}{2}}} \cdot r^\varkappa,$$

we obtain

$$\frac{\partial w}{\partial n_k}\Big|_{\Sigma_k} = m_k \cdot \frac{(1+\varkappa)(1+B) + (1-\varkappa)m^2 m_k^2 + 2m^2}{(1+m_k^2)^{\frac{1+\varkappa}{2}}} \cdot r^\varkappa.$$

Thus,

$$\mathcal{S}_k[w] = \frac{\beta_k(\omega)(B+1-m^2m_k^2) + m_k[a]_{\Sigma_k}\{(1+\varkappa)(1+B) + (1-\varkappa)m^2m_k^2 + 2m^2\}}{(1+m_k^2)^{\frac{1+\varkappa}{2}}} \cdot r^\varkappa.$$

Since $0 < \omega_1 \le \theta_k < \omega_0$ for all $k = 1, \dots, N-1$, then $|m_k| < \tan\frac{\omega_0}{2} = \frac{1}{m}$ for all $k = 1, \dots, N-1$. Now, by virtue of $\beta_k(\omega) \ge \beta_0 > \widetilde{a}\tan\frac{\omega_0}{2}$ on Σ_k, $k = 1, \dots, N-1$ and $\varkappa_0 < m\frac{\beta_0}{a_0} - 1$, for $\varkappa \le \varkappa_0$, we get

$$\mathcal{S}_k[w] \ge \{\beta_0 Bm - a_0\left((1+\varkappa_0)B + 2(1+m^2)\right)\}\frac{m^{\varkappa_0}}{(1+m^2)^{\frac{1+\varkappa_0}{2}}} \cdot r^\varkappa \ge r^\delta \cdot h_0, \tag{4.3.14}$$

if we choose $\varkappa \le \delta \implies r^\varkappa \ge r^\delta$. Hence it follows that

$$B \ge \left\{\frac{h_0(1+m^2)^{\frac{\varkappa_0+1}{2}}}{m^{\varkappa_0}} + 2a_0(1+m^2)\right\} \cdot \frac{1}{m\beta_0 - a_0(1+\varkappa_0)}. \tag{4.3.15}$$

Thus, (4.3.3) is proved.

Now we will show (4.3.4). Let us rewrite the function (4.3.6) in spherical coordinates. Recalling that $m = \cot\frac{\omega_0}{2}$ we obtain

$$w(x;y,x') = (1+B)(r\cos\omega)^{1+\varkappa} - m^2r^2\sin^2\omega(r\cos\omega)^{\varkappa-1}$$
$$= r^{1+\varkappa}\cos^{\varkappa-1}\omega\left(B\cos^2\omega + \frac{\chi(\omega)}{\sin^2\frac{\omega_0}{2}}\right), \quad \forall\omega \in \left[-\frac{\omega_0}{2}; \frac{\omega_0}{2}\right],$$

where $\chi(\omega) = \sin\left(\frac{\omega_0}{2} - \omega\right) \cdot \sin\left(\frac{\omega_0}{2} + \omega\right)$. We find $\chi'(\omega) = -\sin 2\omega$ and $\chi'(\omega) = 0$ for $\omega = 0$. Now we see that $\chi''(0) = -2\cos 0 = -2 < 0$. In this way we have $\max\limits_{\omega\in[-\omega_0/2,\omega_0/2]} \chi(\omega) = \chi(0) = \sin^2\frac{\omega_0}{2}$ and therefore

$$w(x;y,x') \le r^{1+\varkappa}\cos^{\varkappa-1}\omega(B\cos^2\omega + 1) \le r^{1+\varkappa}\cos^{\varkappa+1}\omega\left(B + \frac{1}{\cos^2\omega}\right)$$

$$\leq r^{1+\varkappa}\left(B+\frac{1}{\cos^2\frac{\omega_0}{2}}\right).$$

Hence (4.3.4) follows. Finally, (4.3.5) follows in the same way, by virtue of (4.3.11). □

Now we can estimate $|u(x)|$ for (LN) in a neighborhood of a conical point.

Theorem 4.11. *Let $u(x)$ be a weak solution of the problem (LN) and satisfy assumptions* (a)–(c). *Then there exist numbers $d \in (0,1)$ and $\varkappa > 0$ depending only on $a_*, n, \omega_0, f_0, h_0, g_0, \beta_0, \gamma_0, s$ and the domain G such that*

$$|u(x)-u(0)| \leq C_0|x|^{\varkappa+1}, \quad x \in G_0^d, \tag{4.3.16}$$

where the positive constant C_0 depends only on $a_*, n, \omega_0, f_0, h_0, g_0, \beta_0, \gamma_0, s, M_0$ and the domain G, and does not depend on $u(x)$.

Proof. Without loss of generality we may suppose that $u(0) \geq 0$. Let us take the barrier function $w(x)$ defined by (4.3.6) with $\varkappa \in (0, \varkappa_0)$ and the function $v(x) = u(x) - u(0)$. For them we shall verify Proposition 4.9. Let us calculate operator Q on these functions. Because of the definition $(II)_{loc}$, by Lemma 4.2 and by integrating by parts, we have:

$$Q(Aw,\eta) \equiv \int\limits_{G_0^d}\left\{Aaw_{x_i}\eta_{x_i} + (f(x)+Apw(x))\,\eta(x)\right\}dx - \int\limits_{\Omega_d} Aa\frac{\partial w}{\partial r}\eta(x)d\Omega_d$$
$$-\int\limits_{\Gamma_0^d}\alpha(x)\left(g(x)-\frac{1}{r}\gamma(\omega)Aw(x)\right)\eta(x)ds - \int\limits_{\Sigma_0^d}\left(h(x)-\frac{1}{r}\beta(\omega)Aw(x)\right)\eta(x)ds$$
$$= \int\limits_{G_0^d}(f(x)+Apw(x)-Aa\triangle w)\,\eta(x)dx$$
$$+\int\limits_{\Gamma_0^d}(\mathcal{B}[Aw]-g(x))\,\eta(x)ds + \int\limits_{\Sigma_0^d}(\mathcal{S}[Aw]-h(x))\,\eta(x)ds$$

with any $A > 0$. Hence, by Lemma 4.10 and conditions (b), (c), we obtain

$$Q(Aw,\eta) \geq \int\limits_{G_0^d}\left\{f(x)+Aa_*m^2r^{\varkappa-1}\right\}\eta(x)dx + \int\limits_{\Gamma_0^d}\left(Ag_0r^\delta - g(x)\right)\eta(x)ds$$
$$+\int\limits_{\Sigma_0^d}\left(Ah_0r^\delta - h(x)\right)\eta(x)ds$$

$$\geq \int\limits_{G_0^d} r^{\varkappa-1}\Big\{Aa_*m^2 - f_0 r^{\beta+1-\varkappa}\Big\}\eta(x)dx + +g_0 \int\limits_{\Gamma_0^d} (Ar^\delta - r^{s-1})\eta(x)ds$$

$$+ h_0 \int\limits_{\Sigma_0^d} (Ar^\delta - r^{s-1})\eta(x)ds \geq 0,$$

because of $0 < \varkappa < \varkappa_0$, if numbers $\varkappa_0, \delta, A$ are chosen such that

$$\varkappa_0 \leq \beta + 1,\ \delta \leq s - 1,\ A \geq \max\{1, \frac{f_0}{a_* m^2}\}. \tag{4.3.17}$$

On the other hand, we have

$$Q(v,\eta) \equiv -\int\limits_{G_0^d} pu(0)\eta(x)dx - \int\limits_{\Gamma_0^d} \frac{\gamma(\omega)}{r} u(0)\eta(x)ds - \int\limits_{\Sigma_0^d} \frac{\sigma(\omega)}{r} u(0)\eta(x)ds \leq 0$$

and thus we get

$$Q(v,\eta) \leq Q(Aw,\eta) \tag{4.3.18}$$

for all non-negative $\eta \in \mathbf{C}^0(\overline{G_0^d}) \cap \overset{\circ}{\mathbf{W}}{}^1_0(G_0^d)$.

Now we compare $v(x)$ and $w(x)$ on Ω_d. Since $x^2 \geq h^2 y^2$ in $\overline{K}$ from (4.3.6) we have

$$w(x)\Big|_{r=d} \geq Bd^{1+\varkappa} \cos^{\varkappa+1} \frac{\omega_0}{2}. \tag{4.3.19}$$

On the other hand,

$$v(x)\Big|_{\Omega_d} = (u(x) - u(0))\Big|_{\Omega_d} \leq M_0 \tag{4.3.20}$$

and therefore from (4.3.19)–(4.3.20) we obtain

$$Aw(x)\Big|_{\Omega_d} \geq ABd^{1+\varkappa} \cos^{\varkappa+1} \frac{\omega_0}{2} \geq ABd^{1+\varkappa} \frac{m^{1+\varkappa_0}}{(1+m^2)^{\frac{1+\varkappa_0}{2}}} \geq M_0 \geq v\Big|_{\Omega_d}$$

if we choose A possibly greatest,

$$A \geq \frac{M_0(1+m^2)^{\frac{1+\varkappa_0}{2}}}{B(md)^{1+\varkappa_0}}, \tag{4.3.21}$$

where B satisfies inequalities (4.3.13) and (4.3.15).

Finally, if $\Gamma_0^d \cap \mathcal{D} \neq \emptyset$ and $u(x) = g(x),\ x \in \Gamma_0^d \cap \mathcal{D}$, where $|g(x)| \leq g_0|x|^s$, then we have

$$v(x) = u(x) - u(0) \leq u(x) = g(x) \leq g_0|x|^s,\ x \in \Gamma_0^d \cap \mathcal{D};$$

$$Aw\Big|_{\Gamma_0^d \cap \mathcal{D}} = ABr^{1+\varkappa} \cos^{\varkappa+1} \frac{\omega_0}{2} \geq ABr^{1+\varkappa} \frac{m^{1+\varkappa_0}}{(1+m^2)^{\frac{1+\varkappa_0}{2}}} \geq g_0 r^s \geq v\Big|_{\Gamma_0^d \cap \mathcal{D}}$$

if $\varkappa_0 \le s - 1$ and if we choose A possibly greatest,

$$A \ge \frac{(1+m^2)^{\frac{1+\varkappa_0}{2}}}{Bm^{1+\varkappa_0}} g_0, \tag{4.3.22}$$

where B satisfies inequalities (4.3.13) and (4.3.15). Thus, if we choose large numbers $B > 0,\ A \ge 1$ according to (4.3.13), (4.3.15), (4.3.17), (4.3.21) and (4.3.22), we provide the validity of Proposition 4.9.

Therefore, by the comparison principle (Proposition 4.9), we have:

$$u(x) - u(0) \le Aw(x),\ x \in \overline{G_0^d}. \tag{4.3.23}$$

Similarly, we derive the estimate

$$u(x) - u(0) \ge -Aw(x)$$

if we consider the auxiliary function $v(x) = u(0) - u(x)$. By virtue of (4.3.4), our theorem is proved. □

Now we will estimate the gradient modulus of the problem (LN) solution near a conical point.

Theorem 4.12. *Let $u(x)$ be a weak solution of the problem (LN) and assumptions* (b)–(c) *be satisfied. Let $\varkappa > 0$ be a number defined by Lemma* 4.10 *with $\varkappa_0 \le s-1$. Then there exists a number $d > 0$ such that*

$$|\nabla u(x)| < C_1 |x|^{\varkappa},\ \ x \in G_0^d, \tag{4.3.24}$$

where the constant C_1 does not depend on u, but depends only on $a_*, a^*, n, \omega_0, f_0, h_0, g_0, \beta_0, \gamma_0, s, M_0$ and the domain G.

Proof. Let us consider the set $G_{\varrho/2}^{\varrho} \subset G,\ 0 < \rho < d$. We make the transformation $x = \varrho x'$; $v(x') = \varrho^{-1-\varkappa} u(\varrho x')$. The function $v(x')$ satisfies the problem

$$\begin{cases} a\Delta' v - p\varrho^2 v(x') = \varrho^{1-\varkappa} f(x'), & x \in G_{1/2}^1 \setminus \bigcup\limits_{k=1}^{N-1} (\Sigma_k)_{1/2}^1; \\ \left[a\frac{\partial v}{\partial n'_k}\right]_{(\Sigma_k)_{1/2}^1} + \frac{1}{|x'|}\beta_k(\omega) v(x') = \varrho^{-\varkappa} h_k(\varrho x'), & x' \in (\Sigma_k)_{1/2}^1,\ k = 1, \dots, N-1; \\ \alpha(x') \cdot a\frac{\partial v}{\partial n'} + \frac{1}{|x'|}\gamma(\omega) v(x') = \varrho^{-\varkappa} g(\rho x'), & x' \in \Gamma_{1/2}^1. \end{cases} \tag{LN)$'$}$$

Now we apply Theorem 2.1 §2 [6] for $\alpha = 1$ and Theorem 16.2 chapter III [43] for $\alpha = 0$; according to these Theorems

$$\max_{x' \in G_{1/2}^1} |\nabla' v(x')| \le M_1'. \tag{4.3.25}$$

Returning to the variable x and the function $u(x)$ we obtain from (4.3.25)

$$|\nabla u(x)| \le M_1 \rho^{\varkappa},\ x \in G_{\rho/2}^{\rho},\ 0 < \rho < d.$$

Putting now $|x| = \frac{2}{3}\rho$ we obtain the desired estimate (4.3.24). □

Corollary 4.13. *Let $u(x)$ be a weak solution of problem (LN) and assumptions of Theorem 4.12 be satisfied. Then $u(0) = 0$ and therefore the inequality (4.3.16) takes the form*

$$|u(x)| \le C_0 |x|^{\varkappa+1}, \quad x \in G_0^d. \tag{4.3.26}$$

Proof. From the problem boundary condition it follows that

$$\gamma(\omega)u(x) = |x|g(x) - \alpha(x) \cdot a|x|\frac{\partial u}{\partial n}, \quad x \in \partial G \setminus \mathcal{O}.$$

By the assumption (b)–(c) and the estimate (4.3.24), we obtain

$$\gamma_0|u(x)| \le \gamma(\omega)|u(x)| \le |x||g(x)| + a^*|x||\nabla u| \le g_0|x|^s + C_1 a^*|x|^{\varkappa+1}.$$

By letting $|x|$ tend to 0 we get, because of the continuity of $u(x)$, that $\gamma_0|u(0)| \le 0$ and taking into account $\gamma_0 > 0$, we obtain that $u(0) = 0$. □

4.4 Local estimate at the boundary

Here we formulate a result asserting the local boundedness (near the conical point) of the weak solution of problem (LN). This result is a generalization of Theorem 3.4 in the case of a domain with N different media and it proves verbatim as Theorem 3.4.

Theorem 4.14. *Let $u(x)$ be a weak solution of problem (LN) and assumptions* (a)–(c) *be satisfied. If, in addition, $h(x) \in \mathbf{L}_\infty(\Sigma_0)$, $g(x) \in L_\infty(\partial G)$, then the inequality*

$$\sup_{G_0^{\varkappa\varrho}} |u(x)| \le C\left\{\varrho^{-n/t}\|u\|_{t,G_0^\varrho} + \varrho^{2(1-n/p)}\|f\|_{p/2,G_0^\varrho} + \varrho\left(\|g\|_{\infty,\Gamma_0^\varrho} + \|h\|_{\infty,\Sigma_0^\varrho}\right)\right\} \tag{4.4.1}$$

holds for any $t > 0$, $\varkappa \in (0,1)$ and $\varrho \in (0,d)$, where $C = C(n, a_*, a^*, t, p, \varkappa, G)$.

4.5 Global integral estimates

First we will obtain a global estimate for the Dirichlet integral.

Theorem 4.15. *Let $u(x)$ be a weak solution of problem (LN) and assumptions* (a)–(c) *be satisfied. Suppose, in addition, that $h(x) \in \mathbf{L}_2(\Sigma)$, $g(x) \in \mathbf{L}_2(\partial G)$. Then the inequality*

$$\int_G a|\nabla u|^2 dx + \int_\Sigma \frac{\beta(\omega)}{r} u^2(x) ds + \int_{\partial G} \frac{\gamma(\omega)}{r} u^2(x) ds$$
$$\le C\left\{\int_G \left(u^2(x) + f^2(x)\right) dx + \frac{1}{\beta_0}\int_\Sigma h^2(x) ds + \frac{1}{\gamma_0}\int_{\partial G} g^2(x) ds\right\} \tag{4.5.1}$$

holds, where constant C depends only on p^*, diam G.

Proof. Setting in (II) $\eta(x) = u(x)$ and using the classical Hölder inequality, by assumptions (a), (c), we get

$$\int_G \langle a|\nabla u|^2 + pu^2(x)\rangle\, dx + \int_\Sigma \frac{\beta(\omega)}{r} u^2(x)ds + \int_{\partial G} \frac{\gamma(\omega)}{r} u^2(x)ds$$
$$\leq \int_\Sigma |u||h(x)|ds + \int_{\partial G} |u||g(x)|ds + \int_G |u||f(x)|dx. \quad (4.5.2)$$

By the Cauchy inequality, and by virtue of the assumption (c), we obtain

$$\int_\Sigma |u||h(x)|ds = \int_\Sigma \left(\sqrt{\frac{\beta(\omega)}{r}}|u|\right)\left(\sqrt{\frac{r}{\beta(\omega)}}|h(x)|\right) ds$$
$$\leq \frac{1}{2}\int_\Sigma \frac{\beta(\omega)}{r} u^2(x)ds + \frac{1}{2\beta_0}\int_\Sigma rh^2(x)ds;$$
$$\int_{\partial G} |u||g(x)|ds - \int_{\partial G} \left(\sqrt{\frac{\gamma(\omega)}{r}}|u|\right)\left(\sqrt{\frac{r}{\gamma(\omega)}}|g(x)|\right) ds$$
$$\leq \frac{1}{2}\int_{\partial G} \frac{\gamma(\omega)}{r} u^2(x)ds + \frac{1}{2\gamma_0}\int_{\partial G} rg^2(x)ds;$$
$$\int_G |u||f(x)|dx \leq \frac{1}{2}\int_G |u|^2dx + \frac{1}{2}\int_G |f|^2dx.$$

Hence we get the desired inequality (4.5.1). □

Further, we will obtain a global estimate for the weighted Dirichlet integral.

Theorem 4.16. *Let $u(x)$ be a weak solution of problem (LN) and assumptions* (a)–(c) *be satisfied. Let λ be as above in* (4.1.3), *$\varkappa_0$ be as above in Lemma* 4.10 *about the barrier function. In addition, let $\lambda > 1$ and $0 < \varkappa \leq \min(s-1, \varkappa_0, \lambda - 1)$ as well as*

$$f(x) \in \overset{\circ}{\mathbf{W}}{}^0_\alpha(G), \quad \int_\Sigma r^{\alpha-1}h^2(x)ds < \infty, \quad \int_{\partial G} r^{\alpha-1}g^2(x)ds < \infty,$$

where

$$2 - n - 2\varkappa < \alpha \leq 2. \quad (4.5.3)$$

Then $u(x) \in \overset{\circ}{\mathbf{W}}{}^1_{\alpha-2}(G)$ and the inequality

$$\int_G a\left(r^{\alpha-2}|\nabla u|^2 + r^{\alpha-4}u^2\right)dx + \int_\Sigma r^{\alpha-3}\beta(\omega)u^2(x)ds + \int_{\partial G} r^{\alpha-3}\gamma(\omega)u^2(x)ds$$

$$\leq C\Big\{\int_G \left(u^2 + (1+r^\alpha)f^2(x)\right)dx + \int_\Sigma r^{\alpha-1}h^2(x)ds + \int_{\partial G} r^{\alpha-1}g^2(x)ds\Big\} \quad (4.5.4)$$

holds, where the constant $C > 0$ depends only on $n, a_*, a^*, \alpha, \lambda$ and the domain G.

Proof. Setting in (II) $\eta(x) = r_\varepsilon^{\alpha-2}u(x)$, with regard to $p \geq 0$ and $\eta_{x_i} = r_\varepsilon^{\alpha-2}u_{x_i} + (\alpha-2)r_\varepsilon^{\alpha-3}\frac{x_i-\varepsilon l_i}{r_\varepsilon}u(x)$, we obtain

$$\int_G ar_\varepsilon^{\alpha-2}|\nabla u|^2dx + \int_\Sigma r^{-1}r_\varepsilon^{\alpha-2}\beta(\omega)u^2(x)ds + \int_{\partial G}\alpha(x)r^{-1}r_\varepsilon^{\alpha-2}\gamma(\omega)u^2(x)ds$$

$$= \frac{2-\alpha}{2}\int_G ar_\varepsilon^{\alpha-4}(x_j-\varepsilon l_j)(u^2)_{x_j}dx - \int_G f(x)r_\varepsilon^{\alpha-2}u(x)dx$$

$$+ \int_\Sigma r_\varepsilon^{\alpha-2}u(x)h(x)ds + \int_{\partial G}\alpha(x)r_\varepsilon^{\alpha-2}u(x)g(x)ds. \quad (4.5.5)$$

We transform the first integral on the right-hand side by integrating by parts:

$$\int_G ar_\varepsilon^{\alpha-4}(x_j-\varepsilon l_j)\frac{\partial u^2}{\partial x_j}dx = \sum_{i=1}^N\int_{G_i} a_ir_\varepsilon^{\alpha-4}(x_j-\varepsilon l_j)\frac{\partial u_i^2}{\partial x_j}dx$$

$$= -\int_G au^2\frac{\partial}{\partial x_i}\Big(r_\varepsilon^{\alpha-4}(x_i-\varepsilon l_i)\Big)dx + \sum_{i=1}^N\int_{\partial G_i} a_iu_i^2r_\varepsilon^{\alpha-4}(x_j-\varepsilon l_j)\cos(\overrightarrow{n},x_j)ds$$

$$= -\int_G au^2\frac{\partial}{\partial x_i}\Big(r_\varepsilon^{\alpha-4}(x_i-\varepsilon l_i)\Big)dx + \int_{\partial G} au^2r_\varepsilon^{\alpha-4}(x_i-\varepsilon l_i)\cos(\overrightarrow{n},x_i)ds$$

$$+ \sum_{k=1}^{N-1}[a]_{\Sigma_k}\int_{\Sigma_k} u^2r_\varepsilon^{\alpha-4}(x_i-\varepsilon l_i)\cos(\overrightarrow{n_k},x_i)ds, \quad (4.5.6)$$

because of $[u]_{\Sigma_k} = 0,\ k = 1,\dots,N-1$. We calculate:

1) $\frac{\partial}{\partial x_i}\Big(r_\varepsilon^{\alpha-4}(x_i-\varepsilon l_i)\Big) = nr_\varepsilon^{\alpha-4} + (\alpha-4)(x_i-\varepsilon l_i)r_\varepsilon^{\alpha-5}\frac{x_i-\varepsilon l_i}{r_\varepsilon} = (n+\alpha-4)r_\varepsilon^{\alpha-4}$;

2) by (4.2.1), $(x_i-\varepsilon l_i)\cos(\overrightarrow{n_k},x_i)\Big|_{\Sigma_k} = \varepsilon\cos(\overrightarrow{n_k},x_1\Big|_{\Sigma_k} = \varepsilon\sin\left(\frac{\omega_0}{2}-\theta_k\right),\ k = 1,\dots,N-1$;

3) representing $\partial G = \Gamma_0^d \cup \Gamma_d$ and by (4.2.1),

$$(x_i - \varepsilon l_i)\cos(\overrightarrow{n}, x_i)\Big|_{\Gamma_0^d} = -\varepsilon \sin\frac{\omega_0}{2} \Longrightarrow$$

$$\int\limits_{\partial G} a u^2 r_\varepsilon^{\alpha-4}(x_i - \varepsilon l_i)\cos(\overrightarrow{n}, x_i)ds$$

$$= \int\limits_{\Gamma_d} a u^2 r_\varepsilon^{\alpha-4}(x_i - \varepsilon l_i)\cos(\overrightarrow{n}, x_i)ds - \varepsilon\sin\frac{\omega_0}{2}\int\limits_{\Gamma_0^d} a u^2 r_\varepsilon^{\alpha-4}ds.$$

Hence and from (4.5.6) it follows that

$$\frac{2-\alpha}{2}\int\limits_G a r_\varepsilon^{\alpha-4}(x_i - \varepsilon l_i)\frac{\partial u^2}{\partial x_i}dx = \frac{(2-\alpha)(4-n-\alpha)}{2}\int\limits_G a r_\varepsilon^{\alpha-4}u^2 dx$$

$$-\varepsilon\frac{2-\alpha}{2}\sin\frac{\omega_0}{2}\int\limits_{\Gamma_0^d} a u^2 r_\varepsilon^{\alpha-4}ds + \frac{2-\alpha}{2}\int\limits_{\Gamma_d} a u^2 r_\varepsilon^{\alpha-4}(x_i - \varepsilon l_i)\cos(\overrightarrow{n}, x_i)ds$$

$$+\varepsilon\sum_{k=1}^{N-1}[a]_{\Sigma_k}\int\limits_{\Sigma_k} u^2 r_\varepsilon^{\alpha-4}\sin\left(\frac{\omega_0}{2} - \theta_k\right)ds. \qquad (4.5.7)$$

From (4.5.5) and (4.5.7) we obtain the following equality:

$$\int\limits_G a r_\varepsilon^{\alpha-2}|\nabla u|^2 dx + \int\limits_\Sigma r^{-1} r_\varepsilon^{\alpha-2}\beta(\omega)u^2(x)ds + \int\limits_{\partial G}\alpha(x) r^{-1} r_\varepsilon^{\alpha-2}\gamma(\omega)u^2(x)ds$$

$$+\varepsilon\frac{2-\alpha}{2}\sin\frac{\omega_0}{2}\int\limits_{\Gamma_0^d} a u^2 r_\varepsilon^{\alpha-4}ds$$

$$= \frac{(2-\alpha)(4-n-\alpha)}{2}\int\limits_G a r_\varepsilon^{\alpha-4}u^2 dx + \int\limits_\Sigma r_\varepsilon^{\alpha-2}u(x)h(x)ds$$

$$+\int\limits_{\partial G}\alpha(x) r_\varepsilon^{\alpha-2}u(x)g(x)ds + \frac{2-\alpha}{2}\int\limits_{\Gamma_d} a u^2 r_\varepsilon^{\alpha-4}(x_i - \varepsilon l_i)\cos(\overrightarrow{n}, x_i)ds$$

$$+\varepsilon\sum_{k=1}^{N-1}[a]_{\Sigma_k}\int\limits_{\Sigma_k} u^2 r_\varepsilon^{\alpha-4}\sin\left(\frac{\omega_0}{2} - \theta_k\right)ds - \int\limits_G f(x) r_\varepsilon^{\alpha-2}u(x)dx. \qquad (4.5.8)$$

Now we estimate the integral over Γ_d. Because on Γ_d: $r_\varepsilon \geq hr \geq hd$ and $(\alpha-3)\ln r_\varepsilon \leq (\alpha-3)\ln(hd)$, by $\alpha \leq 2$, we have $r_\varepsilon^{\alpha-3}|_{\Gamma_d} \leq (hd)^{\alpha-3}$ and therefore, by (1.5.12),

$$\frac{2-\alpha}{2}\int\limits_{\Gamma_d} au^2 r_\varepsilon^{\alpha-4}(x_i-\varepsilon l_i)\cos(\vec{n},x_i)ds \le \frac{2-\alpha}{2}\int\limits_{\Gamma_d} a r_\varepsilon^{\alpha-3}u^2 ds$$
$$\le \frac{2-\alpha}{2}(hd)^{\alpha-3}\int\limits_{\Gamma_d} au^2 ds \le c_\delta \int\limits_{G_d} u^2 dx + \delta\int\limits_{G_d}|\nabla u|^2 dx,\ \forall\delta>0. \quad (4.5.9)$$

Further, by the Cauchy inequality and because $\gamma(\omega) \ge \gamma_0 > 0$,

$$ug \le \left(r^{\frac{1}{2}}\frac{1}{\sqrt{\gamma(\omega)}}|g|\right)\left(r^{-\frac{1}{2}}\sqrt{\gamma(\omega)}|u|\right) \le \frac{\delta_1}{2}r^{-1}\gamma(\omega)u^2 + \frac{1}{2\delta_1\gamma_0}rg^2(x),\ \forall\delta_1>0;$$

taking into account property 1) of r_ε we obtain

$$\int\limits_{\partial G} r_\varepsilon^{\alpha-2}|u||g|ds \le \frac{\delta_1}{2}\int\limits_{\partial G} r_\varepsilon^{\alpha-2}\frac{1}{r}\gamma(\omega)u^2 ds + \frac{1}{2\delta_1\gamma_0}\int\limits_{\partial G} r^{\alpha-1}g^2(x)ds,\ \forall\delta_1>0. \quad (4.5.10)$$

Similarly, because $\beta(\omega) \ge \beta_0 > 0$,

$$\int\limits_{\Sigma} r_\varepsilon^{\alpha-2}|u||h(x)|ds \le \frac{\delta_1}{2}\int\limits_{\Sigma} r_\varepsilon^{\alpha-2}\frac{1}{r}\beta(\omega)u^2 ds + \frac{1}{2\delta_1\beta_0}\int\limits_{\Sigma_0} r^{\alpha-1}h^2(x)ds,\ \forall\delta_1>0 \quad (4.5.11)$$

and

$$\int\limits_{G} r_\varepsilon^{\alpha-2}uf(x)dx \le \frac{\delta}{2}\int\limits_{G} ar^{-2}r_\varepsilon^{\alpha-2}u^2 dx + \frac{1}{2a_*\delta}\int\limits_{G} r^{\alpha}f^2(x)dx,\ \forall\delta>0. \quad (4.5.12)$$

Finally, we estimate the integral $\varepsilon\sum\limits_{k=1}^{N-1}[a]_{\Sigma_k}\int\limits_{\Sigma_k} u^2 r_\varepsilon^{\alpha-4}\sin\left(\frac{\omega_0}{2}-\theta_k\right)ds$. At first, by (4.1.2), we get

$$\varepsilon\sum_{k=1}^{N-1}[a]_{\Sigma_k}\int\limits_{\Sigma_k} u^2 r_\varepsilon^{\alpha-4}\sin\left(\frac{\omega_0}{2}-\theta_k\right)ds \le a_0\varepsilon\sum_{k=1}^{N-1}\int\limits_{\Sigma_k} u^2 r_\varepsilon^{\alpha-4}ds$$
$$= a_0\varepsilon\int\limits_{\Sigma} r_\varepsilon^{\alpha-4}u^2(x)ds. \quad (4.5.13)$$

Further, we use the representation $\Sigma = \Sigma_0^\varepsilon \cup \Sigma_\varepsilon$. Then, by property 1) of $r_\varepsilon(x)$ and by virtue of inequality (4.3.26) of Corollary 4.13, we get

$$\varepsilon \int\limits_{\Sigma_0^\varepsilon} r_\varepsilon^{\alpha-4} u^2(x)ds \le \int\limits_{\Sigma_0^\varepsilon} r_\varepsilon^{\alpha-3} u^2(x)ds \le C_0^2 \operatorname{meas}\sigma \cdot \int_0^\varepsilon r^{\alpha+2\varkappa+n-3}dr$$

$$= \frac{C_0^2 \operatorname{meas}\sigma}{\alpha+2\varkappa+n-2}\varepsilon^{\alpha+2\varkappa+n-2}, \quad (4.5.14)$$

because of $\alpha+2\varkappa+n-2>0$, by our assumption (4.5.3). Similarly, for the integral over Σ_ε we obtain

$$\varepsilon \int\limits_{\Sigma_\varepsilon} r_\varepsilon^{\alpha-4} u^2(x)ds \le \varepsilon \cdot \operatorname{meas}\sigma \cdot \int_\varepsilon^R r^{\alpha+2\varkappa+n-4}dr$$

$$= \operatorname{meas}\sigma \cdot \begin{cases} \frac{\varepsilon R^{\alpha+2\varkappa+n-3} - \varepsilon^{\alpha+2\varkappa+n-2}}{\alpha+2\varkappa+n-3}, & \text{if } \alpha+2\varkappa+n-3 \ne 0, \\ \varepsilon \ln \frac{R}{\varepsilon}, & \text{if } \alpha+2\varkappa+n-3 = 0, \end{cases} \quad (4.5.15)$$

where $R = \max\limits_{1\le k\le N-1} \operatorname{diam}\Sigma_k$. Thus, from (4.5.13)–(4.5.15) we get

$$\varepsilon \cdot \sum_{k-1}^{N-1} [a]_{\Sigma_k} \int\limits_{\Sigma_k} u^2 r_\varepsilon^{\alpha-4} \sin\left(\frac{\omega_0}{2} - \theta_k\right) ds \le J(\varepsilon), \quad (4.5.16)$$

where

$$J(\varepsilon) = \varepsilon \cdot a_0 \cdot \operatorname{meas}\sigma \quad (4.5.17)$$

$$\times \begin{cases} \frac{R^{\alpha+2\varkappa+n-3}}{\alpha+2\varkappa+n-3} + \left(\frac{C_0^2}{\alpha+2\varkappa+n-2} - \frac{1}{\alpha+2\varkappa+n-3}\right)\varepsilon^{\alpha+2\varkappa+n-3}, & \text{if } \alpha+2\varkappa+n-3 \ne 0, \\ C_0^2 + \ln\frac{R}{\varepsilon}, & \text{if } \alpha+2\varkappa+n-3 = 0, \end{cases}$$

provided that $\alpha+2\varkappa+n-2>0$. Hence follows

Corollary 4.17.

$$\lim_{\varepsilon\to+0} J(\varepsilon) = 0 \implies \lim_{\varepsilon\to+0} \varepsilon \cdot \sum_{k=1}^{N-1} [a]_{\Sigma_k} \int\limits_{\Sigma_k} u^2 r_\varepsilon^{\alpha-4} \sin\left(\frac{\omega_0}{2} - \theta_k\right) ds = 0.$$

As a result from (4.5.8)–(4.5.16) we obtain:

$$\int\limits_G a r_\varepsilon^{\alpha-2} |\nabla u|^2 dx + \int\limits_\Sigma r^{-1} r_\varepsilon^{\alpha-2} \beta(\omega) u^2(x)ds + \int\limits_{\partial G} r^{-1} r_\varepsilon^{\alpha-2} \gamma(\omega) u^2(x)ds$$

$$+ \varepsilon \frac{2-\alpha}{2} \sin\frac{\omega_0}{2} \int\limits_{\Gamma_0^d} a u^2 r_\varepsilon^{\alpha-4} ds$$

$$
\begin{aligned}
&\le J(\varepsilon) + \frac{(2-\alpha)(4-n-\alpha)}{2}\int\limits_G a r_\varepsilon^{\alpha-4}u^2dx\\
&+\frac{\delta}{2}\int\limits_{G_0^d} a r^{-2} r_\varepsilon^{\alpha-2}u^2dx + c(a_*,\alpha)\int\limits_G \left(a|\nabla u|^2+u^2\right)dx\\
&+\frac{\delta_1}{2}\left\{\int\limits_\Sigma r^{-1}r_\varepsilon^{\alpha-2}\beta(\omega)u^2(x)ds + \int\limits_{\partial G} r^{-1}r_\varepsilon^{\alpha-2}\gamma(\omega)u^2(x)ds\right\} \qquad (4.5.18)\\
&+\frac{1}{2\delta_1\gamma_0}\int\limits_{\partial G} r^{\alpha-1}g^2(x)ds + \frac{1}{2\delta_1\beta_0}\int\limits_\Sigma r^{\alpha-1}h^2(x)ds + \frac{1}{2a_*\delta}\int\limits_G r^\alpha f^2(x)dx,
\end{aligned}
$$

for all $\delta,\delta_1>0$. It is clear that it is sufficient to consider the case $\alpha<4-n$. At first we apply Lemma 4.7. Then from (4.5.18) we get:

$$
\begin{aligned}
&\left(1-\frac{2\,(2-\alpha)\,(4-\alpha-n)}{(4-n-\alpha)^2+4\lambda(\lambda+n-2)}\right)\\
&\times\left\{\int\limits_G a r_\varepsilon^{\alpha-2}|\nabla u|^2dx + \int\limits_\Sigma \frac{1}{r}r_\varepsilon^{\alpha-2}\beta(\omega)u^2(x)ds + \int\limits_{\partial G}\frac{1}{r}r_\varepsilon^{\alpha-2}\gamma(\omega)u^2(x)ds\right\}\\
&\le J(\varepsilon) + \frac{\delta}{2}\int\limits_{G_0^d} a r^{-2}r_\varepsilon^{\alpha-2}u^2dx + c(a_*,\alpha)\int\limits_G\left(a|\nabla u|^2+u^2\right)dx\\
&+\frac{1}{2\delta_1\gamma_0}\int\limits_{\partial G} r^{\alpha-1}g^2(x)ds\\
&+\frac{\delta_1}{2}\left\{\int\limits_\Sigma r^{-1}r_\varepsilon^{\alpha-2}\beta(\omega)u^2(x)ds + \int\limits_{\partial G} r^{-1}r_\varepsilon^{\alpha-2}\gamma(\omega)u^2(x)ds\right\}\\
&+\frac{1}{2\delta_1\beta_0}\int\limits_\Sigma r^{\alpha-1}h^2(x)ds + \frac{1}{2a_*\delta}\int\limits_G r^\alpha f^2(x)dx,\ \forall\delta,\delta_1>0,\ \forall\varepsilon>0. \qquad (4.5.19)
\end{aligned}
$$

However, if $\alpha>4-n-2\lambda$, then

$$
\frac{2\,(2-\alpha)\,(4-\alpha-n)}{(4-n-\alpha)^2+4\lambda(\lambda+n-2)}<1. \qquad (4.5.20)
$$

This inequality is satisfied, by virtue of $\alpha>2-n-2\varkappa$ and $0<\varkappa\le\lambda-1$. Therefore (4.5.20) is true. Let us apply Lemma 4.8 and choose $\delta_1=\frac{\delta}{\lambda(\lambda+n-2)}$. As a result we obtain

$$\left(1 - \frac{2\,(2-\alpha)\,(4-\alpha-n)}{(4-n-\alpha)^2 + 4\lambda(\lambda+n-2)}\right)$$

$$\times \left\{ \int\limits_G a r_\varepsilon^{\alpha-2} |\nabla u|^2 dx + \int\limits_\Sigma \frac{1}{r} r_\varepsilon^{\alpha-2} \beta(\omega) u^2(x) ds + \int\limits_{\partial G} \frac{1}{r} r_\varepsilon^{\alpha-2} \gamma(\omega) u^2(x) ds \right\}$$

$$\le J(\varepsilon) + c(a_*, \alpha) \int\limits_G \left(a|\nabla u|^2 + u^2\right) dx + \frac{1}{2\delta_1 \gamma_0} \int\limits_{\partial G} r^{\alpha-1} g^2(x) ds$$

$$+ \frac{\delta}{\lambda(\lambda+n-2)}$$

$$\times \left\{ \int\limits_G a r_\varepsilon^{\alpha-2} |\nabla u|^2 dx + \int\limits_\Sigma r^{-1} r_\varepsilon^{\alpha-2} \beta(\omega) u^2(x) ds + \int\limits_{\partial G} r^{-1} r_\varepsilon^{\alpha-2} \gamma(\omega) u^2(x) ds \right\}$$

$$+ \frac{1}{2\delta_1 \beta_0} \int\limits_\Sigma r^{\alpha-1} h^2(x) ds + \frac{1}{2a_* \delta} \int\limits_G r^\alpha f^2(x) dx, \ \forall \delta > 0, \ \forall \varepsilon > 0. \tag{4.5.21}$$

We can choose

$$\delta = \frac{\lambda(\lambda+n-2)}{2} \left(1 - \frac{2\,(2-\alpha)\,(4-\alpha-n)}{(4-n-\alpha)^2 + 4\lambda(\lambda+n-2)}\right).$$

Thus we get

$$\int\limits_G a r_\varepsilon^{\alpha-2} |\nabla u|^2 dx + \int\limits_\Sigma \frac{1}{r} r_\varepsilon^{\alpha-2} \beta(\omega) u^2(x) ds + \int\limits_{\partial G} \frac{1}{r} r_\varepsilon^{\alpha-2} \gamma(\omega) u^2(x) ds$$

$$\le C(n, a_*, \alpha, \lambda) \left\{ J(\varepsilon) + \int\limits_G \left(a|\nabla u|^2 + u^2 + r^\alpha f^2(x)\right) dx \right.$$

$$\left. + \frac{1}{\gamma_0} \int\limits_{\partial G} r^{\alpha-1} g^2(x) ds + \frac{1}{\beta_0} \int\limits_\Sigma r^{\alpha-1} h^2(x) ds \right\}, \ \forall \varepsilon > 0. \tag{4.5.22}$$

Now we can perform the passage to the limit as $\varepsilon \to +0$ by the Fatou Theorem. Taking into account Corollary 4.17, it follows that

$$\int\limits_G a r^{\alpha-2} |\nabla u|^2 dx + \int\limits_\Sigma r^{\alpha-3} \beta(\omega) u^2(x) ds + \int\limits_{\partial G} r^{\alpha-3} \gamma(\omega) u^2(x) ds \tag{4.5.23}$$

$$\le C(n, a_*, \alpha, \lambda)$$

$$\times \left\{ \int\limits_G \left(a|\nabla u|^2 + u^2 + r^\alpha f^2(x)\right) dx + \frac{1}{\gamma_0} \int\limits_{\partial G} r^{\alpha-1} g^2(x) ds + \frac{1}{\beta_0} \int\limits_\Sigma r^{\alpha-1} h^2(x) ds \right\}.$$

Applying Theorem 4.15, by the Hardy-Friedrichs-Wirtinger inequality (2.2.3), from (4.5.23) we get the desired estimate (4.5.4). □

4.6 Local integral weighted estimates

Theorem 4.18. *Let $u(x)$ be a weak solution of problem (LN), $\lambda > 1$ be as above in* (4.1.3) *and assumptions* (a)–(c) *be satisfied. Then $u(x) \in \overset{\circ}{\mathbf{W}}{}^{1}_{2-n}(G)$ and there exist $d \in (0,1)$ and a constant $C > 0$ depending only on $n, s, \lambda, a_*, G, \Sigma$ such that the inequality*

$$\int\limits_{G_0^\varrho} a\left(r^{2-n}|\nabla u|^2 + r^{-n}u^2(x)\right)dx + \int\limits_{\Sigma_0^\varrho} r^{1-n}\beta(\omega)u^2(x)ds + \int\limits_{\Gamma_0^\varrho} r^{1-n}\gamma(\omega)u^2(x)ds$$
$$\leq C\Big(\|u\|^2_{2,G} + f_0^2 + \frac{1}{\gamma_0}g_0^2 + \frac{1}{\beta_0}h_0^2\Big)\cdot\begin{cases}\varrho^{2\lambda}, & \text{if } s > \lambda,\\ \varrho^{2\lambda}\ln^2\Big(\frac{1}{\varrho}\Big), & \text{if } s = \lambda,\\ \varrho^{2s}, & \text{if } s < \lambda\end{cases} \tag{4.6.1}$$

holds for almost all $\varrho \in (0,d)$.

Proof. By Theorem 4.16, $u(x)$ belongs to $\overset{\circ}{\mathbf{W}}{}^{1}_{2-n}(G)$, therefore it is enough to prove the estimate (4.6.1). Putting $\eta(x) = r^{2-n}u(x)$ in $(II)_{loc}$ and taking into account the definition (2.4.10), we obtain

$$U(\varrho) = \varrho\int\limits_{\Omega} au(x)\frac{\partial u}{\partial r}\Big|_{r=\varrho}d\Omega + \int\limits_{\Gamma_0^\varrho}\alpha(x)r^{2-n}u(x)g(x)ds + \int\limits_{\Sigma_0^\varrho} r^{2-n}u(x)h(x)ds$$
$$+\int\limits_{G_0^\varrho}\Big\{(n-2)ar^{-n}u(x)x_iu_{x_i} - pr^{2-n}u^2(x) - r^{2-n}u(x)f(x)\Big\}dx. \tag{4.6.2}$$

By the divergence theorem, we find

$$(n-2)\int\limits_{G_0^\varrho} ar^{-n}u(x)x_iu_{x_i}dx = \frac{n-2}{2}\int\limits_{G_0^\varrho} ar^{-n}x_i\frac{\partial u^2}{\partial x_i}dx \tag{4.6.3}$$
$$= \frac{n-2}{2}\Big\{-\int\limits_{G_0^\varrho} au^2(x)\left(nr^{-n} - nr^{-n}\right)dx + \varrho^{-n}\int\limits_{\Omega_\varrho} au^2(x)x_i\cos(r,x_i)d\Omega_\varrho$$
$$+\sum_{k=1}^{N-1}[a]_{\Sigma_k}\int\limits_{(\Sigma_k)_0^\varrho} r^{-n}u^2(x)x_i\cos(n_k,x_i)ds + \int\limits_{\Gamma_0^\varrho}\alpha(x)ar^{-n}u^2(x)x_i\cos(n,x_i)ds\Big\}.$$

By Lemmas 1.1 and 4.4 we have

$$(n-2)\int\limits_{G_0^\varrho} ar^{-n}u(x)x_iu_{x_i}dx = \frac{n-2}{2}\int\limits_{\Omega} au^2(x)d\Omega. \tag{4.6.4}$$

Because of Lemma 2.12 and $p \geq 0$, from (4.6.2)–(4.6.4) it follows that

$$U(\varrho) \leq \frac{\varrho}{2\lambda} U'(\varrho) + \int\limits_{\Gamma_0^\varrho} r^{2-n} |u(x)| \cdot |g(x)| ds + \int\limits_{\Sigma_0^\varrho} r^{2-n} |u(x)| \cdot |h(x)| ds + \int\limits_{G_0^\varrho} r^{2-n} |u(x)| \cdot |f(x)| dx. \quad (4.6.5)$$

We shall obtain upper bounds for each integral on the right. At first, applying the Cauchy and Hardy-Friedrichs-Wirtinger inequality (2.2.3)), we have for any $\forall \delta > 0$:

$$\int\limits_{\Gamma_0^\varrho} r^{2-n} |u||g| ds = \int\limits_{\Gamma_0^\varrho} \left(r^{\frac{1-n}{2}} \sqrt{\gamma(\omega)} |u| \right) \left(r^{\frac{3-n}{2}} \frac{1}{\sqrt{\gamma(\omega)}} |g| \right) ds$$
$$\leq \frac{\delta}{2} \int\limits_{\Gamma_0^\varrho} r^{1-n} \gamma(\omega) |u|^2 ds + \frac{1}{2\delta\gamma_0} \int\limits_{\Gamma_0^\varrho} r^{3-n} |g|^2 ds; \quad (4.6.6)$$

$$\int\limits_{\Sigma_0^\varrho} r^{2-n} |u||h| ds = \int\limits_{\Sigma_0^\varrho} \left(r^{\frac{1-n}{2}} \sqrt{\beta(\omega)} |u| \right) \left(r^{\frac{3-n}{2}} \frac{1}{\sqrt{\beta(\omega)}} |g| \right) ds$$
$$\leq \frac{\delta}{2} \int\limits_{\Sigma_0^\varrho} r^{1-n} \beta(\omega) |u|^2 ds + \frac{1}{2\delta\beta_0} \int\limits_{\Sigma_0^\varrho} r^{3-n} |h|^2 ds; \quad (4.6.7)$$

$$\int\limits_{G_0^\varrho} r^{2-n} |u(x)||f(x)| dx \leq \frac{\delta}{2a_*} \int\limits_{G_0^\varrho} a r^{-n} |u|^2 dx + \frac{1}{2\delta} \int\limits_{G_0^\varrho} r^{4-n} |f|^2 dx$$
$$\leq \frac{\delta}{2a_* \lambda(\lambda + n - 2)} U(\varrho) + \frac{1}{2\delta} \int\limits_{G_0^\varrho} r^{4-n} |f|^2 dx; \quad (4.6.8)$$

Thus, from (4.6.5)–(4.6.8) we get

$$\{1 - c_1(n, \lambda, a_*)\delta\} U(\varrho) \leq \frac{\varrho}{2\lambda} U'(\varrho)$$
$$+ \frac{1}{2\delta} \left\{ \int\limits_{G_0^\varrho} r^{4-n} |f|^2 dx + \frac{1}{\gamma_0} \int\limits_{\Gamma_0^\varrho} r^{3-n} |g|^2 ds + \frac{1}{\beta_0} \int\limits_{\Sigma_0^\varrho} r^{3-n} |h|^2 ds \right\}, \ \forall \delta > 0. \quad (4.6.9)$$

However, by the condition (c),

$$\int\limits_{G_0^\varrho} r^{4-n}|f|^2dx + \frac{1}{\gamma_0}\int\limits_{\Gamma_0^\varrho} r^{3-n}|g|^2ds + \frac{1}{\beta_0}\int\limits_{\Sigma_0^\varrho} r^{3-n}|h|^2ds$$
$$\leq \frac{c_0(G)}{2s}\left(f_0^2 + \frac{1}{\gamma_0}g_0^2 + \frac{1}{\beta_0}h_0^2\right)\cdot \varrho^{2s}.$$

Now, from (4.6.9) we obtain the differential inequality (CP) §1.7 with

$$\begin{aligned} \mathcal{P}(\varrho) &= \frac{2\lambda}{\varrho}\cdot\{1 - c_1(n,\lambda,a_*)\delta\},\ \forall\delta>0;\quad \mathcal{N}(\varrho)\equiv 0;\\ \mathcal{Q}(\varrho) &= \frac{c_0\lambda}{2s}\left(f_0^2 + \frac{1}{\gamma_0}g_0^2 + \frac{1}{\beta_0}h_0^2\right)\cdot\delta^{-1}\varrho^{2s-1},\ \forall\delta>0; \qquad (4.6.10)\\ U_0 &= C\Big\{\int\limits_G \left(u^2 + (1+r^{4-n})f^2(x)\right)dx + \int\limits_\Sigma r^{3-n}h^2(x)ds + \int\limits_{\partial G} r^{3-n}g^2(x)ds\Big\}, \end{aligned}$$

by (4.5.4) with $\alpha = 4-n$.

1) Case $s>\lambda$. Choosing $\delta = \varrho^\varepsilon$, $\forall\varepsilon>0$, we get:

$$\mathcal{P}(\varrho) = \frac{2\lambda}{\varrho} - 2\lambda c_1(n,\lambda,a_*)\varrho^{\varepsilon-1};\quad \mathcal{Q}(\varrho) = \frac{c_0\lambda}{2s}\left(f_1^2 + \frac{1}{\gamma_0}g_1^2 + \frac{1}{\beta_0}h_1^2\right)\cdot\varrho^{2s-1-\varepsilon}.$$

Now for $0<\varrho<\tau<d$,

$$-\int\limits_\varrho^\tau \mathcal{P}(s)ds = -2\lambda\ln\left(\frac{\tau}{\varrho}\right) + 2\lambda c_1\frac{s^\varepsilon}{\varepsilon}\Bigg|_\varrho^\tau \leq \ln\left(\frac{\varrho}{\tau}\right)^{2\lambda} + 2\lambda c_1\frac{d^\varepsilon}{\varepsilon} \Longrightarrow$$

$$\exp\left(-\int\limits_\varrho^d \mathcal{P}(\tau)d\tau\right) \leq \left(\frac{\varrho}{d}\right)^{2\lambda}\exp\left(2\lambda c_1\frac{d^\varepsilon}{\varepsilon}\right) = K_0\left(\frac{\varrho}{d}\right)^{2\lambda};$$

$$\exp\left(-\int\limits_\varrho^\tau \mathcal{P}(\tau)d\tau\right) \leq K_0\left(\frac{\varrho}{\tau}\right)^{2\lambda},$$

where $K_0 = \exp\left\{2\lambda c_1\frac{d^\varepsilon}{\varepsilon}\right\}$. We have also

$$\int\limits_\varrho^d \mathcal{Q}(\tau)\exp\Big(-\int\limits_\varrho^\tau \mathcal{P}(\sigma)d\sigma\Big)d\tau$$
$$\leq \frac{\lambda c_0K_0}{2s}\left(f_0^2 + \frac{1}{\gamma_0}g_0^2 + \frac{1}{\beta_0}h_0^2\right)\varrho^{2\lambda}\int\limits_\varrho^d \tau^{2s-2\lambda-\varepsilon-1}d\tau \leq$$
$$\leq \frac{\lambda c_0K_0}{2s}\left(f_0^2 + \frac{1}{\gamma_0}g_0^2 + \frac{1}{\beta_0}h_0^2\right)\cdot\frac{d^{s-\lambda}}{s-\lambda}\varrho^{2\lambda},$$

since $s > \lambda$ and we can choose $\varepsilon = s - \lambda$. Therefore we have in our case $K_0 = \exp\left\{2\lambda c_1 \frac{d^{s-\lambda}}{s-\lambda}\right\}$.

Now we apply Theorem 1.21. Then from (1.7.1), by virtue of the deduced inequalities and with regard to (2.2.3) for $\alpha = 4 - n$, we obtain the statement of (4.6.1) for $s > \lambda$.

2) Case $s = \lambda$. Taking in (4.6.10) any function $\delta(\varrho) > 0$ instead of $c_1\delta > 0$, we obtain the problem (CP) with

$$\mathcal{P}(\varrho) = \frac{2\lambda(1-\delta(\varrho))}{\varrho}; \ \mathcal{N}(\varrho) = 0; \ \mathcal{Q}(\varrho) = \frac{c_0}{2}\left(f_0^2 + \frac{1}{\gamma_0}g_0^2 + \frac{1}{\beta_0}h_0^2\right)\cdot \delta^{-1}(\varrho)\varrho^{2\lambda-1}.$$

We choose $\delta(\varrho) = \frac{1}{2\lambda \ln\left(\frac{ed}{\varrho}\right)}$, $0 < \varrho < d$, where e is the Euler number. Then we obtain

$$-\int_\varrho^\tau \mathcal{P}(\sigma)d\sigma \le \ln\left(\frac{\varrho}{\tau}\right)^{2\lambda} + \int_\varrho^\tau \frac{d\sigma}{\sigma\ln\left(\frac{ed}{\sigma}\right)} = \ln\left(\frac{\varrho}{\tau}\right)^{2\lambda} + \ln\left(\frac{\ln\left(\frac{ed}{\varrho}\right)}{\ln\left(\frac{ed}{\tau}\right)}\right) \Longrightarrow$$

$$\exp\left(-\int_\varrho^d \mathcal{P}(\tau)d\tau\right) \le \left(\frac{\varrho}{d}\right)^{2\lambda}\ln\left(\frac{ed}{\varrho}\right), \quad \exp\left(-\int_\varrho^\tau \mathcal{P}(\sigma)d\sigma\right) \le \left(\frac{\varrho}{\tau}\right)^{2\lambda}\cdot\frac{\ln\left(\frac{ed}{\varrho}\right)}{\ln\left(\frac{ed}{\tau}\right)}.$$

In this case we also have

$$\begin{aligned}
&\int_\varrho^d \mathcal{Q}(\tau)\exp\left(-\int_\varrho^\tau \mathcal{P}(\sigma)d\sigma\right)d\tau \\
&\le c_2\left(f_1^2 + \frac{1}{\nu_0}g_1^2 + \frac{1}{\nu_0}h_1^2\right)\varrho^{2\lambda}\ln\left(\frac{ed}{\varrho}\right)\cdot\int_\varrho^d \frac{d\tau}{\tau\delta(\tau)\ln\left(\frac{ed}{\tau}\right)} \\
&\le 2\lambda c_2\left(f_1^2 + \frac{1}{\nu_0}g_1^2 + \frac{1}{\nu_0}h_1^2\right)\cdot\varrho^{2\lambda}\ln^2\left(\frac{ed}{\varrho}\right).
\end{aligned}$$

Now we apply Theorem 1.21, and from (1.7.1), by virtue of the deduced inequalities, we obtain

$$U(\varrho) \le c_3(U_0 + f_0^2 + \frac{1}{\gamma_0}g_0^2 + \frac{1}{\beta_0}h_0^2)\varrho^{2\lambda}\ln^2\frac{1}{\varrho}, \quad 0 < \varrho < d < \frac{1}{e}.$$

Thus we have proved the statement of (4.6.1) for $s = \lambda$.

3) Case $0 < s < \lambda$. Analogously to case 1) taking into account (4.6.10) we have $\exp\left(-\int_\varrho^d \mathcal{P}(\tau)d\tau\right) \le \left(\frac{\varrho}{d}\right)^{2\lambda(1-c_1\delta)}$. In this case we also have

$$\int_\varrho^d \mathcal{Q}(\tau)\exp\Big(-\int_\varrho^\tau \mathcal{P}(\sigma)d\sigma\Big)d\tau \le \frac{\lambda c_0}{2s\delta}\left(f_0^2+\frac{1}{\gamma_0}g_0^2+\frac{1}{\beta_0}h_0^2\right)\cdot\varrho^{2\lambda(1-c_1\delta)}$$
$$\times\int_\varrho^d \tau^{2s-2\lambda(1-c_1\delta)-1}d\tau \le c_0c_4(n,\lambda,a_*,s)\left(f_0^2+\frac{1}{\gamma_0}g_0^2+\frac{1}{\beta_0}h_0^2\right)\cdot\varrho^{2s}.$$

Now we apply Theorem 1.21, and then from (1.7.1), by virtue of the deduced inequalities, we obtain

$$U(\varrho)\le c_5\left(U_0\varrho^{2\lambda(1-c_1\delta)}+\left(f_0^2+\frac{1}{\gamma_0}g_0^2+\frac{1}{\beta_0}h_0^2\right)\cdot\varrho^{2s}\right)$$
$$\le c_6(U_0+f_0^2+\frac{1}{\gamma_0}g_0^2+\frac{1}{\beta_0}h_0^2)\varrho^{2s}.$$

Thus we have proved the statement of (4.6.1) for $s<\lambda$. □

4.7 The power modulus of continuity at the conical point for weak solutions

Proof of Theorem 4.3. We define the function

$$\psi(\varrho)=\begin{cases}\varrho^\lambda, & \text{if } s>\lambda,\\ \varrho^\lambda\ln\left(\frac{1}{\varrho}\right), & \text{if } s=\lambda,\\ \varrho^s, & \text{if } s<\lambda\end{cases} \tag{4.7.1}$$

for $0<\varrho<d$.

By Theorem 4.14 about the local bound of the weak solution modulus we have

$$\sup_{G_0^{\varrho/2}}|u(x)|\le C\left\{\varrho^{-n/2}\|u\|_{2,G_0^\varrho}+\varrho^{2(1-n/p)}\|f\|_{p/2,G_0^\varrho}+\varrho\left(\|g\|_{\infty,\Gamma_0^\varrho}+\|h\|_{\infty,\Sigma_0^\varrho}\right)\right\} \tag{4.7.2}$$

where $C=C(n,a_*,a^*)$ and $p>n$. By Theorem 4.18, we have

$$\varrho^{-n/2}\|u\|_{2,G_0^\varrho}\le 2^{n/2}\Bigg(\int_{G_0^\varrho} r^{-n}u^2(x)dx\Bigg)^{1/2} \tag{4.7.3}$$
$$\le C\Big(\|u\|_{2,G}+\|f\|_{2,G}+\|g\|_{2,\partial G}+\|h\|_{2,\Sigma_0}+f_0+\frac{1}{\sqrt{\gamma_0}}g_0+\frac{1}{\sqrt{\beta_0}}h_0\Big)\psi(\varrho).$$

By the assumption (c), we obtain

$$\varrho^{2(1-n/p)}\|f\|_{p/2,G_0^\varrho}+\varrho\left(\|g\|_{\infty,\Gamma_0^\varrho}+\|h\|_{\infty,\Sigma_0^\varrho}\right)\le c\left(f_0+\frac{1}{\sqrt{\gamma_0}}g_0+\frac{1}{\sqrt{\beta_0}}h_0\right)\psi(\varrho). \tag{4.7.4}$$

From (4.7.2)–(4.7.4) it follows that

$$\sup_{G_{\varrho/4}^{\varrho/2}}|u(x)|\le C\Big(\|u\|_{2,G}+\|f\|_{2,G}+\|g\|_{2,\partial G}+\|h\|_{2,\Sigma_0}+f_0+\frac{1}{\sqrt{\gamma_0}}g_0+\frac{1}{\sqrt{\beta_0}}h_0\Big)\psi(\varrho).$$

Putting now $|x|=\frac{1}{3}\varrho$ we obtain finally the desired estimate (4.1.4).

Let us consider two sets $G_{\varrho/4}^{2\varrho}$ and $G_{\varrho/2}^{\varrho}\subset G_{\varrho/4}^{2\varrho}$, $\varrho>0$. We perform the change of variables $x=\varrho x'$ and $u(\varrho x')=\psi(\varrho)v(x')$. Then the function $v(x')$ satisfies the problem

$$\begin{cases} a\triangle' v-p\varrho^2 v(x')=\frac{\varrho^2}{\psi(\varrho)}f(\varrho x'), & x'\in G_{1/4}^2;\\ [v(x')]_{(\Sigma_k)_{1/4}^2}=0, & k=1,\dots,N-1;\\ \left[a\frac{\partial v}{\partial n_k'}\right]_{(\Sigma_k)_{1/4}^2}+\frac{1}{|x'|}\beta_k(\omega)v(x')=\frac{\varrho}{\psi(\varrho)}h_k(\varrho x'), & x'\in(\Sigma_k)_{1/4}^2,\ k=1,\dots,N-1;\\ \alpha(x')\cdot a\frac{\partial v}{\partial n'}+\frac{1}{|x'|}\gamma(\omega)v(x')=\frac{\varrho}{\psi(\varrho)}g(\rho x'), & x'\in\Gamma_{1/4}^2. \end{cases} \tag{$(LN)''$}$$

By the Sobolev Imbedding Theorem 1.19,

$$\sup_{x'\in G_{1/2}^1}|\nabla' v(x')|\le c\|v\|_{\mathbf{W}^{2,p}(G_{1/2}^1)},\quad p>n. \tag{4.7.5}$$

By virtue of the local L^p *a priori* estimate [66, 67], for the solution to the equation of problem $(LN)''$ inside the domain $G_{1/4}^2$ and near the smooth portions of the boundaries $\Sigma_{1/4}^2$ and $\Gamma_{1/4}^2$, we have

$$\|v\|_{\mathbf{W}^{2,p}(G_{1/2}^1)}\le c\frac{\varrho}{\psi(\varrho)}\left\{\varrho\|f\|_{\mathbf{L}^p(G_{1/4}^2)}+\|h\|_{\mathbf{W}^{1-1/p,p}(\Sigma_{1/4}^2)}+\|g\|_{\mathbf{W}^{1-1/p,p}(\Gamma_{1/4}^2)}\right\}$$
$$+c\|v\|_{\mathbf{L}^p(G_{1/4}^2)}. \tag{4.7.6}$$

Returning to the variables x, from (4.7.5) and (4.7.6), it follows that

$$\sup_{G_{\varrho/2}^{\varrho}}|\nabla u|$$
$$\le c\varrho^{2-n/p}\left\{\varrho^{-3}\|u\|_{\mathbf{L}^p(G_{\varrho/4}^{2\varrho})}+\|f\|_{p,G_{\varrho/4}^{2\varrho}}+\|g\|_{\mathbf{V}_{p,0}^{1-1/p}(\Gamma_{\varrho/4}^{2\varrho})}+\|h\|_{\mathbf{V}_{p,0}^{1-1/p}(\Sigma_{\varrho/4}^{2\varrho})}\right\}$$

and

$$\varrho^{2-n/p}\|u\|_{\mathbf{V}^2_{p,0}(G^{\varrho}_{\varrho/2})}$$
$$\le c\varrho^{2-n/p}\Big\{\varrho^{-2}\|u\|_{\mathbf{L}^p(G^{2\varrho}_{\varrho/4})} + \|f\|_{p,G^{2\varrho}_{\varrho/4}} + \|g\|_{\mathbf{V}^{1-1/p}_{p,0}(\Gamma^{2\varrho}_{\varrho/4})} + \|h\|_{\mathbf{V}^{1-1/p}_{p,0}(\Sigma^{2\varrho}_{\varrho/4})}\Big\}$$

or

$$\sup_{G^{\varrho}_{\varrho/2}} |\nabla u| \le c\varrho^{-1}\{|u|_{0,G^{2\varrho}_{\varrho/4}} + \|f\|_{\mathbf{V}^0_{p,2p-N}(G^{2\varrho}_{\varrho/4})} + \|g\|_{\mathbf{V}^{1-1/p}_{p,2p-n}(\Gamma^{2\varrho}_{\varrho/4})} + \|h\|_{\mathbf{V}^{1-1/p}_{p,2p-n}(\Sigma^{2\varrho}_{\varrho/4})}\}$$

and

$$\|u\|_{\mathbf{V}^2_{p,2p-n}(G^{\varrho}_{\varrho/2})}$$
$$\le c\{|u|_{0,G^{2\varrho}_{\varrho/4}} + \|f\|_{\mathbf{V}^0_{p,2p-N}(G^{2\varrho}_{\varrho/4})} + \|g\|_{\mathbf{V}^{1-1/p}_{p,2p-n}(\Gamma^{2\varrho}_{\varrho/4})} + \|h\|_{\mathbf{V}^{1-1/p}_{p,2p-n}(\Sigma^{2\varrho}_{\varrho/4})}\}.$$

Hence, because of (4.1.4), (4.1.5) and the assumption (c), there follow the required results (4.1.6) and (4.1.7).

Now we can make more precise the statement of Theorem 4.16. In fact, estimate (4.1.4) proved above allows us to consider in Theorem 4.16 the value $\varkappa = \lambda - 1$. As a result we obtain the last statement of our theorem and the estimate (4.1.9) with the best possible exponent that satisfies the inequality (4.1.8). □

4.8 Appendix: Eigenvalue transmission problem in a composite plane domain with an angular point

We consider the eigenvalue transmission boundary value problem for (LN) in a composite plane domain with an angular point.

Let $G \subset \mathbb{R}^2$ be a bounded domain with boundary ∂G that is a smooth curve everywhere except at the origin $\mathcal{O} \in \partial G$. Near the point $\mathcal{O}$ it is a fan that consists of N corners with vertexes at $\mathcal{O}$. Thus

$$G = \bigcup_{i=1}^{N} G_i; \quad \partial G = \bigcup_{j=0}^{N+1} \Gamma_j; \quad \Sigma = \bigcup_{k=1}^{N-1} \Sigma_k.$$

Here Σ_k, $k = 1, \dots, N-1$ are the rays that divide G into angular domains G_i, $i = 1, \dots, N$. Let ω_i be apertures at the vertex $\mathcal{O}$ in domains G_i, $i = 1, \dots, N$. We define the value $\theta_k = \omega_1 + \omega_2 + \dots + \omega_k$. Let $\Gamma = \bigcup_{j=1}^{N} \Gamma_j$ be the curvilinear portion of the boundary ∂G. In this case we have $\vartheta = \lambda^2$.

We also assume that $\Gamma_0 = \{(r,\omega) \big| r > 0,\ \omega = 0\}$; $\Gamma_{N+1} = \{(r,\omega) \big| r > 0,\ \omega = \theta_N\}$; $\beta_k\big|_{\Sigma_k} = \beta_k(\theta_k) = \beta_k = const$; $\gamma(0) = \gamma_1 = const$, $\gamma(\omega_0) = \gamma_N = const$.

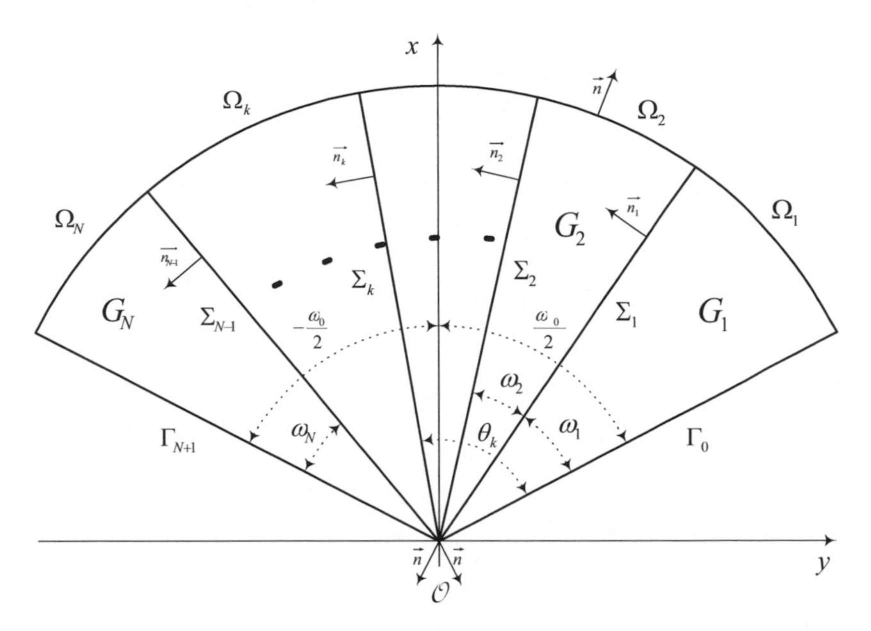

Figure 7

Eigenvalue problem $(EVPN)$ in this case has the form

$$\begin{cases} \psi_i'' + \lambda^2 \psi_i(\omega) = 0, \quad \omega \in \Omega_i = \{\omega_{i-1} < \omega < \omega_i\}, & i = 1, \ldots, N; \\ \psi_i(\theta_i) = \psi_{i+1}(\theta_i), & i = 1, \ldots, N-1; \\ a_i \psi_i'(\theta_i) - a_{i+1} \psi_{i+1}'(\theta_i) + \beta_i \psi_i(\theta_i) = 0, & i = 1, \ldots, N-1; \\ \alpha_1 a_1 \psi_1'(0) + \gamma_1 \psi_1(0) = 0, & \\ \alpha_N a_N \psi_N'(\omega_0) + \gamma_N \psi_N(\omega_0) = 0, & \end{cases}$$

where $a_1, \alpha_N \in \{0, 1\}$.

By direct calculation, we get $\psi_i(\omega) = A_i \cos(\lambda\omega) + B_i \sin(\lambda\omega)$, $i = 1, \ldots, N$, where constants $A_1, \ldots A_N$; $B_1, \ldots B_N$ are determined from the algebraic homogeneous system

$$\begin{cases} \lambda\alpha_1 a_1 B_1 + \gamma_1 = 0, \\ A_{i+1} = \Big(\cos^2(\lambda\theta_i) + \frac{a_i}{a_{i+1}}\sin^2(\lambda\theta_i) - \frac{\beta_i}{\lambda a_{i+1}}\sin(\lambda\theta_i)\cos(\lambda\theta_i)\Big)\cdot A_i \\ \qquad + \Big(\sin(\lambda\theta_i)\cos(\lambda\theta_i)\Big(1 - \frac{a_i}{a_{i+1}}\Big) - \frac{\beta_i}{\lambda a_{i+1}}\sin^2(\lambda\theta_i)\Big)\cdot B_i, \\ B_{i+1} = \Big(\sin(\lambda\theta_i)\cos(\lambda\theta_i)\Big(1 - \frac{a_i}{a_{i+1}}\Big) + \frac{\beta_i}{\lambda a_{i+1}}\cos^2(\lambda\theta_i)\Big)\cdot A_i \\ \qquad + \Big(\sin^2(\lambda\theta_i) + \frac{a_i}{a_{i+1}}\cos^2(\lambda\theta_i) + \frac{\beta_i}{\lambda a_{i+1}}\sin(\lambda\theta_i)\cos(\lambda\theta_i)\Big)\cdot B_i, \\ (\gamma_N\cos(\lambda\omega_0) - \lambda\alpha_N a_N\sin(\lambda\omega_0))\cdot A_N \\ + (\gamma_N\sin(\lambda\omega_0) + \lambda\alpha_N a_N\cos(\lambda\omega_0))\cdot B_N = 0, \end{cases}$$

$i = 1, \ldots, N-1$. The least positive eigenvalue λ is defined from the vanishing of the determinant of this system.

4.8.1 Four-media transmission problem

Our goal is the derivation of the eigenvalues equation that corresponds to our transmission problem for the case $N = 4$. Let S^1 be the unit circle in $\mathbb{R}^2$ centered at $\mathcal{O}$. We write: $\Omega_i = G_i \cap S^1$; $i = 1, 2, 3, 4$. The eigenvalue problem is the following:

$$\begin{cases} \psi_i'' + \lambda^2\psi_i(\omega) = 0, \ \omega \in \Omega_i; \ i = 1, 2, 3, 4; \\ \psi_2(\omega_1) = \psi_1(\omega_1); \ \psi_3(\theta_2) = \psi_2(\theta_2); \ \psi_4(\theta_3) = \psi_3(\theta_3); \\ a_1\psi_1'(\omega_1) - a_2\psi_2'(\omega_1) + \beta_1\psi_1(\omega_1) = 0; \\ a_2\psi_2'(\theta_2) - a_3\psi_3'(\theta_2) + \beta_2\psi_2(\theta_2) = 0; \\ a_3\psi_3'(\theta_3) - a_4\psi_4'(\theta_3) + \beta_3\psi_3(\theta_3) = 0; \\ \alpha_1 a_1\psi_1'(0) + \gamma_1\psi_1(0) = 0; \\ \alpha_4 a_4\psi_4'(\theta_4) + \gamma_4\psi_4(\theta_4) = 0, \end{cases} \tag{4.8.1}$$

where $\alpha_1 = \alpha\big|_{\Gamma_0} = \alpha\big|_{\omega=0}$, $\alpha_4 = \alpha\big|_{\Gamma_5} = \alpha\big|_{\omega=\theta_4}$, $\gamma_1 = \gamma(0)$, $\gamma_4 = \gamma(\theta_4)$; $\alpha_{1,4} \in \{0, 1\}$.

We find the general solution of equation (4.8.1):

$$\psi_i(\omega) = A_i\cos(\lambda\omega) + B_i\sin(\lambda\omega) \Longrightarrow \psi_i'(\omega) = -\lambda A_i\sin(\lambda\omega) + \lambda B_i\cos(\lambda\omega);$$

for $i = 1, 2, 3, 4$, where A_i, B_i $(i = 1, 2, 3, 4)$ are arbitrary constants. From the boundary condition of (4.8.1) we obtain the homogenous algebraic system of eight linear equations for the finding of A_i, B_i $(i = 1, 2, 3, 4)$:

$$\begin{cases} A_2\cos\lambda\omega_1 + B_2\sin\omega_1 - A_1\cos\lambda\omega_1 - B_1\sin\lambda\omega_1 = 0,\\ A_3\cos\lambda\theta_2 + B_3\sin\theta_2 - A_2\cos\lambda\theta_2 - B_2\sin\lambda\theta_2 = 0,\\ A_4\cos\lambda\theta_3 + B_4\sin\theta_3 - A_3\cos\lambda\theta_3 - B_3\sin\lambda\theta_3 = 0,\\ \lambda a_2A_2\sin\lambda\omega_1 - \lambda a_2B_2\cos\lambda\omega_1 - \lambda a_1A_1\sin\lambda\omega_1\\ +\lambda a_1B_1\cos\lambda\omega_1 + \beta_1A_1\cos\lambda\omega_1 + \beta_1B_1\sin\lambda\omega_1 = 0,\\ \lambda a_3A_3\sin\lambda\theta_2 - \lambda a_3B_3\cos\lambda\theta_2 - \lambda a_2A_2\sin\lambda\theta_2\\ +\lambda a_2B_2\cos\lambda\theta_2 + \beta_2A_2\cos\lambda\theta_2 + \beta_2B_2\sin\lambda\theta_2 = 0,\\ \lambda a_4A_4\sin\lambda\theta_3 - \lambda a_4B_4\cos\lambda\theta_3 - \lambda a_3A_3\sin\lambda\theta_3\\ +\lambda a_3B_3\cos\lambda\theta_3 + \beta_3A_3\cos\lambda\theta_3 + \beta_3B_3\sin\lambda\theta_3 = 0,\\ \alpha_1a_1\lambda B_1 + \gamma_1A_1 = 0,\\ \alpha_4a_4\lambda A_4\sin\lambda\theta_4 - \alpha_4a_4\lambda B_4\cos\lambda\theta_4 - \gamma_4A_4\cos\lambda\theta_4 - \gamma_4B_4\sin\lambda\theta_4 = 0. \end{cases}$$

The determinant of this system has to be equal to zero for the nontrivial solution of this system to exist. The latter gives us the required equation for eigenvalues λ:

$$\begin{aligned}
&[\lambda^4(\alpha_1\alpha_4\beta_2a_1^2a_4^2 - \alpha_4\gamma_1a_2^2a_4^2 + \alpha_1\gamma_4a_1^2a_3^2) + \lambda^2(\alpha_4\beta_1\beta_2a_4^2 - \alpha_1\beta_2\beta_3\gamma_4a_1^2\\
&+ \beta_3\gamma_1\gamma_4a_2^2 + \gamma_1\gamma_4\beta_1a_3^2) - \beta_1\beta_2\beta_3\gamma_1\gamma_4]\cdot\sin\lambda\omega_1\sin\lambda\omega_2\sin\lambda\omega_3\sin\lambda\omega_4\\
&+ \lambda a_2[\alpha_4\lambda^2(\beta_2\gamma_1a_4^2 + \beta_1\gamma_1a_4^2 + \lambda^2\alpha_1a_1^2a_4^2) - \gamma_1\gamma_4\beta_2\beta_3 - \gamma_1\gamma_4\beta_1\beta_3\\
&- \lambda^2\alpha_1\beta_3\gamma_4a_1^2 + \lambda^2\gamma_1\gamma_4a_3^2]\cdot\sin\lambda\omega_1\cos\lambda\omega_2\sin\lambda\omega_3\sin\lambda\omega_4\\
&- \lambda a_3[\gamma_4(\lambda^2\alpha_1\beta_2a_1^2 + \gamma_1\beta_1\beta_2 - \lambda^2\gamma_1a_2^2) - \lambda^2\alpha_4\beta_1\gamma_1a_4^2 - \lambda^4\alpha_1\alpha_4a_1^2a_4^2\\
&+ \gamma_1\gamma_4\beta_1\beta_3 + \lambda^2\alpha_1\beta_3\gamma_4a_1^2]\cdot\sin\lambda\omega_1\sin\lambda\omega_2\cos\lambda\omega_3\sin\lambda\omega_4\\
&- \lambda a_4[\lambda^2\alpha_1\beta_2\gamma_4a_1^2 + \beta_1\beta_2\gamma_1\gamma_4 - \lambda^2\gamma_1\gamma_4a_2^2 + \lambda^2\alpha_1\alpha_4\beta_2\beta_3a_1^2 + \alpha_4\beta_1\beta_2\beta_3\gamma_1\\
&- \lambda^2\alpha_4\beta_3\gamma_1a_2^2 - \lambda^2\alpha_4a_3^2(\gamma_1\beta_1 + \lambda^2\alpha_1a_1^2)]\cdot\sin\lambda\omega_1\sin\lambda\omega_2\sin\lambda\omega_3\cos\lambda\omega_4\\
&- \lambda a_1[\lambda^2\alpha_1\alpha_4\beta_1\beta_2a_4^2 - \lambda^2\alpha_4\beta_2\gamma_1a_4^2 - \lambda^4\alpha_1\alpha_4a_2^2a_4^2 - \alpha_1\beta_1\beta_2\beta_3\gamma_4 + \gamma_1\gamma_4\beta_2\beta_3\\
&+ \lambda^2\alpha_1\beta_3\gamma_4a_2^2 - \lambda^2\gamma_4a_3^2(\gamma_1 - \alpha_1\beta_1)]\cdot\cos\lambda\omega_1\sin\lambda\omega_2\sin\lambda\omega_3\sin\lambda\omega_4\\
&+ \lambda^2a_3a_4[\alpha_4(\lambda^2\alpha_1\beta_2a_1^2 + \gamma_1\beta_1\beta_2 - \lambda^2\gamma_1a_2^2) + \alpha_4\beta_1\beta_3\gamma_1 + \lambda^2\alpha_1\alpha_4\beta_3a_1^2\\
&+ \beta_1\gamma_1\gamma_4 + \lambda^2\alpha_1\gamma_4a_1^2]\cdot\sin\lambda\omega_1\sin\lambda\omega_2\cos\lambda\omega_3\cos\lambda\omega_4\\
&+ \lambda^2a_2a_3[\gamma_1(\lambda^2\alpha_4a_4^2 - \gamma_4\beta_3) - \gamma_4(\gamma_1\beta_2 + \beta_1\gamma_1 + \lambda^2\alpha_1a_1^2)]\\
&\quad\times\sin\lambda\omega_1\cos\lambda\omega_2\cos\lambda\omega_3\sin\lambda\omega_4\\
&+ \lambda^2a_2a_4[\lambda^2\alpha_4\gamma_1a_3^2 - \gamma_1\beta_2\gamma_4 - \beta_1\gamma_1\gamma_4 - \lambda^2\alpha_1\gamma_4a_1^2 - \alpha_4\beta_2\beta_3\gamma_1\\
&- \alpha_4\beta_1\beta_3\gamma_1 - \lambda^2\alpha_1\alpha_4\beta_3a_1^2]\cdot\sin\lambda\omega_1\cos\lambda\omega_2\sin\lambda\omega_3\cos\lambda\omega_4\\
&+ \lambda^2a_1a_2[\lambda^2\alpha_4\gamma_1a_4^2 - \lambda^2\alpha_1\alpha_4\beta_2a_4^2 - \lambda^2\alpha_1\alpha_4\beta_1a_4^2 - \beta_3\gamma_1\gamma_4 + \alpha_1\beta_2\beta_3\gamma_4\\
&+ \alpha_1\beta_1\beta_3\gamma_4 - \lambda^2\alpha_1\gamma_4a_3^2]\cdot\cos\lambda\omega_1\cos\lambda\omega_2\sin\lambda\omega_3\sin\lambda\omega_4\\
&+ \lambda^2a_1a_4[\alpha_1\beta_1\beta_2\gamma_4 - \beta_2\gamma_1\gamma_4 - \lambda^2\alpha_1\gamma_4a_2^2 + \alpha_1\alpha_4\beta_1\beta_2\beta_3 - \alpha_4\beta_2\beta_3\gamma_1\\
&- \lambda^2\alpha_1\alpha_4\beta_3a_2^2 + \lambda^2\alpha_4a_3^2(\gamma_1 - \alpha_1\beta_1)]\cdot\cos\lambda\omega_1\sin\lambda\omega_2\sin\lambda\omega_3\cos\lambda\omega_4\\
&+ \lambda^2a_1a_3[\gamma_4(\alpha_1\beta_1\beta_2 - \gamma_1\beta_2 - \lambda^2\alpha_1a_2^2) + \lambda^2\alpha_4\gamma_1a_4^2 - \lambda^2\alpha_1\alpha_4\beta_1a_4^2
\end{aligned}$$

$$
\begin{aligned}
&- \gamma_1\gamma_4\beta_3 + \alpha_1\beta_1\beta_3\gamma_4] \cdot \cos\lambda\omega_1 \sin\lambda\omega_2 \cos\lambda\omega_3 \sin\lambda\omega_4 \\
&- \lambda^3 a_2 a_3 a_4 [\alpha_4(\beta_2\gamma_1 + \beta_1\gamma_1 + \lambda^2\alpha_1 a_1^2) + \gamma_1(\alpha_4\beta_3 + \gamma_4)] \\
&\quad \times \sin\lambda\omega_1 \cos\lambda\omega_2 \cos\lambda\omega_3 \cos\lambda\omega_4 \\
&- \lambda^3 a_1 a_3 a_4 [\alpha_4\beta_3\gamma_1 - \alpha_1\alpha_4\beta_1\beta_3 + \gamma_1\gamma_4 - \alpha_1\beta_1\gamma_4 \\
&- \alpha_4(\alpha_1\beta_1\beta_2 - \gamma_1\beta_2 - \lambda^2\alpha_1 a_2^2)] \cdot \cos\lambda\omega_1 \sin\lambda\omega_2 \cos\lambda\omega_3 \cos\lambda\omega_4 \\
&- \lambda^3 a_1 a_2 a_4 [\gamma_1\gamma_4 - \alpha_1\beta_2\gamma_4 - \alpha_1\beta_1\gamma_4 + \alpha_4\beta_3\gamma_1 - \alpha_1\alpha_4\beta_2\beta_3 \\
&- \alpha_1\alpha_4\beta_1\beta_3 + \lambda^2\alpha_1\alpha_4 a_3^2] \cdot \cos\lambda\omega_1 \cos\lambda\omega_2 \sin\lambda\omega_3 \cos\lambda\omega_4 \\
&- \lambda^3 a_1 a_2 a_3 [\gamma_4(\gamma_1 - \alpha_1\beta_2 - \alpha_1\beta_1) + \alpha_1(\lambda^2\alpha_4 a_4^2 - \gamma_4\beta_1)] \\
&\quad \times \cos\lambda\omega_1 \cos\lambda\omega_2 \cos\lambda\omega_3 \sin\lambda\omega_4 \\
&+ \lambda^4 a_1 a_2 a_3 a_4 [\alpha_1(\alpha_4\beta_3 + \gamma_4) - \alpha_4(\gamma_1 - \alpha_1\beta_2 - \alpha_1\beta_1)] \\
&\quad \times \cos\lambda\omega_1 \cos\lambda\omega_2 \cos\lambda\omega_3 \cos\lambda\omega_4 \\
&= 0.
\end{aligned}
$$

We consider the following particular cases of boundary conditions.

1) *The Dirichlet problem:* $\alpha_1 = \alpha_4 = \beta_1 = \beta_2 = \beta_3 = 0$; $\gamma_1 = \gamma_4 = 1$.

$$
\begin{aligned}
&\lambda^3 a_2 a_3^2 \sin\lambda\omega_1 \cos\lambda\omega_2 \sin\lambda\omega_3 \sin\lambda\omega_4 + a_2^2 a_3 \sin\lambda\omega_1 \sin\lambda\omega_2 \cos\lambda\omega_3 \sin\lambda\omega_4 \\
&+ a_2^2 a_4 \sin\lambda\omega_1 \sin\lambda\omega_2 \sin\lambda\omega_3 \cos\lambda\omega_4 + a_1 a_3^2 \cos\lambda\omega_1 \sin\lambda\omega_2 \sin\lambda\omega_3 \sin\lambda\omega_4 \\
&- a_2 a_3 a_4 \sin\lambda\omega_1 \cos\lambda\omega_2 \cos\lambda\omega_3 \cos\lambda\omega_4 - a_1 a_3 a_4 \cos\lambda\omega_1 \sin\lambda\omega_2 \cos\lambda\omega_3 \cos\lambda\omega_4 \\
&- a_1 a_2 a_4 \cos\lambda\omega_1 \cos\lambda\omega_2 \sin\lambda\omega_3 \cos\lambda\omega_4 - a_1 a_2 a_3 \cos\lambda\omega_1 \cos\lambda\omega_2 \cos\lambda\omega_3 \sin\lambda\omega_4 \\
&= 0.
\end{aligned}
$$

In the isotropic case ($a_1 = a_2 = a_3 = a_4$) we hence obtain the following well-known result: $\sin(\lambda\theta_4) = 0 \Rightarrow \lambda_n = \frac{\pi n}{\theta_4}$, $n = 1, 2 \dots$.

Corollary. $\lambda = \frac{\pi}{\omega_0} > 1$, if $\omega_0 < \pi$.

2) *The Neumann problem:* $\alpha_1 = \alpha_4 = 1$; $\beta_1 = \beta_2 = \beta_3 = 0$; $\gamma_1 = \gamma_4 = 0$.

$$
\begin{aligned}
&- a_1^2 a_2 a_4^2 \sin\lambda\omega_1 \cos\lambda\omega_2 \sin\lambda\omega_3 \sin\lambda\omega_4 - a_1^2 a_3 a_4^2 \sin\lambda\omega_1 \sin\lambda\omega_2 \cos\lambda\omega_3 \sin\lambda\omega_4 \\
&- a_1^2 a_3^2 a_4 \sin\lambda\omega_1 \sin\lambda\omega_2 \sin\lambda\omega_3 \cos\lambda\omega_4 - a_1 a_2^2 a_4^2 \cos\lambda\omega_1 \sin\lambda\omega_2 \sin\lambda\omega_3 \sin\lambda\omega_4 \\
&+ a_1^2 a_2 a_3 a_4 \sin\lambda\omega_1 \cos\lambda\omega_2 \cos\lambda\omega_3 \cos\lambda\omega_4 \\
&+ a_1 a_2^2 a_3 a_4 \cos\lambda\omega_1 \sin\lambda\omega_2 \cos\lambda\omega_3 \cos\lambda\omega_4 \\
&+ a_1 a_2 a_3^2 a_4 \cos\lambda\omega_1 \cos\lambda\omega_2 \sin\lambda\omega_3 \cos\lambda\omega_4 \\
&+ a_1 a_2 a_3 a_4^2 \cos\lambda\omega_1 \cos\lambda\omega_2 \cos\lambda\omega_3 \sin\lambda\omega_4 = 0.
\end{aligned}
$$

In the isotropic case ($a_1 = a_2 = a_3 = a_4$) we hence obtain the following well-known result: $\sin(\lambda\theta_3) = 0 \Rightarrow \lambda_n = \frac{\pi n}{\theta_4}$, $n = 0, 1, 2 \dots$.

Corollary. $\lambda = \frac{\pi}{\omega_0} > 1$, if $\omega_0 < \pi$.

3) *The mixed problem:* $\alpha_1 = \gamma_4 = 1$, $\alpha_4 = \beta_1 = \beta_2 = \beta_3 = 0$; $\gamma_1 = 0$.

$$\begin{aligned}
&a_1^2 a_3^2 \sin\lambda\omega_1 \sin\lambda\omega_2 \sin\lambda\omega_3 \sin\lambda\omega_4 - a_1^2 a_3 a_4 \sin\lambda\omega_1 \sin\lambda\omega_2 \cos\lambda\omega_3 \cos\lambda\omega_4\\
&- a_1^2 a_2 a_3 \sin\lambda\omega_1 \cos\lambda\omega_2 \cos\lambda\omega_3 \sin\lambda\omega_4 - a_1^2 a_2 a_4 \sin\lambda\omega_1 \cos\lambda\omega_2 \sin\lambda\omega_3 \cos\lambda\omega_4\\
&- a_1 a_2 a_3^2 \cos\lambda\omega_1 \cos\lambda\omega_2 \sin\lambda\omega_3 \sin\lambda\omega_4 - a_1 a_2^2 a_4 \cos\lambda\omega_1 \sin\lambda\omega_2 \sin\lambda\omega_3 \cos\lambda\omega_4\\
&- a_1 a_2^2 a_4 \cos\lambda\omega_1 \sin\lambda\omega_2 \cos\lambda\omega_3 \sin\lambda\omega_4\\
&+ a_1 a_2 a_3 a_4 \cos\lambda\omega_1 \cos\lambda\omega_2 \cos\lambda\omega_3 \cos\lambda\omega_4 = 0.
\end{aligned}$$

In the isotropic case ($a_1 = a_2 = a_3 = a_4$) we hence obtain the following well-known result: $\cos(\lambda\theta_4) = 0 \Rightarrow \lambda_n = \frac{\pi(2n-1)}{2\theta_4}$, $n = 1, 2 \ldots$.

Corollary. $\lambda = \frac{\pi}{2\omega_0} > 1$, if $\omega_0 < \frac{\pi}{2}$.

4) *The Robin problem:* $\alpha_1 = \alpha_4 = 1$.

In the isotropic case ($a_1 = a_2 = a_3 = a_4 = 1$; $\beta_1 = \beta_2 = \beta_3 = 0$) we obtain: $\tan(\lambda\omega_0) = \frac{\lambda(\gamma_4 - \gamma_1)}{\lambda^2 + \gamma_1\gamma_4}$.

4.8.2 Three-media transmission problem

In this subsection we consider the eigenvalue problem corresponding to our transmission problem for the case $N = 3$. Let S^1 be the unit circle in $\mathbb{R}^2$ centered at $\mathcal{O}$. We write: $\Omega_i = G_i \cap S^1$; $i = 1, 2, 3$. The eigenvalue problem is the following:

$$\begin{cases}
\psi_i'' + \lambda^2\psi_i(\omega) = 0, \ \omega \in \Omega_i; \ (i = 1, 2, 3);\\
\psi_1(\omega_1) = \psi_2(\omega_1); \ \psi_3(\theta_2) = \psi_2(\theta_2);\\
a_2\psi_2'(\omega_1) - a_1\psi_1'(\omega_1) + \beta_1\psi_1(\omega_1) = 0;\\
a_3\psi_3'(\theta_2) - a_2\psi_2'(\theta_2) + \beta_2\psi_2(\theta_2) = 0;\\
\alpha_1 a_1\psi_1'(0) + \gamma_1\psi_1(0) = 0;\\
\alpha_3 a_3\psi_3'(\theta_3) + \gamma_3\psi_3(\theta_3) = 0.
\end{cases} \tag{4.8.2}$$

We find the general solution of equation (4.8.2):

$$\psi_i(\omega) = A_i\cos(\lambda\omega) + B_i\sin(\lambda\omega) \Longrightarrow \psi_i'(\omega) = -\lambda A_i\sin(\lambda\omega) + \lambda B_i\cos(\lambda\omega);$$

($i = 1, 2, 3$), where A_i, B_i ($i = 1, 2, 3$) are arbitrary constants. From the boundary condition of (4.8.2) we obtain a homogenous algebraic system of six linear equations for the determination of A_i, B_i ($i = 1, 2, 3$). The determinant of the system must be equal to zero for the nontrivial solution of this system to exist. The latter gives us the required equation for eigenvalues λ:

$$\begin{aligned}
&[\lambda^2\alpha_3 a_3^2(\beta_1\gamma_1 + \lambda^2\alpha_1 a_1^2) - \gamma_3(\beta_1\beta_2\gamma_1 + \lambda^2\alpha_1\beta_2 a_1^2 - \lambda^2\gamma_1 a_2^2)] \cdot \sin(\lambda\omega_1)\sin(\lambda\omega_2)\sin(\lambda\omega_3)\\
&+\lambda a_1 \cdot [\lambda^2\alpha_3 a_3^2(\gamma_1 - \beta_1\alpha_1) + \gamma_3(\beta_1\beta_2\alpha_1 - \gamma_1\beta_2 - \lambda^2\alpha_1 a_2^2)] \cdot \cos(\lambda\omega_1)\sin(\lambda\omega_2)\sin(\lambda\omega_3)\\
&-\lambda a_3 \cdot [\gamma_3(\beta_1\gamma_1 + \lambda^2\alpha_1 a_1^2) + \alpha_3(\beta_1\beta_2\gamma_1 + \lambda^2\alpha_1\beta_2 a_1^2 - \lambda^2\gamma_1 a_2^2)] \cdot \sin(\lambda\omega_1)\sin(\lambda\omega_2)\cos(\lambda\omega_3)
\end{aligned}$$

$$+\lambda^2 a_1 a_3 \cdot [\gamma_3(\beta_1\alpha_1 - \gamma_1) + \alpha_3(\beta_1\beta_2\alpha_1 - \gamma_1\beta_2 - \lambda^2\alpha_1 a_2^2)] \cdot \cos(\lambda\omega_1)\sin(\lambda\omega_2)\cos(\lambda\omega_3)$$
$$-\lambda a_2 \cdot [\gamma_3(\beta_2\gamma_1 + \lambda^2\alpha_1 a_1^2 + \beta_1\gamma_1) - \lambda^2\alpha_3\gamma_1 a_3^2] \cdot \sin(\lambda\omega_1)\cos(\lambda\omega_2)\sin(\lambda\omega_3)$$
$$+\lambda^2 a_1 a_2 \cdot [\gamma_3(\beta_2\alpha_1 + \alpha_1\beta_1 - \gamma_1) - \lambda^2\alpha_3\alpha_1 a_3^2] \cdot \cos(\lambda\omega_1)\cos(\lambda\omega_2)\sin(\lambda\omega_3)$$
$$-\lambda^2 a_2 a_3 \cdot [\gamma_1\gamma_3 + \alpha_3(\beta_2\gamma_1 + \lambda^2\alpha_1 a_1^2 + \beta_1\gamma_1)] \cdot \sin(\lambda\omega_1)\cos(\lambda\omega_2)\cos(\lambda\omega_3)$$
$$+\lambda^3 a_1 a_2 a_3 \cdot [\alpha_1\gamma_3 + \alpha_3(\beta_2\alpha_1 + \alpha_1\beta_1 - \gamma_1)] \cdot \cos(\lambda\omega_1)\cos(\lambda\omega_2)\cos(\lambda\omega_3)$$
$$= 0. \tag{4.8.3}$$

We consider the following particular cases of boundary conditions.

1) *The Dirichlet problem:* $\alpha_1 = \alpha_3 = \beta_1 = \beta_2 = 0$; $\gamma_1 = \gamma_3 = 1$.

$$a_1 a_3 \cdot \cos(\lambda\omega_1)\sin(\lambda\omega_2)\cos(\lambda\omega_3) + a_1 a_2 \cdot \cos(\lambda\omega_1)\cos(\lambda\omega_2)\sin(\lambda\omega_3)$$
$$+ a_2 a_3 \cdot \sin(\lambda\omega_1)\cos(\lambda\omega_2)\cos(\lambda\omega_3) - a_2^2 \cdot \sin(\lambda\omega_1)\sin(\lambda\omega_2)\sin(\lambda\omega_3) = 0.$$

In the isotropic case ($a_1 = a_2 = a_3$) we hence obtain the following well-known result: $\sin(\lambda\theta_3) = 0 \Longrightarrow \lambda_n = \frac{\pi n}{\theta_3}$, $n = 1, 2, \ldots$.

Corollary. $\lambda = \frac{\pi}{\theta_3} > 1$, if $\omega_1 + \omega_2 + \omega_3 < \pi$.

2) *The Neumann problem:* $\alpha_0 = \alpha_3 = 1$; $\beta_1 = \beta_2 = \gamma_0 = \gamma_3 = 0$.

$$a_2^2 \cdot \cos(\lambda\omega_1)\sin(\lambda\omega_2)\cos(\lambda\omega_3) + a_2 a_3 \cdot \cos(\lambda\omega_1)\cos(\lambda\omega_2)\sin(\lambda\omega_3)$$
$$+ a_1 a_2 \cdot \sin(\lambda\omega_1)\cos(\lambda\omega_2)\cos(\lambda\omega_3) - a_1 a_3 \cdot \sin(\lambda\omega_1)\sin(\lambda\omega_2)\sin(\lambda\omega_3) = 0.$$

In the isotropic case ($a_1 = a_2 = a_3$) we hence obtain the following well-known result: $\sin(\lambda\theta_3) = 0 \Longrightarrow \lambda_n = \frac{\pi n}{\theta_3}$, $n = 0, 1, 2, \ldots$.

Corollary. $\lambda = \frac{\pi}{\theta_3} > 1$, if $\omega_1 + \omega_2 + \omega_3 < \pi$.

3) *The mixed problem:* $\alpha_0 = \gamma_3 = 1$; $\alpha_3 = \beta_1 = \beta_2 = \gamma_0 = 0$.

$$a_2^2 \cdot \cos(\lambda\omega_1)\sin(\lambda\omega_2)\sin(\lambda\omega_3) + a_1 a_3 \cdot \sin(\lambda\omega_1)\sin(\lambda\omega_2)\cos(\lambda\omega_3)$$
$$+ a_1 a_2 \cdot \sin(\lambda\omega_1)\cos(\lambda\omega_2)\sin(\lambda\omega_3) - a_2 a_3 \cdot \cos(\lambda\omega_1)\cos(\lambda\omega_2)\cos(\lambda\omega_3) = 0.$$

In the isotropic case ($a_1 = a_2 = a_3$) we hence obtain the following well-known result: $\cos(\lambda\theta_3) = 0 \Longrightarrow \lambda_n = \frac{\pi(2n-1)}{2\theta_3}$, $n = 1, 2, \ldots$.

Corollary. $\lambda = \frac{\pi}{2\theta_3} > 1$, if $\omega_1 + \omega_2 + \omega_3 < \frac{\pi}{2}$.

4) *The Robin problem:* $\alpha_1 = 1$, $\alpha_3 = 1$; $\beta_1 = \beta_2 = 0$.

$$(\lambda^2 a_1^2 a_3^2 + \gamma_1\gamma_3 a_2^2) \cdot \sin(\lambda\omega_1)\sin(\lambda\omega_2)\sin(\lambda\omega_3)$$
$$- \lambda \cdot (\gamma_3 a_1 a_2^2 - \gamma_1 a_1 a_3^2) \cdot \cos(\lambda\omega_1)\sin(\lambda\omega_2)\sin(\lambda\omega_3)$$
$$- \lambda a_3 \cdot (\gamma_3 a_1^2 - \gamma_1 a_2^2) \cdot \sin(\lambda\omega_1)\sin(\lambda\omega_2)\cos(\lambda\omega_3)$$
$$- a_1 a_3(\gamma_1\gamma_3 + \lambda^2 a_2^2) \cdot \cos(\lambda\omega_1)\sin(\lambda\omega_2)\cos(\lambda\omega_3)$$
$$- \lambda a_2 \cdot (\gamma_3 a_1^2 - \gamma_1 a_3^2) \cdot \sin(\lambda\omega_1)\cos(\lambda\omega_2)\sin(\lambda\omega_3)$$
$$- a_1 a_2 \cdot (\gamma_1\gamma_3 + \lambda^2 a_3^2) \cdot \cos(\lambda\omega_1)\cos(\lambda\omega_2)\sin(\lambda\omega_3)$$

$$
\begin{aligned}
&- a_2 a_3 \cdot (\gamma_1 \gamma_3 + \lambda^2 a_1^2) \cdot \sin(\lambda\omega_1) \cos(\lambda\omega_2) \cos(\lambda\omega_3) \\
&+ \lambda a_1 a_2 a_3 \cdot (\gamma_3 - \gamma_1) \cdot \cos(\lambda\omega_1) \cos(\lambda\omega_2) \cos(\lambda\omega_3) = 0.
\end{aligned}
$$

In the isotropic case ($a_1 = a_2 = a_3 = 1$) we hence obtain the following well-known result (see Example 1 in §10.1.7 of [14]): $\tan(\lambda\theta_3) = \frac{\lambda(\gamma_3-\gamma_1)}{\lambda^2+\gamma_1\gamma_3}$.

4.8.3 Two-media transmission problem

The two-media transmission problem was considered in detail in Chapter 3.

Chapter 5

Transmission problem for weak quasi-linear elliptic equations in a conical domain

5.1 Introduction

In this chapter we investigate the behavior of weak solutions to the transmission problem for weak nonlinear equations

$$\begin{cases} -\dfrac{d}{dx_i}\left(|u|^q a^{ij}(x)u_{x_j}\right) + b(x,u,\nabla u) = 0, \quad q \geq 0, & x \in G \setminus \Sigma_0; \\ [u]_{\overline{\Sigma_0}} = 0, \quad \mathcal{S}[u] \equiv \left[\frac{\partial u}{\partial \nu}\right]_{\Sigma_0} + \frac{1}{|x|}\sigma\left(\frac{x}{|x|}\right) u \cdot |u|^q = h(x,u), & x \in \Sigma_0; \\ \mathcal{B}[u] \equiv \frac{\partial u}{\partial \nu} + \frac{1}{|x|}\gamma\left(\frac{x}{|x|}\right) u \cdot |u|^q = g(x,u), & x \in \partial G \setminus \{\Sigma_0 \cup \mathcal{O}\}; \end{cases} \qquad (WQL)$$

(summation over repeated indices from 1 to n is understood); here:

- $\frac{\partial}{\partial \nu} = |u|^q a^{ij}(x) n_i \frac{\partial}{\partial x_j}$,
- $\left[\frac{\partial u}{\partial \nu}\right]_{\Sigma_0}$ denotes the saltus of the co-normal derivative of the function $u(x)$ on crossing Σ_0, i.e.,

$$\left[\frac{\partial u}{\partial \nu}\right]_{\Sigma_0} = |u|^q a_+^{ij}(x)\frac{\partial u_+}{\partial x_j} n_i\Big|_{\Sigma_0} - |u|^q a_-^{ij}(x)\frac{\partial u_-}{\partial x_j} n_i\Big|_{\Sigma_0}.$$

Definition 5.1. The function $u(x)$ is called a *weak* solution of the problem (WQL) provided that $u(x) \in \mathbf{C}^0(\overline{G}) \cap \overset{\circ}{\mathbf{W}}{}^1_0(G)$ and satisfies the integral identity

M. Borsuk, *Transmission Problems for Elliptic Second-Order Equations in Non-Smooth Domains*, Frontiers in Mathematics, DOI 10.1007/978-3-0346-0477-2_6, © Springer Basel AG 2010

$$\int\limits_G \{|u|^q a^{ij}(x)u_{x_j}\eta_{x_i} + b(x,u,u_x)\eta(x)\}\,dx + \int\limits_{\Sigma_0} \frac{\sigma(\omega)}{r} u|u|^q \eta(x)ds$$

$$+ \int\limits_{\partial G} \frac{\gamma(\omega)}{r} u|u|^q \eta(x)ds = \int\limits_{\partial G} g(x,u)\eta(x)ds + \int\limits_{\Sigma_0} h(x,u)\eta(x)ds \quad (II)$$

for all functions $\eta(x) \in \mathbf{C}^0(\overline{G}) \cap \overset{\circ}{\mathbf{W}}{}^1_0(G)$.

Lemma 5.2. *Let $u(x)$ be a weak solution of (WQL). For any function $\eta(x) \in \mathbf{C}^0(\overline{G}) \cap \overset{\circ}{\mathbf{W}}{}^1_0(G)$ the equality*

$$\int\limits_{G_0^\varrho} \Big\{|u|^q a^{ij}(x)u_{x_j}\eta_{x_i} + b(x,u,u_x)\eta(x)\Big\}dx = \int\limits_{\Omega_\varrho} |u|^q a^{ij}(x)u_{x_j}\cos(r,x_i)\eta(x)d\Omega_\varrho$$

$$+ \int\limits_{\Gamma_0^\varrho} \left(g(x,u) - \frac{\gamma(\omega)}{r}u|u|^q\right)\eta(x)ds + \int\limits_{\Sigma_0^\varrho} \left(h(x,u) - \frac{\sigma(\omega)}{r}u|u|^q\right)\eta(x)ds \quad (II)_{loc}$$

holds for a.e. $\varrho \in (0,d)$.

Proof. The proof is analogous to the proof of Lemma 3.2 Chapter 3. □

Assumptions. *Let $q \geq 0$, $0 \leq \mu < q+1$, $s > 1$, $f_1 \geq 0$, $g_1 \geq 0$, $h_1 \geq 0$ be given;*

(a) *the condition of the uniform ellipticity:*

$$a_\pm \xi^2 \leq a_\pm^{ij}(x)\xi_i\xi_j \leq A_\pm \xi^2, \quad \forall x \in \overline{G_\pm}, \quad \forall \xi \in \mathbb{R}^n; \quad a_\pm, A_\pm = const > 0,$$

$$a^{ij}(0) = a\delta_i^j, \text{ where } \delta_i^j \text{ is the Kronecker symbol; } a = \begin{cases} a_+, & x \in G_+, \\ a_-, & x \in G_-; \end{cases} \text{ we write}$$

$$a_* = \min\{a_+, a_-\} > 0, \quad a^* = \max\{a_+, a_-\} > 0, \quad A^* = \max(A_-, A_+);$$

(b) *$a^{ij}(x) \in \mathbf{C}^0(\overline{G})$ and the inequality*

$$\left(\sum_{i,j=1}^n |a_\pm^{ij}(x) - a_\pm^{ij}(y)|^2\right)^{\frac{1}{2}} \leq \mathcal{A}(|x-y|)$$

holds for $x, y \in \overline{G}$, where $\mathcal{A}(r)$ is a monotonically increasing, non-negative function, continuous at 0, $\mathcal{A}(0) = 0$;

(c) *$|b(x,u,u_x)| \leq a\mu|u|^{q-1}|\nabla u|^2 + b_0(x)$; $b_0(x) \in L_{p/2}(G)$, $n < p < 2n$;*

(d) *$\sigma(\omega) \geq \nu_0 > 0$ on σ_0; $\gamma(\omega) \geq \nu_0 > 0$ on ∂G;*

(e) $\frac{\partial h(x,u)}{\partial u} \leq 0$, $\frac{\partial g(x,u)}{\partial u} \leq 0$;

(f) $|b_0(x)| \leq f_1|x|^{s-2}$, $|g(x,0)| \leq g_1|x|^{s-1}$, $|h(x,0)| \leq h_1|x|^{s-1}$.

We make the function change

$$u = v|v|^{\varsigma-1} \text{ with } \varsigma = \frac{1}{q+1}. \tag{5.1.1}$$

Then identities (II) and $(II)_{loc}$ can be presented in the form

$$\int_G \Big\langle \varsigma a^{ij}(x) v_{x_j}\eta_{x_i} + \mathcal{B}(x,v,v_x)\eta\Big\rangle dx + \int_{\partial G} \frac{\gamma(\omega)}{r} v\eta(x)ds + \int_{\Sigma_0} \frac{\sigma(\omega)}{r} v\eta(x)ds$$
$$= \int_{\partial G} \mathcal{G}(x,v)\eta(x)ds + \int_{\Sigma_0} \mathcal{H}(x,v)\eta(x)ds; \tag{$\widetilde{II}$}$$

$$\int_{G_0^\varrho} \Big\langle \varsigma a^{ij}(x) v_{x_j}\eta_{x_i} + \mathcal{B}(x,v,v_x)\eta\Big\rangle dx + \int_{\Gamma_0^\varrho} \frac{\gamma(\omega)}{r} v\eta(x)ds + \int_{\Sigma_0^\varrho} \frac{\sigma(\omega)}{r} v\eta(x)ds$$
$$= \int_{\Omega_\varrho} \varsigma a^{ij}(x) v_{x_j} \cos(r,x_i)\eta(x)d\Omega_\varrho + \int_{\Gamma_0^\varrho} \mathcal{G}(x,v)\eta(x)ds + \int_{\Sigma_0^\varrho} \mathcal{H}(x,v)\eta(x)ds \tag{$\widetilde{II}_{loc}$}$$

for a.e. $\varrho \in (0,d)$, $v(x) \in \mathbf{C}^0(\overline{G}) \cap \overset{\circ}{\mathbf{W}}{}^1_0(G)$ and any $\eta(x) \in \mathbf{C}^0(\overline{G}) \cap \overset{\circ}{\mathbf{W}}{}^1_0(G)$, where

$$\mathcal{B}(x,v,v_x) \equiv b(x, v|v|^{\varsigma-1}, \varsigma|v|^{\varsigma-1}v_x), \quad \mathcal{G}(x,v) \equiv g(x, v|v|^{\varsigma-1}),$$
$$\mathcal{H}(x,v) \equiv h(x, v|v|^{\varsigma-1}). \tag{5.1.2}$$

We assume without loss of generality that there exists $d > 0$ such that G_0^d is *a rotational cone* with the vertex at $\mathcal{O}$ and the aperture ω_0, thus

$$\Gamma_0^d = \left\{ (r,\omega) \Big| x_1^2 = \cot^2\frac{\omega_0}{2}\sum_{i=2}^n x_i^2;\ r \in (0,d),\ \omega_1 = \frac{\omega_0}{2},\ \omega_0 \in (0,2\pi) \right\}. \tag{5.1.3}$$

Theorem 5.3. *Let u be a weak solution of the problem (WQL) and assumptions* (a)–(f) *be satisfied where $\mathcal{A}(r)$ is Dini-continuous at zero. Let us assume that $M_0 = \max_{x\in\overline{G}} |u(x)|$ is known and λ is as above in* (2.2.1). *Then there exist $d \in (0,1)$ and a constant $C_0 > 0$ depending only on $n, a_*, A^*, p, q, \lambda, \mu, f_1, h_1, g_1, \nu_0, s, M_0$, meas G, diam G and on the quantity $\int_0^1 \frac{\mathcal{A}(r)}{r}dr$ such that the inequality*

$$|u(x)| \leq C_0\Big(\|u\|_{2(q+1),G} + f_1 + g_1 + h_1\Big) \cdot \begin{cases} |x|^{\frac{\lambda(1+q-\mu)}{(q+1)^2}}, & \text{if } s > \lambda\frac{1+q-\mu}{1+q}, \\ |x|^{\frac{\lambda(1+q-\mu)}{(q+1)^2}} \ln^{\frac{1}{q+1}}\left(\frac{1}{|x|}\right), & \text{if } s = \lambda\frac{1+q-\mu}{1+q}, \\ |x|^{\frac{s}{q+1}}, & \text{if } s < \lambda\frac{1+q-\mu}{1+q} \end{cases} \tag{5.1.4}$$

holds for all $x \in G_0^d$.

In addition, if coefficients of the problem (WQL) satisfy such conditions, which guarantee the local a-priori estimate $|\nabla u|_{0,G'} \le M_1$ *for any smooth* $G' \subset\subset \overline{G} \setminus \{\mathcal{O}\}$ *(see for example §4 in [6] or [64], [47]), then the inequality*

$$|\nabla u(x)| \le C_1 \cdot \begin{cases} |x|^{\frac{\lambda(1+q-\mu)}{(q+1)^2}-1}, & \text{if } s > \lambda\frac{1+q-\mu}{1+q}, \\ |x|^{\frac{\lambda(1+q-\mu)}{(q+1)^2}-1} \ln^{\frac{1}{q+1}}\left(\frac{1}{|x|}\right), & \text{if } s = \lambda\frac{1+q-\mu}{1+q}, \\ |x|^{\frac{s}{q+1}-1}, & \text{if } s < \lambda\frac{1+q-\mu}{1+q} \end{cases} \tag{5.1.5}$$

holds for all $x \in G_0^d$ *with* $C_1 = c_1\big(\|u\|_{2(q+1),G} + f_1 + g_1 + h_1\big)$, *where* c_1 *depends on* M_0, M_1 *and* C_0 *from above.*

5.2 Local estimate at the boundary

We formulate here a result asserting the local boundedness (near the conical point) of the weak solution of problem (WQL).

Theorem 5.4. *Let* $u(x)$ *be a weak solution of the problem (WQL). Let assumptions* (a), (c)–(e) *be satisfied. Suppose, in addition, that* $h(x,0) \in L_\infty(\Sigma_0)$, $g(x,0) \in L_\infty(\partial G)$. *Then the inequality*

$$\begin{aligned} \sup_{G_0^{\varkappa\varrho}} |u(x)| \le C\Big\{ & \varrho^{-n/t(q+1)} \|u\|_{t(q+1),G_0^\varrho} + \varrho^{\frac{2}{q+1}(1-n/p)} \|b_0\|_{p/2,G_0^\varrho}^{\frac{1}{q+1}} \\ & + \varrho^{\frac{1}{q+1}} \left(\|g(x,0)\|_{\infty,\Gamma_0^\varrho}^{\frac{1}{q+1}} + \|h(x,0)\|_{\infty,\Sigma_0^\varrho}^{\frac{1}{q+1}} \right) \Big\} \end{aligned} \tag{5.2.1}$$

holds for any $t > 0$, $\varkappa \in (0,1)$ *and* $\varrho \in (0,d)$, *where* $C = const(n, a_*, A^*, t, p, q, \varkappa, \mu, G)$ *and* $d \in (0,1)$.

Proof. See the proof of Theorem 3.4 or Theorem 6.6 for $m = 2$. □

5.3 Global integral estimate

In this section we estimate the weighted Dirichlet integral.

Theorem 5.5. [1] *Let* u *be a weak solution of the problem* (WQL) *and assumptions* (a)–(e) *are satisfied where* $\mathcal{A}(r)$ *is Dini-continuous at zero. In addition, let us satisfy*

$$b_0(x) \in \overset{\circ}{\mathbf{W}}{}^0_\alpha(G), \quad \int_{\Sigma_0} r^{\alpha-1} h^2(x,0)ds < \infty, \quad \int_{\partial G} r^{\alpha-1} g^2(x,0)ds < \infty, \quad 4-n \le \alpha \le 2.$$

[1] See also Subsection 5.5.2.

Then $|u(x)|^{q+1} \in \overset{\circ}{\mathbf{W}}{}^{1}_{\alpha-2}(G)$ *and there exists a constant* $C > 0$ *depending only on* $a_*, \alpha, \mu, q, n, \lambda$ *from* (2.2.1) *and the domain* G *such that the inequality*

$$\int_G a\left(r^{\alpha-2}|u|^{2q}|\nabla u|^2 + r^{\alpha-4}|u|^{2(q+1)}\right)dx + \int_{\Sigma_0} r^{\alpha-3}\sigma(\omega)|u|^{2(q+1)}ds$$
$$+ \int_{\partial G} r^{\alpha-3}\gamma(\omega)|u|^{2(q+1)}ds \tag{5.3.1}$$
$$\le C\Big\{\int_G \left(|u|^{2(q+1)} + (1+r^\alpha)b_0^2(x)\right)dx + \int_{\Sigma_0} r^{\alpha-1}h^2(x,0)ds + \int_{\partial G} r^{\alpha-1}g^2(x,0)ds\Big\}$$

holds.

Proof. At first we make the function change (5.1.1) and consider the integral identity $(\widetilde{II})$ for the function $v(x)$. Putting in this identity $\eta(x) = r_\varepsilon^{\alpha-2}v(x)$ we obtain

$$\varsigma\int_G ar_\varepsilon^{\alpha-2}|\nabla v|^2dx + \int_{\Sigma_0} r^{-1}r_\varepsilon^{\alpha-2}\sigma(\omega)v^2(x)ds + \int_{\partial G} r^{-1}r_\varepsilon^{\alpha-2}\gamma(\omega)v^2(x)ds$$
$$= \varsigma\frac{2-\alpha}{2}\int_G ar_\varepsilon^{\alpha-4}(x_i - \varepsilon l_i)(v^2)_{x_i}dx$$
$$+ \varsigma(2-\alpha)\int_G \left(a^{ij}(x) - a^{ij}(0)\right)r_\varepsilon^{\alpha-4}(x_i - \varepsilon l_i)v_{x_j}v(x)dx$$
$$- \varsigma\int_G \left(a^{ij}(x) - a^{ij}(0)\right)r_\varepsilon^{\alpha-2}v_{x_i}v_{x_j}dx - \int_G \mathcal{B}(x,v,v_x)r_\varepsilon^{\alpha-2}v(x)dx$$
$$+ \int_{\Sigma_0} r_\varepsilon^{\alpha-2}v(x)\mathcal{H}(x,v)ds + \int_{\partial G} r_\varepsilon^{\alpha-2}v(x)\mathcal{G}(x,v)ds. \tag{5.3.2}$$

Integrating by parts, we have

$$\int_G ar_\varepsilon^{\alpha-4}(x_i - \varepsilon l_i)\frac{\partial v^2}{\partial x_i}dx = \int_{G_+} a_+r_\varepsilon^{\alpha-4}(x_i - \varepsilon l_i)\frac{\partial v_+^2}{\partial x_i}dx$$
$$+ \int_{G_-} a_-r_\varepsilon^{\alpha-4}(x_i - \varepsilon l_i)\frac{\partial v_-^2}{\partial x_i}dx \tag{5.3.3}$$
$$= -\int_G av^2\frac{\partial}{\partial x_i}\left(r_\varepsilon^{\alpha-4}(x_i - \varepsilon l_i)\right)dx + \int_{\partial G_+} a_+v_+^2r_\varepsilon^{\alpha-4}(x_i - \varepsilon l_i)\cos(\overrightarrow{n}, x_i)ds$$

$$+\int\limits_{\partial G_-} a_- v_-^2 r_\varepsilon^{\alpha-4}(x_i-\varepsilon l_i)\cos(\vec{n},x_i)ds = -\int\limits_G av^2\frac{\partial}{\partial x_i}\Big(r_\varepsilon^{\alpha-4}(x_i-\varepsilon l_i)\Big)dx$$

$$+\int\limits_{\partial G} av^2 r_\varepsilon^{\alpha-4}(x_i-\varepsilon l_i)\cos(\vec{n},x_i)ds + [a]_{\Sigma_0}\int\limits_{\Sigma_0} v^2 r_\varepsilon^{\alpha-4}(x_i-\varepsilon l_i)\cos(\vec{n},x_i)ds,$$

because of $[v]_{\Sigma_0}=0$. Now we calculate:

1) $\frac{\partial}{\partial x_i}\left(r_\varepsilon^{\alpha-4}(x_i-\varepsilon l_i)\right) = nr_\varepsilon^{\alpha-4}+(\alpha-4)(x_i-\varepsilon l_i)r_\varepsilon^{\alpha-5}\frac{x_i-\varepsilon l_i}{r_\varepsilon} = (n+\alpha-4)r_\varepsilon^{\alpha-4}$;

2) because of $\cos(\vec{n},x_i)\Big|_{\Sigma_0} = \cos(x_n,x_i) = \delta_i^n$,

$$(x_i-\varepsilon l_i)\cos(\vec{n},x_i)\Big|_{\Sigma_0} = \delta_i^n(x_i-\varepsilon l_i)\Big|_{\Sigma_0} = (x_n-\varepsilon l_n)\Big|_{\Sigma_0} = x_n\Big|_{\Sigma_0} = 0,$$

since $\Sigma_0=\{x_n=0\}\cap G$ and $l_n=0$;

3) from the representation $\partial G = \Gamma_0^d\cup\Gamma_d$ and by (4.2.1),

$$(x_i-\varepsilon l_i)\cos(\vec{n},x_i)\Big|_{\Gamma_0^d} = -\varepsilon\sin\frac{\omega_0}{2} \implies$$

$$\int\limits_{\partial G} av^2 r_\varepsilon^{\alpha-4}(x_i-\varepsilon l_i)\cos(\vec{n},x_i)ds$$

$$= -\varepsilon\sin\frac{\omega_0}{2}\int\limits_{\Gamma_0^d} av^2 r_\varepsilon^{\alpha-4}ds + \int\limits_{\Gamma_d} av^2 r_\varepsilon^{\alpha-4}(x_i-\varepsilon l_i)\cos(\vec{n},x_i)ds.$$

Then from (5.3.3) it follows that

$$\frac{2-\alpha}{2}\int\limits_G ar_\varepsilon^{\alpha-4}(x_i-\varepsilon l_i)\frac{\partial v^2}{\partial x_i}dx = \frac{(2-\alpha)(4-n-\alpha)}{2}\int\limits_G ar_\varepsilon^{\alpha-4}v^2dx \tag{5.3.4}$$

$$-\varepsilon\frac{2-\alpha}{2}\sin\frac{\omega_0}{2}\int\limits_{\Gamma_0^d} av^2r_\varepsilon^{\alpha-4}ds + \frac{2-\alpha}{2}\int\limits_{\Gamma_d} av^2 r_\varepsilon^{\alpha-4}(x_i-\varepsilon l_i)\cos(\vec{n},x_i)ds.$$

Then we can rewrite (5.3.2) in the following form:

$$\varsigma\int\limits_G ar_\varepsilon^{\alpha-2}|\nabla v|^2dx + \varepsilon\varsigma\frac{2-\alpha}{2}\sin\frac{\omega_0}{2}\int\limits_{\Gamma_0^d} av^2r_\varepsilon^{\alpha-4}ds + \int\limits_{\Sigma_0} r^{-1}r_\varepsilon^{\alpha-2}\sigma(\omega)v^2(x)ds$$

$$+\int\limits_{\partial G} r^{-1}r_\varepsilon^{\alpha-2}\gamma(\omega)v^2(x)ds = \frac{2-\alpha}{2}\varsigma\int\limits_{\Gamma_d} av^2r_\varepsilon^{\alpha-4}(x_i-\varepsilon l_i)\cos(\vec{n},x_i)ds$$

$$+\varsigma\frac{(2-\alpha)(4-n-\alpha)}{2}\int\limits_G ar_\varepsilon^{\alpha-4}v^2dx$$

$$+\varsigma(2-\alpha)\int\limits_G \big(a^{ij}(x)-a^{ij}(0)\big)r_\varepsilon^{\alpha-4}(x_i-\varepsilon l_i)v_{x_j}v(x)dx$$

$$-\varsigma\int\limits_G \big(a^{ij}(x)-a^{ij}(0)\big)r_\varepsilon^{\alpha-2}v_{x_i}v_{x_j}dx-\int\limits_G \mathcal{B}(x,v,v_x)r_\varepsilon^{\alpha-2}v(x)dx$$

$$+\int\limits_{\Sigma_0} r_\varepsilon^{\alpha-2}v(x)\mathcal{H}(x,v)ds+\int\limits_{\partial G} r_\varepsilon^{\alpha-2}v(x)\mathcal{G}(x,v)ds. \quad (5.3.5)$$

Now we estimate the integral over Γ_d. Because on Γ_d:

$$r_\varepsilon\ge hr\ge hd \;\Rightarrow\; (\alpha-3)\ln r_\varepsilon\le(\alpha-3)\ln(hd),$$

by $\alpha\le 2$, we have $r_\varepsilon^{\alpha-3}|_{\Gamma_d}\le(hd)^{\alpha-3}$ and therefore:

$$\frac{2-\alpha}{2}\varsigma\int\limits_{\Gamma_d} av^2r_\varepsilon^{\alpha-4}(x_i-\varepsilon l_i)\cos(\vec{n},x_i)ds\le\frac{2-\alpha}{2}\varsigma\int\limits_{\Gamma_d} ar_\varepsilon^{\alpha-3}v^2ds$$

$$\le\frac{2-\alpha}{2}\varsigma\,(hd)^{\alpha-3}\int\limits_{\Gamma_d} av^2ds\le c\int\limits_{G_d}(v^2+|\nabla v|^2)dx, \quad (5.3.6)$$

by (1.5.12). By virtue of assumption (c) and the function change (5.1.1), we have

$$|v\cdot\mathcal{B}(x,v,v_x)|\le a\mu\varsigma^2|\nabla v|^2+b_0(x)|v|. \quad (5.3.7)$$

Using the Cauchy inequality we deduce the following:

$$\int\limits_G \mathcal{B}(x,v,v_x)r_\varepsilon^{\alpha-2}v(x)dx\le\mu\varsigma^2\int\limits_G ar_\varepsilon^{\alpha-2}|\nabla v|^2dx+\int\limits_G r_\varepsilon^{\alpha-2}|v|b_0(x)dx \quad (5.3.8)$$

$$\le\mu\varsigma^2\int\limits_G ar_\varepsilon^{\alpha-2}|\nabla v|^2dx+\frac{\delta}{2}\int\limits_G ar^{-2}r_\varepsilon^{\alpha-2}v^2dx+\frac{1}{2\delta a_*}\int\limits_G r^2r_\varepsilon^{\alpha-2}b_0^2(x)dx,\ \forall\delta>0.$$

Now we use the representation $G=G_0^d\cup G_d$. For the estimating integrals over G_0^d, by assumption (b) and the Cauchy inequality, we obtain

$$\int\limits_{G_0^d}\Big\{\big(a^{ij}(x)-a^{ij}(0)\big)\left(r_\varepsilon^{\alpha-2}v_{x_i}v_{x_j}+r_\varepsilon^{\alpha-4}(x_i-\varepsilon l_i)v(x)v_{x_j}\right)dx \quad (5.3.9)$$

$$\le\mathcal{A}(d)\int\limits_{G_0^d} a\Big(r_\varepsilon^{\alpha-2}|\nabla v|^2+r_\varepsilon^{\alpha-3}|\nabla v|\cdot|v(x)|\Big)dx$$

$$\leq \frac{3}{2}\mathcal{A}(d) \int\limits_{G_0^d} a \left(r_\varepsilon^{\alpha-2}|\nabla v|^2 + r_\varepsilon^{\alpha-4}v^2\right) dx.$$

Now we estimate integrals over G_d. By assumptions (a), the Cauchy inequality and taking into account that $r_\varepsilon \geq hd$ for $r \geq d$, we get

$$\int\limits_{G_d} \Big\{\left(a^{ij}(x) - a^{ij}(0)\right)\left(r_\varepsilon^{\alpha-2}v_{x_i}v_{x_j} + r_\varepsilon^{\alpha-4}(x_i - \varepsilon l_i)v(x)v_{x_j}\right)\Big\}dx \tag{5.3.10}$$

$$\leq A^* \int\limits_{G_d} \left(\frac{3}{2}r_\varepsilon^{\alpha-2}|\nabla v|^2 + r_\varepsilon^{\alpha-4}|v|^2\right) dx \leq C(A^*, h, \alpha, d) \int\limits_{G_d} \left(|\nabla v|^2 + v^2\right) dx.$$

Further, because of assumption (e),

$$v\mathcal{G}(x, v) = v\mathcal{G}(x, 0) + v^2 \cdot \int\limits_0^1 \frac{\partial \mathcal{G}(x, \tau v)}{\partial(\tau v)} d\tau \leq |g(x, 0)| \cdot |v|. \tag{5.3.11}$$

But, by the Cauchy inequality and $\gamma(\omega) \geq \nu_0 > 0$,

$$|g(x, 0)| \cdot |v| = \left(r^{\frac{1}{2}} \frac{1}{\sqrt{\gamma(\omega)}}|g(x, 0)|\right)\left(r^{-\frac{1}{2}}\sqrt{\gamma(\omega)}|v|\right)$$
$$\leq \frac{\delta}{2}r^{-1}\gamma(\omega)v^2 + \frac{1}{2\delta\nu_0}rg^2(x, 0),$$

for all $\delta > 0$; taking into account that $r_\varepsilon \geq hr$ (see Subsection 1.3) we obtain

$$\int\limits_{\partial G} r_\varepsilon^{\alpha-2}v\mathcal{G}(x, v)ds \leq \frac{\delta}{2}\int\limits_{\partial G} r_\varepsilon^{\alpha-2}\frac{1}{r}\gamma(\omega)v^2 ds + \frac{1}{2\delta\nu_0}\int\limits_{\partial G} r^{\alpha-1}g^2(x, 0)ds, \tag{5.3.12}$$

for all $\delta > 0$. Similarly,

$$\int\limits_{\Sigma_0} r_\varepsilon^{\alpha-2}v\mathcal{H}(x, v)ds \leq \frac{\delta}{2}\int\limits_{\Sigma_0} r_\varepsilon^{\alpha-2}\frac{1}{r}\sigma(\omega)v^2 ds + \frac{1}{2\delta\nu_0}\int\limits_{\Sigma_0} r^{\alpha-1}h^2(x, 0)ds, \tag{5.3.13}$$

for all $\delta > 0$. As a result, with regard to $\varsigma \leq 1$, $\mu\varsigma < 1$ and $4 - n \leq \alpha \leq 2$ from (5.3.5)–(5.3.13), we obtain

$$\varsigma(1 - \mu\varsigma)\int\limits_G ar_\varepsilon^{\alpha-2}|\nabla v|^2 dx + \int\limits_{\Sigma_0} \frac{1}{r}r_\varepsilon^{\alpha-2}\sigma(\omega)v^2(x)ds + \int\limits_{\partial G} \frac{1}{r}r_\varepsilon^{\alpha-2}\gamma(\omega)v^2(x)ds$$
$$\leq \frac{3}{2}\mathcal{A}(d)\int\limits_{G_0^d} a\left(r_\varepsilon^{\alpha-2}|\nabla v|^2 + r_\varepsilon^{\alpha-4}v^2\right) dx + \frac{\delta}{2}\int\limits_G ar^{-2}r_\varepsilon^{\alpha-2}v^2 dx$$

$$
\begin{aligned}
&+C\int\limits_{G_d}\left(|\nabla v|^2+v^2\right)dx+\frac{1}{2a_*\delta}\int\limits_G r^\alpha b_0^2(x)dx\\
&+\frac{1}{2\delta\nu_0}\left\{\int\limits_{\partial G} r^{\alpha-1}g^2(x,0)ds+\int\limits_{\Sigma_0} r^{\alpha-1}h^2(x,0)ds\right\}\\
&+\frac{\delta}{2}\left\{\int\limits_{\Sigma_0} r_\varepsilon^{\alpha-2}\frac{1}{r}\sigma(\omega)v^2ds+\int\limits_{\partial G} r_\varepsilon^{\alpha-2}\frac{1}{r}\gamma(\omega)v^2ds\right\},
\end{aligned}
\tag{5.3.14}
$$

for all $\delta>0$. By the inequality $r_\varepsilon\ge hr$, we have $r_\varepsilon^{\alpha-4}\le h^{-2}r^{-2}r_\varepsilon^{\alpha-2}$. Hence, by Lemma 2.5, from (5.3.14) it follows that

$$
\begin{aligned}
&\varsigma(1-\mu\varsigma)\left\{\int\limits_G ar_\varepsilon^{\alpha-2}|\nabla v|^2dx+\int\limits_{\Sigma_0}\frac{1}{r}r_\varepsilon^{\alpha-2}\sigma(\omega)v^2(x)ds+\int\limits_{\partial G}\frac{1}{r}r_\varepsilon^{\alpha-2}\gamma(\omega)v^2(x)ds\right\}\\
&\le c(\lambda,\omega_0)\left(\delta+\mathcal{A}(d)\right)\left\{\int\limits_G ar_\varepsilon^{\alpha-2}|\nabla v|^2dx+\int\limits_{\Sigma_0} r^{-1}r_\varepsilon^{\alpha-2}\sigma(\omega)v^2(x)ds\right.\\
&\left.+\int\limits_{\partial G} r^{-1}r_\varepsilon^{\alpha-2}\gamma(\omega)v^2ds\right\}+C\int\limits_G\left(|\nabla v|^2+v^2\right)dx+\frac{1}{2a_*\delta}\int\limits_G r^\alpha b_0^2(x)dx\\
&+\frac{1}{2\delta\nu_0}\int\limits_{\partial G} r^{\alpha-1}g^2(x,0)ds+\frac{1}{2\delta\nu_0}\int\limits_{\Sigma_0} r^{\alpha-1}h^2(x,0)ds,\quad\forall\delta>0,\ \forall\varepsilon>0.
\end{aligned}
\tag{5.3.15}
$$

Because of $0\le\mu<1+q$, we can choose $\delta=\frac{\varsigma}{4c(\lambda,\omega_0)}(1-\mu\varsigma)$ and next $d>0$ such that, by the continuity of $\mathcal{A}(r)$ at zero, $c(\lambda,\omega_0)\mathcal{A}(d)\le\frac{\varsigma}{4}(1-\mu\varsigma)$. Thus, from (5.3.15) we get

$$
\begin{aligned}
&\int\limits_G ar_\varepsilon^{\alpha-2}|\nabla v|^2dx+\int\limits_{\Sigma_0}\frac{1}{r}r_\varepsilon^{\alpha-2}\sigma(\omega)v^2(x)ds+\int\limits_{\partial G}\frac{1}{r}r_\varepsilon^{\alpha-2}\gamma(\omega)v^2(x)ds\\
&\le C(a_*,\alpha,\lambda,\mu,q,n,d)\left\{\int\limits_G(|\nabla v|^2+v^2)dx+\int\limits_G r^\alpha b_0^2(x)dx\right.\\
&\left.+\frac{1}{\nu_0}\int\limits_{\partial G} r^{\alpha-1}g^2(x,0)ds+\frac{1}{\nu_0}\int\limits_{\Sigma_0} r^{\alpha-1}h^2(x,0)ds\right\},\ \forall\varepsilon>0.
\end{aligned}
\tag{5.3.16}
$$

We can observe that the right-hand side of (5.3.16) does not depend on ε. Therefore, performing the passage to the limit as $\varepsilon\to+0$, by the Fatou Theorem, we derive $v(x)\in\overset{\circ}{\mathbf{W}}{}^1_{\alpha-2}(G)$ and

$$
\int\limits_G ar^{\alpha-2}|\nabla v|^2dx+\int\limits_{\Sigma_0} r^{\alpha-3}\sigma(\omega)v^2(x)ds+\int\limits_{\partial G} r^{\alpha-3}\gamma(\omega)v^2(x)ds
$$

$$\leq C(a_*, \alpha, \lambda, \mu, q, n, d)\Big\{\int\limits_G (|\nabla v|^2 + v^2)dx + \int\limits_G r^\alpha b_0^2(x)dx$$
$$+\frac{1}{\nu_0}\int\limits_{\partial G} r^{\alpha-1} g^2(x,0)ds + \frac{1}{\nu_0}\int\limits_{\Sigma_0} r^{\alpha-1} h^2(x,0)ds\Big\}. \tag{5.3.17}$$

Returning to the integral identity $(\widetilde{II})$ and setting in it $\eta(x) = v(x)$, we get

$$\int\limits_G \Big\langle \varsigma a^{ij}(x) v_{x_j} v_{x_i} + \mathcal{B}(x, v, v_x)v\Big\rangle dx + \int\limits_{\partial G} \frac{\gamma(\omega)}{r} v^2 ds + \int\limits_{\Sigma_0} \frac{\sigma(\omega)}{r} v^2 ds$$
$$= \int\limits_{\partial G} \mathcal{G}(x, v)vds + \int\limits_{\Sigma_0} \mathcal{H}(x, v)vds.$$

From the ellipticity condition (a), inequalities (5.3.7)and (5.3.12)–(5.3.13) for $\alpha = 2$, $\delta = 2$, it follows that

$$\varsigma(1-\mu\varsigma)\int\limits_G a|\nabla v|^2 dx \tag{5.3.18}$$
$$\leq c(a_*, \nu_0, diamG)\Big\{\int\limits_G \left(|v|^2 + b_0^2(x)\right) dx + \int\limits_{\partial G} r^{\alpha-1} g^2(x,0)ds + \int\limits_{\Sigma_0} r^{\alpha-1} h^2(x,0)ds\Big\}.$$

Now, using the inequality (2.2.3) and returning to the function $u(x)$, by means of the function change (5.1.1), from (5.3.17)–(5.3.18) we get the desired estimate (5.3.1). □

5.4 Local integral weighted estimates

In this section we will derive a local estimate for the weighted Dirichlet integral.

Theorem 5.6. *Let u be a weak solution of the problem (WQL), λ be as above in* (2.2.1) *and assumptions* (a)–(f) *be satisfied where $\mathcal{A}(r)$ is Dini-continuous at zero. Then $|u(x)|^{q+1} \in \overset{\circ}{\mathbf{W}}{}^1_{2-n}(G)$ and there exist $d \in (0,1)$ and a constant $C > 0$ depending only on $n, s, \lambda, q, \mu, \nu_0, a_*, G, \Sigma_0$ and on $\int_0^1 \frac{\mathcal{A}(r)}{r}dr$ such that the inequality*

$$\int\limits_{G_0^\varrho} a\left(r^{2-n}|u|^{2q}|\nabla u|^2 + r^{-n}|u|^{2(q+1)}\right) dx + \int\limits_{\Sigma_0^\varrho} r^{1-n}\sigma(\omega)|u|^{2(q+1)}ds$$
$$+\int\limits_{\Gamma_0^\varrho} r^{1-n}\gamma(\omega)|u|^{2(q+1)}ds$$

$$\le C\Big(\int\limits_G |u|^{2(q+1)}dx+f_1^2+\frac{1}{\nu_0}g_1^2+\frac{1}{\nu_0}h_1^2\Big)\cdot\begin{cases}\varrho^{2\lambda(1-\mu\varsigma)}, & \text{if } s>\lambda(1-\mu\varsigma),\\ \varrho^{2\lambda(1-\mu\varsigma)}\ln^2\big(\frac{1}{\varrho}\big), & \text{if } s=\lambda(1-\mu\varsigma),\\ \varrho^{2s}, & \text{if } s<\lambda(1-\mu\varsigma)\end{cases} \tag{5.4.1}$$

and $\varsigma=\frac{1}{1+q}$ *holds for all* $\varrho\in(0,d)$.

Proof. Performing the function change (5.1.1) we consider the integral identity $(\widetilde{II})_{loc}$ for the function $v(x)$. From Theorem 5.5 it follows that $v(x)$ belongs to $\overset{\circ}{\mathbf{W}}{}^{1}_{2-n}(G)$, so it is enough to derive the estimate (5.4.1). Using the function $V(\varrho)$ that is defined by (2.4.10) and setting $\eta(x)=r^{2-n}v(x)$ in the integral identity $(\widetilde{II})_{loc}$, we obtain

$$\varsigma V(\varrho)\le\varsigma\varrho\int\limits_{\Omega}av(x)\frac{\partial v}{\partial r}\Big|_{r=\varrho}d\Omega+\varsigma\int\limits_{\Omega_\varrho}r^{2-n}v(x)\left(a^{ij}(x)-a^{ij}(0)\right)v_{x_j}\cos(r,x_i)d\Omega_\varrho$$

$$+\int\limits_{\Gamma_0^\varrho}r^{2-n}v(x)\mathcal{G}(x,v)ds+\int\limits_{\Sigma_0^\varrho}r^{2-n}v(x)\mathcal{H}(x,v)ds+\varsigma(n-2)\int\limits_{G_0^\varrho}ar^{-n}x_ivv_{x_i}dx$$

$$+\int\limits_{G_0^\varrho}\Big\{-\varsigma r^{2-n}\left(a^{ij}(x)-a^{ij}(0)\right)v_{x_i}v_{x_j}+\varsigma(n-2)r^{-n}v(x)\left(a^{ij}(x)-a^{ij}(0)\right)x_iv_{x_j}$$

$$-r^{2-n}v(x)\mathcal{B}(x,v,v_x)\Big\}dx. \tag{5.4.2}$$

We transform some integrals on the right. By the divergence theorem,

$$(n-2)\int\limits_{G_0^\varrho}ar^{-n}v(x)x_iv_{x_i}dx=\frac{n-2}{2}\int\limits_{G_0^\varrho}\frac{ax_i}{r^n}\frac{\partial v^2}{\partial x_i}dx$$

$$=\frac{n-2}{2}\Big\{\frac{1}{\varrho^n}\int\limits_{\Omega_\varrho}av^2(x)x_i\cos(r,x_i)d\Omega_\varrho$$

$$+[a]_{\Sigma_0}\int\limits_{\Sigma_0^\varrho}r^{-n}v^2(x)x_i\cos(n,x_i)ds+\int\limits_{\Gamma_0^\varrho}ar^{-n}v^2(x)x_i\cos(n,x_i)ds\Big\}.$$

Since $x_i\cos(n,x_i)\Big|_{\Gamma_0^\varrho}=0$, $x_i\cos(r,x_i)\Big|_{\Omega_\varrho}=\varrho$; $x_i\cos(n,x_i)\Big|_{\Sigma_0}=x_i\cos(x_n,x_i)\Big|_{\Sigma_0}=x_n\Big|_{\Sigma_0}=0$, we have from above

$$(n-2)\int\limits_{G_0^\varrho}ar^{-n}v(x)x_iv_{x_i}dx=\frac{n-2}{2}\int\limits_{\Omega}av^2(x)d\Omega. \tag{5.4.3}$$

Under Lemma 2.12, from (5.4.2)–(5.4.3) it follows that

$$V(\varrho) \le \frac{\varrho}{2\lambda} V'(\varrho) + \varrho \int\limits_{\Omega} v(x) \left(a^{ij}(x) - a^{ij}(0)\right) v_{x_j} \cos(r, x_i) d\Omega$$

$$+ \frac{1}{\varsigma} \int\limits_{\Gamma_0^\varrho} r^{2-n} v(x) \mathcal{G}(x, v) ds + \frac{1}{\varsigma} \int\limits_{\Sigma_0^\varrho} r^{2-n} v(x) \mathcal{H}(x, v) ds$$

$$+ \int\limits_{G_0^\varrho} \Big\{ -r^{2-n} \left(a^{ij}(x) - a^{ij}(0)\right) v_{x_i} v_{x_j} + (n-2) r^{-n} v(x) \left(a^{ij}(x) - a^{ij}(0)\right) x_i v_{x_j}$$

$$- \frac{1}{\varsigma} r^{2-n} v(x) \mathcal{B}(x, v, v_x) \Big\} dx. \tag{5.4.4}$$

By virtue of assumptions (b)–(d) together with inequalities (5.3.7) and (5.3.11), from (5.4.4) it follows that

$$(1 - \mu\varsigma) V(\varrho) \le \frac{\varrho}{2\lambda} V'(\varrho) + \varrho \mathcal{A}(\varrho) \int\limits_{\Omega} a |v| |\nabla v| d\Omega + \frac{1}{\varsigma} \int\limits_{\Gamma_0^\varrho} r^{2-n} |v(x)| |g(x, 0)| ds$$

$$+ \frac{1}{\varsigma} \int\limits_{\Sigma_0^\varrho} r^{2-n} |v(x)| |h(x, 0)| ds + c_1(n) \mathcal{A}(\varrho) \int\limits_{G_0^\varrho} a \left(r^{2-n} |\nabla v|^2 + r^{1-n} |v| |\nabla v|\right) dx$$

$$+ \frac{1}{\varsigma} \int\limits_{G_0^\varrho} r^{2-n} |v(x)| |b_0(x)| dx. \tag{5.4.5}$$

Applying the Cauchy and Friedrichs-Wirtinger inequalities we have (see $(W)_2$, (2.4.12))

$$\int\limits_{\Omega} a \varrho |v| |\nabla v| d\Omega \le \frac{1}{2} \int\limits_{\Omega} a \left(\varrho^2 |\nabla v|^2 + |v|^2\right) d\Omega \le c_2(\lambda) \varrho V'(\varrho) \tag{5.4.6}$$

as well as, by virtue of the inequality (2.2.3) with $\alpha = 4 - n$,

$$\int\limits_{G_0^\varrho} a r^{1-n} |v| |\nabla v| dx \le \int\limits_{G_0^\varrho} a \left(r^{2-n} |\nabla v|^2 + r^{-n} |v|^2\right) dx \le c_3(\lambda) V(\varrho). \tag{5.4.7}$$

Further, by the Cauchy inequality with $\forall \delta > 0$,

$$\int\limits_{\Gamma_0^\varrho} r^{2-n} |v| |g(x, 0)| ds = \int\limits_{\Gamma_0^\varrho} \left(r^{\frac{1-n}{2}} \sqrt{\gamma(\omega)} |v|\right) \left(r^{\frac{3-n}{2}} \frac{1}{\sqrt{\gamma(\omega)}} |g(x, 0)|\right) ds$$

$$\le \frac{\delta}{2} \int\limits_{\Gamma_0^\varrho} r^{1-n} \gamma(\omega) |v|^2 ds + \frac{1}{2\delta\nu_0} \int\limits_{\Gamma_0^\varrho} r^{3-n} |g(x, 0)|^2 ds; \tag{5.4.8}$$

$$\int\limits_{\Sigma_0^\varrho} r^{2-n}|v||h(x,0)|ds = \int\limits_{\Sigma_0^\varrho} \left(r^{\frac{1-n}{2}}\sqrt{\sigma(\omega)}|v|\right)\left(r^{\frac{3-n}{2}}\frac{1}{\sqrt{\sigma(\omega)}}|h(x,0)|\right)ds$$
$$\leq \frac{\delta}{2}\int\limits_{\Sigma_0^\varrho} r^{1-n}\sigma(\omega)|v|^2 ds + \frac{1}{2\delta\nu_0}\int\limits_{\Sigma_0^\varrho} r^{3-n}|h(x,0)|^2 ds; \quad (5.4.9)$$

$$\int\limits_{G_0^\varrho} r^{2-n}|v(x)||b_0(x)|dx \leq \frac{\delta}{2a_*}\int\limits_{G_0^\varrho} ar^{-n}|v|^2dx + \frac{1}{2\delta}\int\limits_{G_0^\varrho} r^{4-n}|b_0|^2 dx$$
$$\leq \frac{\delta}{2a_*}c_4(\lambda)V(\varrho) + \frac{1}{2\delta}\int\limits_{G_0^\varrho} r^{4-n}|b_0|^2 dx \quad (5.4.10)$$

because of the inequality (2.2.3) with $\alpha = 4-n$. Thus, from (5.4.5)–(5.4.10) we get

$$\{(1-\mu\varsigma) - c_5(n,\lambda,q,a_*)(\delta + \mathcal{A}(\varrho))\}\,V(\varrho) \leq \frac{\varrho}{2\lambda}\left(1 + c_6(\lambda)\mathcal{A}(\varrho)\right)V'(\varrho) \quad (5.4.11)$$
$$+\frac{1}{2\delta}\Bigg\{\int\limits_{G_0^\varrho} r^{4-n}|b_0|^2dx + \frac{1}{\nu_0}\int\limits_{\Gamma_0^\varrho} r^{3-n}|g(x,0)|^2ds + \frac{1}{\nu_0}\int\limits_{\Sigma_0^\varrho} r^{3-n}|h(x,0)|^2ds\Bigg\}, \ \forall\delta>0.$$

However, by the condition (f),

$$\int\limits_{G_0^\varrho} r^{4-n}|b_0|^2dx + \frac{1}{\nu_0}\int\limits_{\Gamma_0^\varrho} r^{3-n}|g(x,0)|^2ds + \frac{1}{\nu_0}\int\limits_{\Sigma_0^\varrho} r^{3-n}|h(x,0)|^2ds$$
$$\leq \frac{c_0(G)}{2s}\left(f_1^2 + \frac{1}{\nu_0}g_1^2 + \frac{1}{\nu_0}h_1^2\right)\cdot\varrho^{2s}.$$

From (5.4.11) we obtain the differential inequality (CP) §2.4 with

$$\mathcal{P}(\varrho) = \frac{2\lambda}{\varrho}\cdot\{(1-\mu\varsigma) - c_5(n,\lambda,a_*)(\delta+\mathcal{A}(\varrho))\}, \ \forall\delta>0; \quad \mathcal{N}(\varrho)\equiv 0;$$
$$\mathcal{Q}(\varrho) = \frac{\lambda}{s}c_0(G)\left(f_1^2 + \frac{1}{\nu_0}g_1^2 + \frac{1}{\nu_0}h_1^2\right)\cdot\delta^{-1}\varrho^{2s-1}, \ \forall\delta>0; \quad (5.4.12)$$
$$V_0 = C\Bigg\{\int\limits_G \left(v^2 + (1+r^{4-n})b_0^2(x)\right)dx + \int\limits_{\Sigma_0} r^{3-n}h^2(x,0)ds + \int\limits_{\partial G} r^{3-n}g^2(x,0)ds\Bigg\},$$

by (5.3.1) with $\alpha = 4-n$.

1) Case $s > \lambda(1-\mu\varsigma)$. Choosing $\delta = \varrho^\varepsilon$, $\forall\varepsilon>0$, we have

$$\mathcal{P}(\varrho) = \frac{2\lambda}{\varrho}\cdot\{(1-\mu\varsigma) - c_5(n,\lambda,a_*)(\varrho^\varepsilon + \mathcal{A}(\varrho))\};$$

$$\mathcal{Q}(\varrho) = \frac{\lambda}{s} c_0(G) \left(f_1^2 + \frac{1}{\nu_0} g_1^2 + \frac{1}{\nu_0} h_1^2 \right) \cdot \varrho^{2s-1-\varepsilon}.$$

We can represent $\mathcal{P}(\varrho) = \frac{2\lambda(1-\mu\varsigma)}{\varrho} - \frac{\mathcal{K}(\varrho)}{\varrho}$, where $\mathcal{K}(\varrho)$ satisfies the Dini condition at zero. Therefore

$$-\int_{\varrho}^{\tau} \mathcal{P}(s)ds = -2\lambda(1-\mu\varsigma)\ln\left(\frac{\tau}{\varrho}\right) + \int_{\varrho}^{\tau} \frac{\mathcal{K}(s)}{s} ds \le \ln\left(\frac{\varrho}{\tau}\right)^{2\lambda(1-\mu)} + \int_0^d \frac{\mathcal{K}(r)}{r} dr \implies$$

$$\exp\left(-\int_{\varrho}^{d} \mathcal{P}(\tau)d\tau \right) \le \left(\frac{\varrho}{d}\right)^{2\lambda(1-\mu\varsigma)} \exp\left(\int_0^d \frac{\mathcal{K}(\tau)}{\tau} d\tau \right) = K_0 \left(\frac{\varrho}{d}\right)^{2\lambda(1-\mu\varsigma)};$$

$$\exp\left(-\int_{\varrho}^{\tau} \mathcal{P}(\tau)d\tau \right) \le \left(\frac{\varrho}{\tau}\right)^{2\lambda(1-\mu\varsigma)} \exp\left(\int_0^d \frac{\mathcal{K}(\tau)}{\tau} d\tau \right) = K_0 \left(\frac{\varrho}{\tau}\right)^{2\lambda(1-\mu\varsigma)}.$$

We have as well:

$$\begin{aligned} &\int_{\varrho}^{d} \mathcal{Q}(\tau) \exp\Big(-\int_{\varrho}^{\tau} \mathcal{P}(\sigma)d\sigma \Big) d\tau \\ &\le \frac{\lambda c_0 K_0}{s} \left(f_1^2 + g_1^2 + h_1^2\right) \varrho^{2\lambda(1-\mu\varsigma\varsigma)} \int_{\varrho}^{d} \tau^{2s-2\lambda(1-\mu\varsigma)-\varepsilon-1} d\tau \\ &\le \frac{\lambda c_0 K_0}{s} \left(f_1^2 + g_1^2 + h_1^2\right) \cdot \frac{d^{s-\lambda(1-\mu\varsigma)}}{s-\lambda(1-\mu\varsigma)} \varrho^{2\lambda(1-\mu\varsigma)}, \end{aligned}$$

since $s > \lambda(1-\mu\varsigma)$ and we can choose $\varepsilon = s - \lambda(1-\mu\varsigma)$.

Now we apply Theorem 1.21: then from (1.7.1), by virtue of the deduced inequalities and taking into account (2.2.3) for $\alpha = 4-n$, we get

$$\begin{aligned} &\int_{G_0^{\varrho}} a\left(r^{2-n}|\nabla v|^2 + r^{-n}v^2\right) dx + \int_{\Sigma_0^{\varrho}} r^{1-n}\sigma(\omega)v^2(x)ds + \int_{\Gamma_0^{\varrho}} r^{1-n}\gamma(\omega)v^2(x)ds \\ &\le C\left(\|v\|_{2,G}^2 + f_1^2 + g_1^2 + h_1^2\right) \varrho^{2\lambda(1-\mu\varsigma)}, \end{aligned} \tag{5.4.13}$$

where $C = const(n, s, q, \lambda, \mu, \nu_0, G)$. Returning to the function $u(x)$ by means of the function change (5.1.1), we obtain the statement of (5.4.1) for $s > \lambda(1-\mu)$.

2) Case $s = \lambda(1-\mu\varsigma)$. Substituting in (5.4.12) an arbitrary function $\delta(\varrho) > 0$ instead of $\delta > 0$, we have the problem (CP) with

$$\mathcal{P}(\varrho) = \frac{2\lambda\{(1-\mu\varsigma) - c_5\delta(\varrho)\}}{\varrho} - c_5\frac{\mathcal{A}(\varrho)}{\varrho}; \quad \mathcal{N}(\varrho) = 0;$$

$$\mathcal{Q}(\varrho) = c_0(G)\left(f_1^2 + \frac{1}{\nu_0}g_1^2 + \frac{1}{\nu_0}h_1^2\right)\cdot\delta^{-1}(\varrho)\varrho^{2\lambda(1-\mu\varsigma)-1}.$$

Choosing $\delta(\varrho) = \frac{1}{2\lambda c_5 \ln\left(\frac{ed}{\varrho}\right)}$, $0 < \varrho < d$, where e is the Euler number, we obtain

$$-\int_\varrho^\tau \mathcal{P}(\sigma)d\sigma \leq \ln\left(\frac{\varrho}{\tau}\right)^{2\lambda(1-\mu\varsigma)} + \int_\varrho^\tau \frac{d\sigma}{\sigma\ln\left(\frac{ed}{\sigma}\right)} + c_5\int_0^d \frac{\mathcal{A}(\tau)}{\tau}d\tau$$
$$= \ln\left(\frac{\varrho}{\tau}\right)^{2\lambda(1-\mu\varsigma)} + \ln\left(\frac{\ln\left(\frac{ed}{\varrho}\right)}{\ln\left(\frac{ed}{\tau}\right)}\right) + c_5\int_0^d \frac{\mathcal{A}(\tau)}{\tau}d\tau \quad\Longrightarrow$$
$$\exp\left(-\int_\varrho^\tau \mathcal{P}(\sigma)d\sigma\right) \leq \left(\frac{\varrho}{\tau}\right)^{2\lambda(1-\mu\varsigma)}\cdot\frac{\ln\left(\frac{ed}{\varrho}\right)}{\ln\left(\frac{ed}{\tau}\right)}\exp\left(c_5\int_0^d \frac{\mathcal{A}(\tau)}{\tau}d\tau\right),$$
$$\exp\left(-\int_\varrho^d \mathcal{P}(\tau)d\tau\right) \leq \left(\frac{\varrho}{d}\right)^{2\lambda(1-\mu\varsigma)}\ln\left(\frac{ed}{\varrho}\right)\exp\left(c_5\int_0^d \frac{\mathcal{A}(\tau)}{\tau}d\tau\right).$$

We also have

$$\int_\varrho^d \mathcal{Q}(\tau)\exp\left(-\int_\varrho^\tau \mathcal{P}(\sigma)d\sigma\right)d\tau$$
$$\leq c_6\left(f_1^2 + \frac{1}{\nu_0}g_1^2 + \frac{1}{\nu_0}h_1^2\right)\varrho^{2\lambda(1-\mu\varsigma)}\ln\left(\frac{ed}{\varrho}\right)\cdot\int_\varrho^d \frac{d\tau}{\tau\delta(\tau)\ln\left(\frac{ed}{\tau}\right)}$$
$$\leq 2\lambda c_5 c_6\left(f_1^2 + \frac{1}{\nu_0}g_1^2 + \frac{1}{\nu_0}h_1^2\right)\cdot\varrho^{2\lambda(1-\mu\varsigma)}\ln^2\left(\frac{ed}{\varrho}\right).$$

Now we apply Theorem 1.21: then from (1.7.1), by virtue of the deduced inequalities and with regard to (2.2.3) for $\alpha = 4 - n$, we get

$$\int_{G_0^\varrho} a\left(r^{2-n}|\nabla v|^2 + r^{-n}v^2\right)dx + \int_{\Sigma_0^\varrho} r^{1-n}\sigma(\omega)v^2(x)ds + \int_{\Gamma_0^\varrho} r^{1-n}\gamma(\omega)v^2(x)ds$$
$$\leq C\left(\|v\|_{2,G}^2 + f_1^2 + g_1^2 + h_1^2\right)\varrho^{2\lambda(1-\mu\varsigma)}\ln^2\left(\frac{ed}{\varrho}\right), \tag{5.4.14}$$

where $C = const(n, s, q, \lambda, \mu, \nu_0, G)$. Returning to the function $u(x)$ by means of the function change (5.1.1) we obtain the statement of (5.4.1) for $s = \lambda(1 - \mu\varsigma)$.

3) Case $0 < s < \lambda(1-\mu\varsigma)$. Analogously to case 1), taking into account (5.4.12), we have

$$\exp\Big(-\int_{\varrho}^{d}\mathcal{P}(\tau)d\tau\Big) \le \Big(\frac{\varrho}{d}\Big)^{2\lambda(1-\mu\varsigma-\delta)}\exp\Big(c_5\int_0^d\frac{\mathcal{A}(\tau)}{\tau}d\tau\Big) = c_8\Big(\frac{\varrho}{d}\Big)^{2\lambda(1-\mu\varsigma-\delta)};$$

$$\int_{\varrho}^{d}\mathcal{Q}(\tau)\exp\Big(-\int_{\varrho}^{\tau}\mathcal{P}(\sigma)d\sigma\Big)d\tau$$

$$\le \frac{c_0(\nu_0,G)}{\delta}\left(f_1^2+g_1^2+h_1^2\right)\varrho^{2\lambda(1-\mu\varsigma-\delta)}\int_{\varrho}^{d}\tau^{2s-2\lambda(1-\mu\varsigma-\delta)-1}d\tau$$

$$\le c_9\left(f_1^2+\frac{1}{\nu_0}g_1^2+\frac{1}{\nu_0}h_1^2\right)\cdot\varrho^{2s},$$

if we choose $\delta\in(0,1-\mu\varsigma-\frac{s}{\lambda})$.

Now we apply Theorem 1.21: then from (1.7.1), by virtue of the deduced inequalities and taking into account (2.2.3) for $\alpha=4-n$, we get

$$\int_{G_0^{\varrho}} a\left(r^{2-n}|\nabla v|^2+r^{-n}v^2\right)dx+\int_{\Sigma_0^{\varrho}} r^{1-n}\sigma(\omega)v^2(x)ds+\int_{\Gamma_0^{\varrho}} r^{1-n}\gamma(\omega)v^2(x)ds$$
$$\le C\left(\|v\|^2_{2,G}+f_1^2+g_1^2+h_1^2\right)\varrho^{2s}, \tag{5.4.15}$$

where $C=const(n,s,q,\lambda,\mu,\nu_0,G)$. Returning to the function $u(x)$ by means of the function change (5.1.1) we obtain the statement of (5.4.1) for $s<\lambda(1-\mu\varsigma)$. □

5.5 The power modulus of continuity at the conical point for weak solutions

5.5.1 Proof of Theorem 5.3.

Let us introduce the function

$$\psi(\varrho)=\begin{cases}\varrho^{\lambda(1-\mu\varsigma)}, & \text{if } s>\lambda(1-\mu\varsigma),\\ \varrho^{\lambda(1-\mu\varsigma)}\ln\Big(\frac{1}{\varrho}\Big), & \text{if } s=\lambda(1-\mu\varsigma),\\ \varrho^{s}, & \text{if } s<\lambda(1-\mu\varsigma)\end{cases} \tag{5.5.1}$$

for $0<\varrho<d$.

We perform the function change (5.1.1) and consider the function $v(x)$. For it, by Theorem 5.4 about the local bound of the weak solution modulus, we have

$$\sup_{G_0^{\varrho/2}} |v(x)| \le C\Big\{\varrho^{-n/2}\|v\|_{2,G_0^\varrho} + \varrho^{2(1-n/p)}\|b_0\|_{p/2,G_0^\varrho}$$
$$+ \varrho\Big(\|g(x,0)\|_{\infty,\Gamma_0^\varrho} + \|h(x,0)\|_{\infty,\Sigma_0^\varrho}\Big)\Big\},\ \varrho \in (0,d), \quad (5.5.2)$$

where $C = C(n, a_*, A^*, p, \mu, G)$ and $n < p < 2n$. Further, by Theorem 5.6 (see (5.4.13)–(5.4.15)), we have

$$\varrho^{-n/2}\|v\|_{2,G_0^\varrho} \le \Big(\int_{G_0^\varrho} r^{-n}v^2(x)dx\Big)^{1/2} \le C\Big(\|v\|_{2,G} + f_1 + g_1 + h_1\Big)\psi(\varrho). \quad (5.5.3)$$

Now, by the assumption (f),

$$\varrho^{2(1-n/p)}\|b_0\|_{p/2,G_0^\varrho} + \varrho\left(\|g(x,0)\|_{\infty,\Gamma_0^\varrho} + \|h(x,0)\|_{\infty,\Sigma_0^\varrho}\right) \le c\,(f_1 + g_1 + h_1)\,\psi(\varrho). \quad (5.5.4)$$

From (5.5.2)–(5.5.4) it follows that

$$\sup_{G_{\varrho/4}^{\varrho/2}} |v(x)| \le C\Big(\|v\|_{2,G} + f_1 + g_1 + h_1\Big)\psi(\varrho).$$

Setting now $|x| = \frac{1}{3}\varrho$ and returning to the function $u(x)$ by (5.1.1) we obtain finally the desired estimate (5.1.4).

As well we shall estimate the gradient modulus of the problem (WQL) solution near a conical point. Let us introduce two sets $G_{\varrho/4}^{2\varrho}$ and $G_{\varrho/2}^{\varrho} \subset G_{\varrho/4}^{2\varrho}$, $\varrho > 0$. We apply the change of variables $x = \varrho x'$ and $v(\varrho x') = \psi(\varrho)z(x')$. Then the function $z(x')$ satisfies in the weak sense the problem

$$\begin{cases} -\varsigma\frac{d}{dx_i'}\left(a^{ij}(\varrho x')z_{x_j'}\right) + \frac{\varrho^2}{\psi(\varrho)}\mathcal{B}(\varrho x', \psi(\varrho)z(x'), \frac{\psi(\varrho)}{\varrho}z_{x'}) = 0, & x \in G_{1/4}^2, \\ [z(x')]_{\Sigma_{1/4}^2} = 0,\ \varsigma\left[\frac{\partial z}{\partial\nu'}\right]_{\Sigma_{1/4}^2} + \frac{\sigma(\omega)}{|x'|}z(x') = \frac{\varrho}{\psi(\varrho)}\mathcal{H}(\varrho x', \psi(\varrho)z(x')), & x \in \Sigma_{1/4}^2, \\ \varsigma\frac{\partial z}{\partial\nu'} + \frac{\gamma(\omega)}{|x'|}z(x') = \frac{\varrho}{\psi(\varrho)}\mathcal{G}(\varrho x', \psi(\varrho)z(x')), & x \in \Gamma_{1/4}^2. \end{cases} \quad (WQL)'$$

Now we apply our assumption about a priori estimate of the gradient modulus of the problem $(WQL)'$ solution

$$\max_{x' \in G_{1/2}^1} |\nabla' z(x')| \le M_1' \quad (5.5.5)$$

(see §4 Theorem 4.4 in [6] or §§3, 5 from [64] as well as [47]). Returning to the variable x and the function $v(x)$ we obtain from (5.5.5)

$$|\nabla v(x)| \le M_1'\varrho^{-1}\psi(\varrho),\ x \in G_{\varrho/2}^{\varrho},\ 0 < \varrho < d.$$

Setting now $|x| = \frac{2}{3}\varrho$ and returning to the function $u(x)$ by (5.1.1), we obtain desired estimate (5.1.5).

5.5.2 Remark to Theorem 5.5

Now we can state that Theorem 5.5 is true for $\alpha \in (4-n-2\lambda, 2]$, if a neighborhood of the conic point is convex and the inequality

$$0 \le \mu < \frac{(1+q)(\alpha+n+2\lambda-4)(n+2\lambda-\alpha)}{4\lambda(\lambda+n-2)+(4-n-\alpha)^2} \tag{5.5.6}$$

is satisfied. In fact, from estimate (5.1.4) we obtain $u(0)=0$ and therefore we can apply Lemma 2.10 to the proof of Theorem 5.5 for $\alpha \in (4-n-2\lambda, 4-n)$. In this case the equality (5.3.4) can be rewritten, by virtue of (2.3.8), in the form of the inequality

$$\frac{2-\alpha}{2}\int\limits_G a r_\varepsilon^{\alpha-4}(x_i-\varepsilon l_i)\frac{\partial v^2}{\partial x_i}dx \le \frac{2-\alpha}{2}\int\limits_{\Gamma_d} a v^2 r_\varepsilon^{\alpha-4}(x_i-\varepsilon l_i)\cos(\overrightarrow{n}, x_i)ds$$
$$+\frac{(2-\alpha)(4-n-\alpha)}{2}H(\lambda,n,\alpha)\Big\{\int\limits_{G_0^d} r_\varepsilon^{\alpha-2}|\nabla v|^2dx + \int\limits_{\Sigma_0^d} r_\varepsilon^{\alpha-3}\sigma(\omega)v^2(x)ds$$
$$+\int\limits_{\Gamma_0^d} r_\varepsilon^{\alpha-3}\gamma(\omega)v^2(x)ds\Big\},$$

where $H(\lambda,n,\alpha)$ is determined by (2.3.3). By virtue of the convexity of G_0^d and the first property of r_ε (see §1.3), we have $r_\varepsilon \ge r$. Therefore (5.3.14)–(5.3.15) take the form

$$\varsigma\left(1-\frac{(2-\alpha)(4-n-\alpha)}{2}H(\lambda,n,\alpha)-\mu\varsigma\right)\Big\{\int\limits_G a r_\varepsilon^{\alpha-2}|\nabla v|^2dx$$
$$+\int\limits_{\Sigma_0}\frac{1}{r}r_\varepsilon^{\alpha-2}\sigma(\omega)v^2(x)ds+\int\limits_{\partial G}\frac{1}{r}r_\varepsilon^{\alpha-2}\gamma(\omega)v^2(x)ds\Big\}$$
$$\le c(\lambda,\omega_0)\left(\delta+\mathcal{A}(d)\right)\Big\{\int\limits_G a r_\varepsilon^{\alpha-2}|\nabla v|^2dx+\int\limits_{\Sigma_0} r^{-1}r_\varepsilon^{\alpha-2}\sigma(\omega)v^2(x)ds$$
$$+\int\limits_{\partial G} r^{-1}r_\varepsilon^{\alpha-2}\gamma(\omega)v^2ds\Big\}+C\int\limits_G\left(|\nabla v|^2+v^2\right)dx+\frac{1}{2a_*\delta}\int\limits_G r^\alpha b_0^2(x)dx$$
$$+\frac{1}{2\delta\nu_0}\int\limits_{\partial G} r^{\alpha-1}g^2(x)ds+\frac{1}{2\delta\nu_0}\int\limits_{\Sigma_0} r^{\alpha-1}h^2(x)ds,\ \forall\delta>0, \forall\varepsilon>0.$$

But, by (2.3.3) and $4-n-2\lambda<\alpha<4-n$, we verify that $1-\frac{(2-\alpha)(4-n-\alpha)}{2}H(\lambda,n,\alpha) > 0$. Then, by virtue of (5.5.6), we obtain that $1-\frac{(2-\alpha)(4-n-\alpha)}{2}H(\lambda,n,\alpha)-\mu\varsigma>0$. Now, choosing first $\delta>0$ and next $d>0$, together with $\mathcal{A}(d)$ appropriately small, we guarantee the realization of (5.3.16). Further, just as in the proof of Theorem 5.5 we obtain the required statement.

5.6 Example

In this section we consider the two-dimensional transmission problem for the Laplace operator with absorbtion term in an angular domain and investigate the corresponding eigenvalue problem. Let $n = 2$, the domain G lie inside the corner

$$G_0 = \{(r,\omega)\,|r > 0; \ -\frac{\omega_0}{2} < \omega < \frac{\omega_0}{2}\}, \quad \omega_0 \in]0, 2\pi[; \ \mathcal{O} \in \partial G$$

and in some neighborhood of $\mathcal{O}$ the boundary ∂G coincide with the sides of the corner $\omega = -\frac{\omega_0}{2}$ and $\omega = \frac{\omega_0}{2}$. We write

$$\Gamma_\pm = \{(r,\omega) \mid r > 0; \ \omega = \pm\frac{\omega_0}{2}\}, \quad \Sigma_0 = \{(r,\omega) \mid r > 0; \ \omega = 0\}$$

and set $\sigma(\omega)\Big|_{\Sigma_0} = \sigma(0) = \sigma = const > 0$, $\gamma(\omega)\Big|_{\omega=\pm\frac{\omega_0}{2}} = \gamma_\pm = const > 0$. Let us consider the problem:

$$\begin{cases} \dfrac{d}{dx_i}(|u|^q u_{x_i}) = a_0 r^{-2} u|u|^q - \mu u|u|^{q-2}|\nabla u|^2, & x \in G_0 \setminus \Sigma_0; \\ [u]_{\Sigma_0} = 0, \quad \left[a|u|^q \frac{\partial u}{\partial n}\right]_{\Sigma_0} + \frac{1}{|x|}\sigma(0) u|u|^q = 0, & x \in \Sigma_0; \\ \alpha_\pm a_\pm |u_\pm|^q \frac{\partial u_\pm}{\partial n} + \frac{1}{|x|}\gamma_\pm u_\pm |u_\pm|^q = 0, & x \in \Gamma_\pm \setminus \mathcal{O} \end{cases} \tag{AL}$$

where $a_0 \geq 0$, $0 \leq \mu < 1+q$, $q \geq 0$; $\alpha_\pm \in \{0; 1\}$; $a_\pm > 0$. We perform the function change (5.1.1); then the problem for the function $v(x)$ has the following form:

$$\begin{cases} \Delta v + \mu\varsigma v^{-1}|\nabla v|^2 = a_0(1+q)r^{-2}v; \quad \varsigma = \frac{1}{1+q}, & x \in G_0 \setminus \Sigma_0; \\ [v]_{\Sigma_0} = 0, \quad \left[a\frac{\partial v}{\partial n}\right]_{\Sigma_0} + (1+q)\sigma(0)\frac{v(x)}{|x|} = 0, & x \in \Sigma_0; \\ \alpha_\pm a_\pm \frac{\partial v_\pm}{\partial n} + (1+q)\gamma_\pm \frac{v_\pm(x)}{|x|} = 0, & x \in \Gamma_\pm \setminus \mathcal{O}. \end{cases}$$

We want to find the exact solution of this problem in the form $v(r,\omega) = r^{\varkappa}\psi(\omega)$. For $\psi(\omega)$ we obtain the problem

$$\begin{cases} \psi''(\omega) + \frac{\mu\varsigma}{\psi(\omega)}\psi'^2(\omega) + \{(1+\mu\varsigma)\varkappa^2 - a_0(1+q)\}\cdot\psi(\omega) = 0, \\ \qquad\qquad\qquad\qquad \omega \in \left(-\frac{\omega_0}{2}, 0\right) \cup \left(0, \frac{\omega_0}{2}\right); \\ [\psi]_{\omega=0} = 0, \quad [a\psi'(0)] = (1+q)\sigma(0)\psi(0); \\ \pm\alpha_\pm a_\pm \psi'_\pm\left(\pm\frac{\omega_0}{2}\right) + (1+q)\gamma_\pm\psi_\pm\left(\pm\frac{\omega_0}{2}\right) = 0. \end{cases}$$

Let us assume $\varkappa^2 > a_0\frac{(1+q)^2}{1+q+\mu}$ and define the value

$$\Upsilon = \sqrt{\varkappa^2 - a_0\frac{(1+q)^2}{1+q+\mu}}. \tag{5.6.1}$$

We consider separately two cases: $\mu = 0$ and $\mu \neq 0$.

$$\boxed{Case \quad \mu = 0.}$$

We have

$$\psi_\pm(\omega) = A\cos(\Upsilon\omega) + B_\pm\sin(\Upsilon\omega), \tag{5.6.2}$$

where constants $A, B_\pm$ should be determined from the conjunction and boundary conditions; namely, they satisfy the system

$$\begin{cases} (1+q)\sigma(0)\cdot A - a_+\Upsilon\cdot B_+ + a_-\Upsilon\cdot B_- & = 0 \\ \left\{(1+q)\gamma_+\cos\left(\Upsilon\frac{\omega_0}{2}\right) - \alpha_+a_+\Upsilon\sin\left(\Upsilon\frac{\omega_0}{2}\right)\right\}\cdot A & \\ +\left\{(1+q)\gamma_+\sin\left(\Upsilon\frac{\omega_0}{2}\right) + \alpha_+a_+\Upsilon\cos\left(\Upsilon\frac{\omega_0}{2}\right)\right\}\cdot B_+ & = 0 \\ \left\{(1+q)\gamma_-\cos\left(\Upsilon\frac{\omega_0}{2}\right) - \alpha_-a_-\Upsilon\sin\left(\Upsilon\frac{\omega_0}{2}\right)\right\}\cdot A & \\ -\left\{(1+q)\gamma_-\sin\left(\Upsilon\frac{\omega_0}{2}\right) + \alpha_-a_-\Upsilon\cos\left(\Upsilon\frac{\omega_0}{2}\right)\right\}\cdot B_- & = 0. \end{cases}$$

The Dirichlet problem: $\alpha_\pm = 0$, $\gamma_\pm \neq 0$.

Direct calculations give

$$\psi_\pm(\omega) = \cos(\Upsilon\omega) \mp \cot\left(\Upsilon\frac{\omega_0}{2}\right)\cdot\sin(\Upsilon\omega), \quad \Upsilon = \begin{cases} \frac{\pi}{\omega_0}, & \text{if } \sigma(0) = 0; \\ \Upsilon^*, & \text{if } \sigma(0) \neq 0, \end{cases}$$

where Υ^* is the least positive root of the transcendence equation

$$\Upsilon\cdot\cot\left(\Upsilon\frac{\omega_0}{2}\right) = -\frac{1+q}{a_+ + a_-}\sigma(0)$$

and from the graphic solution (see Figure 2 §3.6) we obtain $\frac{\pi}{\omega_0} < \Upsilon^* < \frac{2\pi}{\omega_0}$.

The corresponding eigenfunctions are

$$\psi_{\pm}(\omega) = \begin{cases} \cos\left(\frac{\pi\omega}{\omega_0}\right), & \text{if } \sigma(0) = 0; \\ \cos(\Upsilon^*\omega) \pm \frac{1+q}{a_+ + a_-} \cdot \frac{\sigma(0)}{\Upsilon^*} \cdot \sin(\Upsilon^*\omega), & \text{if } \sigma(0) \neq 0. \end{cases}$$

The Neumann problem: $\alpha_{\pm} = 1$, $\gamma_{\pm} = 0$.

Direct calculations will give:

$$\Upsilon = \begin{cases} \frac{\pi}{\omega_0}, & \text{if } \sigma(0) = 0; \\ \Upsilon^*, & \text{if } \sigma(0) \neq 0, \end{cases}$$

where Υ^* is the least positive root of the transcendence equation

$$\Upsilon \cdot \tan\left(\Upsilon \frac{\omega_0}{2}\right) = \frac{1+q}{a_+ + a_-}\sigma(0)$$

and from the graphic solution (see Figure 3 §3.6) we obtain $0 < \Upsilon^* < \frac{\pi}{\omega_0}$.

The corresponding eigenfunctions are

$$\psi_{\pm}(\omega) = \begin{cases} a_+ \sin\left(\frac{\pi\omega}{\omega_0}\right), & \text{if } \sigma(0) = 0; \\ \cos(\Upsilon^*\omega) \pm \frac{1+q}{a_+ + a_-} \cdot \frac{\sigma(0)}{\Upsilon^*} \cdot \sin(\Upsilon^*\omega), & \text{if } \sigma(0) \neq 0. \end{cases}$$

Mixed problem: $\alpha_+ = 1$, $\alpha_- = 0$; $\gamma_+ = 0$, $\gamma_- = 1$.

Direct calculations will give:

$$\Upsilon = \begin{cases} \frac{2}{\omega_0} \arctan\sqrt{\frac{a_-}{a_+}}, & \text{if } \sigma(0) = 0; \\ \Upsilon^*, & \text{if } \sigma(0) \neq 0, \end{cases}$$

where Υ^* is the least positive root of the transcendence equation

$$a_+ \tan\left(\Upsilon \frac{\omega_0}{2}\right) - a_- \cot\left(\Upsilon \frac{\omega_0}{2}\right) = \frac{1+q}{\Upsilon}\sigma(0)$$

and from the graphic solution (see Figure 8) we obtain

$$\frac{2}{\omega_0} \arctan\sqrt{\frac{a_-}{a_+}} < \Upsilon^* < \frac{\pi}{\omega_0}.$$

The corresponding eigenfunctions are

$$\begin{cases} \psi_{\pm}(\omega) = \cos(\Upsilon^*\omega) + \sqrt{\left(\frac{a_-}{a_+}\right)^{\pm 1}} \cdot \sin(\Upsilon^*\omega), & \text{if } \sigma(0) = 0; \\ \psi_{\pm}(\omega) = \cos(\Upsilon^*\omega) + T^{\pm 1} \cdot \sin(\Upsilon^*\omega), \quad \omega \in \left[0, \frac{\omega_0}{2}\right], & \text{if } \sigma(0) \neq 0, \end{cases}$$

$$a \tan x - b \cot x = \frac{c}{x}; \quad a, b, c > 0;\ a \neq b;\ x > 0$$

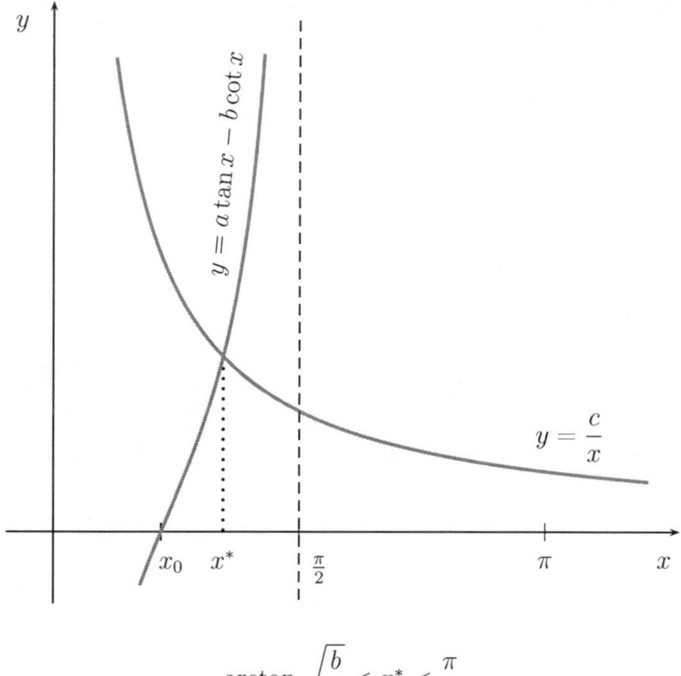

$$\arctan \sqrt{\frac{b}{a}} < x^* < \frac{\pi}{2}$$

Figure 8: The graphic solution

where

$$T = \frac{(1+q)\sigma(0) + \sqrt{(1+q)^2\sigma^2(0) + 4(\Upsilon^*)^2 a_+ a_-}}{2\Upsilon^* a_+}.$$

The Robin problem: $\alpha_\pm = 1$, $\gamma_\pm \neq 0$.

Direct calculations of the above system give:

1) $\dfrac{\gamma_+}{\gamma_-} = \dfrac{a_+}{a_-}$. In this case we get either

$$\psi_\pm(\omega) = a_\mp \sin(\Upsilon^* \omega),$$

where Υ^* is the least positive root of the transcendence equation

$$\Upsilon \cdot \cot\left(\Upsilon \frac{\omega_0}{2}\right) = -(1+q)\frac{\gamma_+}{a_+}$$

and from the graphic solution (see Figure 2 §3.6) we obtain

$$\frac{\pi}{\omega_0} < \Upsilon^* < \frac{2\pi}{\omega_0},$$

or

$$\psi_\pm(\omega) = \cos(\Upsilon^*\omega) \pm \frac{1+q}{a_+ + a_-} \cdot \frac{\sigma(0)}{\Upsilon^*} \sin(\Upsilon^*\omega),$$

where Υ^* is the least positive root of the transcendence equation

$$\tan\left(\Upsilon\frac{\omega_0}{2}\right) = (1+q)\Upsilon\frac{a_+\sigma(0) + \gamma_+(a_+ + a_-)}{a_+(a_+ + a_-)\Upsilon^2 - (1+q)^2\gamma_+\sigma(0)}. \tag{5.6.3}$$

In particular, for $\sigma(0) = 0$ we have

$$\psi_\pm(\omega) = \cos(\Upsilon\omega), \quad \tan\left(\Upsilon\frac{\omega_0}{2}\right) = \frac{1+q}{\Upsilon} \cdot \frac{\gamma_+}{a_+}$$

and from the graphic solution (see Figure 3 §3.6) $0 < \Upsilon^* < \frac{\pi}{\omega_0}$.

We rewrite equation (5.6.3) in the form

$$\begin{cases} \tan x = \frac{x}{a^2x^2 - b^2}, \\ a^2 = \frac{2}{\omega_0} \cdot \frac{a_+(a_+ + a_-)}{(1+q)\{a_+\sigma(0) + \gamma_+(a_+ + a_-)\}}, \quad b^2 = \frac{\omega_0}{2} \cdot \frac{(1+q)\gamma_+\sigma(0)}{a_+\sigma(0) + \gamma_+(a_+ + a_-)} \quad \Longrightarrow \\ \frac{b}{a} = (1+q)\frac{\omega_0}{2} \cdot \sqrt{\frac{\gamma_+\sigma(0)}{a_+(a_+ + a_-)}}. \end{cases}$$

The graphic solution. Now we consider three possible situations.

a) $(1+q) \cdot \sqrt{\frac{\gamma_+\sigma(0)}{a_+(a_+ + a_-)}} < \frac{\pi}{\omega_0}$ (see below Figure 9). In this case we obtain

$$(1+q) \cdot \sqrt{\frac{\gamma_+\sigma(0)}{a_+(a_+ + a_-)}} < \Upsilon^* < \frac{\pi}{\omega_0}.$$

b) $\frac{\pi}{\omega_0} < (1+q) \cdot \sqrt{\frac{\gamma_+\sigma(0)}{a_+(a_+ + a_-)}} < \frac{2\pi}{\omega_0}$ (see below Figure 10). In this case we obtain

$$\frac{\pi}{\omega_0} < \Upsilon^* < (1+q) \cdot \sqrt{\frac{\gamma_+\sigma(0)}{a_+(a_+ + a_-)}}.$$

c) $(1+q) \cdot \sqrt{\frac{\gamma_+\sigma(0)}{a_+(a_+ + a_-)}} = \frac{\pi}{\omega_0}$ (see below Figure 11). In this case we obtain

$$\frac{2\pi}{\omega_0} < \Upsilon^* < \frac{3\pi}{\omega_0}.$$

$$\tan x = \frac{x}{a^2x^2 - b^2}; \quad a, b > 0; \ \frac{b}{a} < \frac{\pi}{2}; \ x > 0$$

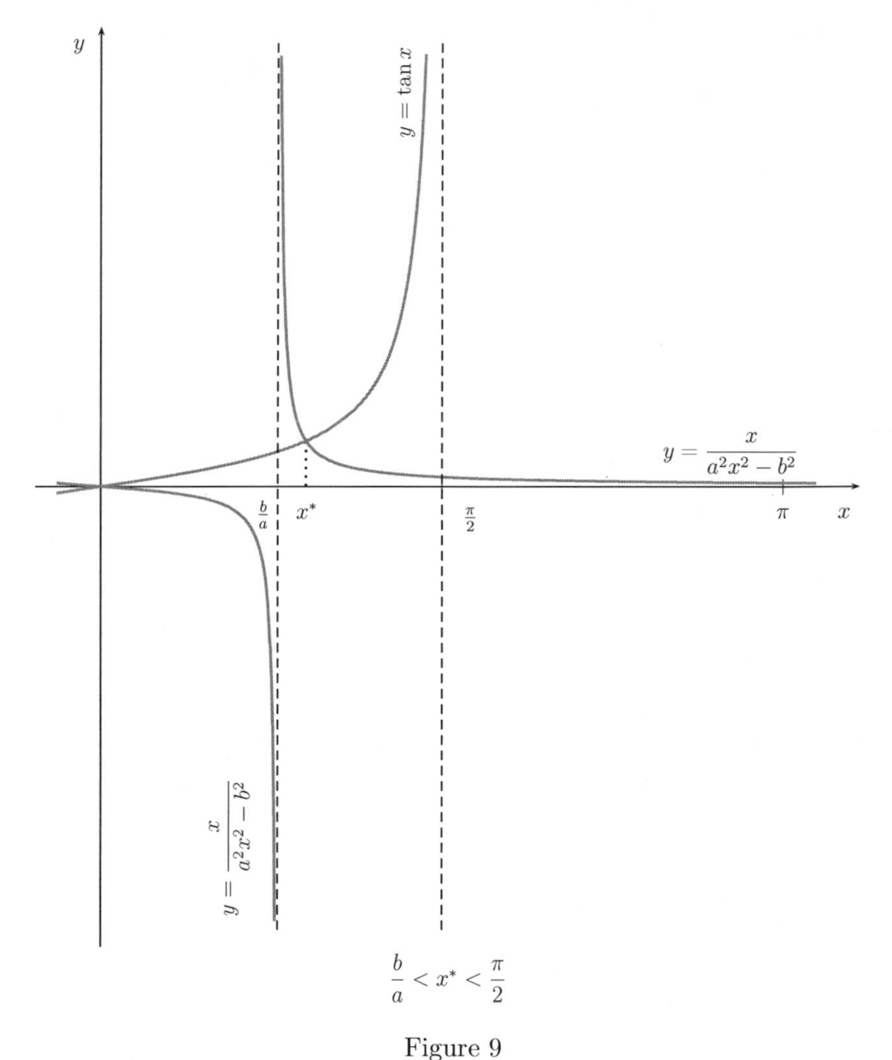

$$\frac{b}{a} < x^* < \frac{\pi}{2}$$

Figure 9

2) $\dfrac{\gamma_+}{\gamma_-} \neq \dfrac{a_+}{a_-}$. In this case we get $A \neq 0$ and from (5.6.2) it follows that $\psi_\pm(0) \neq 0$; further see below the general case $\mu \neq 0$.

Case $\mu \neq 0$.

We observe that in this case $\psi(0) \neq 0$. By setting $y(\omega) = \frac{\psi'(\omega)}{\psi(\omega)}$, we arrive at the

$$\tan x = \frac{x}{a^2x^2 - b^2}; \quad a, b > 0; \ \frac{\pi}{2} < \frac{b}{a} < \pi; \ x > 0$$

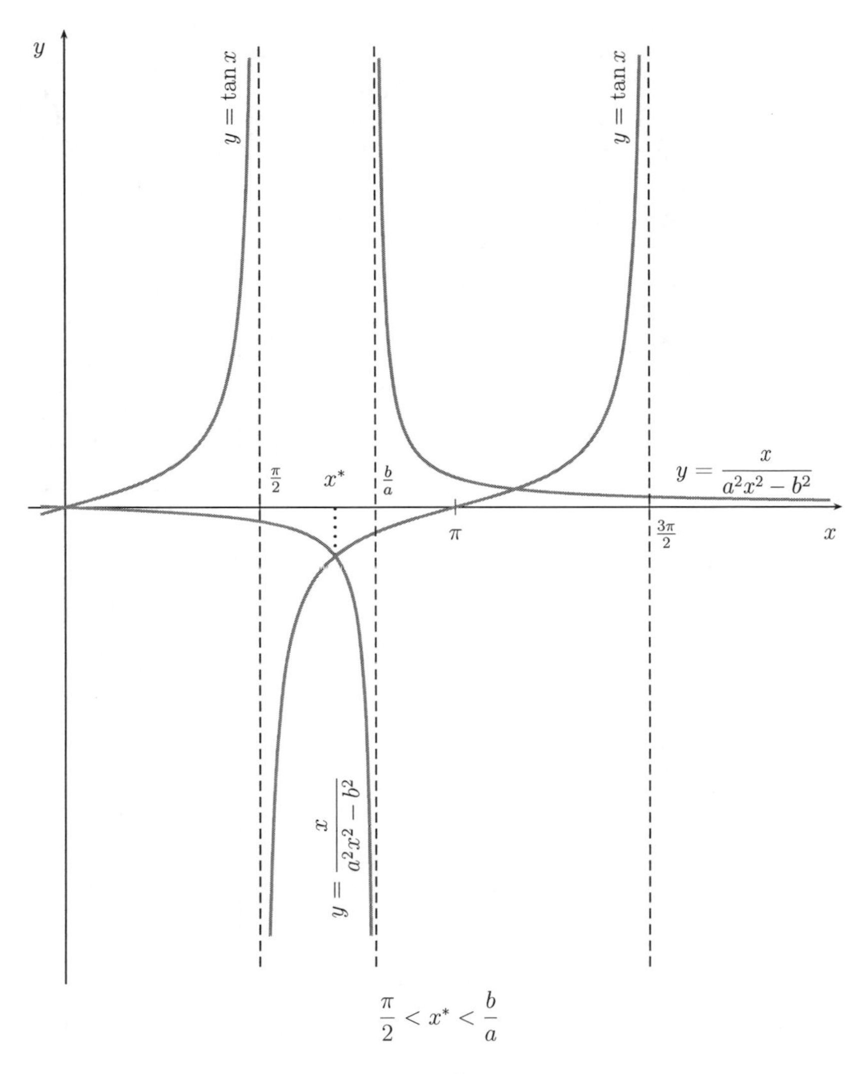

$$\frac{\pi}{2} < x^* < \frac{b}{a}$$

Figure 10

problem for $y(\omega)$:

$$\begin{cases} y' + (1 + \mu\varsigma)y^2(\omega) + (1 + \mu\varsigma)\varkappa^2 - a_0(1 + q) = 0, \quad \omega \in \left(-\frac{\omega_0}{2}, 0\right) \cup \left(0, \frac{\omega_0}{2}\right); \\ a_+y_+(0) - a_-y_-(0) = (1 + q)\sigma(0); \\ \pm\alpha_\pm a_\pm y_\pm\left(\pm\frac{\omega_0}{2}\right) + (1 + q)\gamma_\pm = 0. \end{cases}$$

$$\tan x = \frac{x}{a^2x^2 - b^2}; \quad a, b > 0; \ \frac{b}{a} = \frac{\pi}{2}; \ x > 0$$

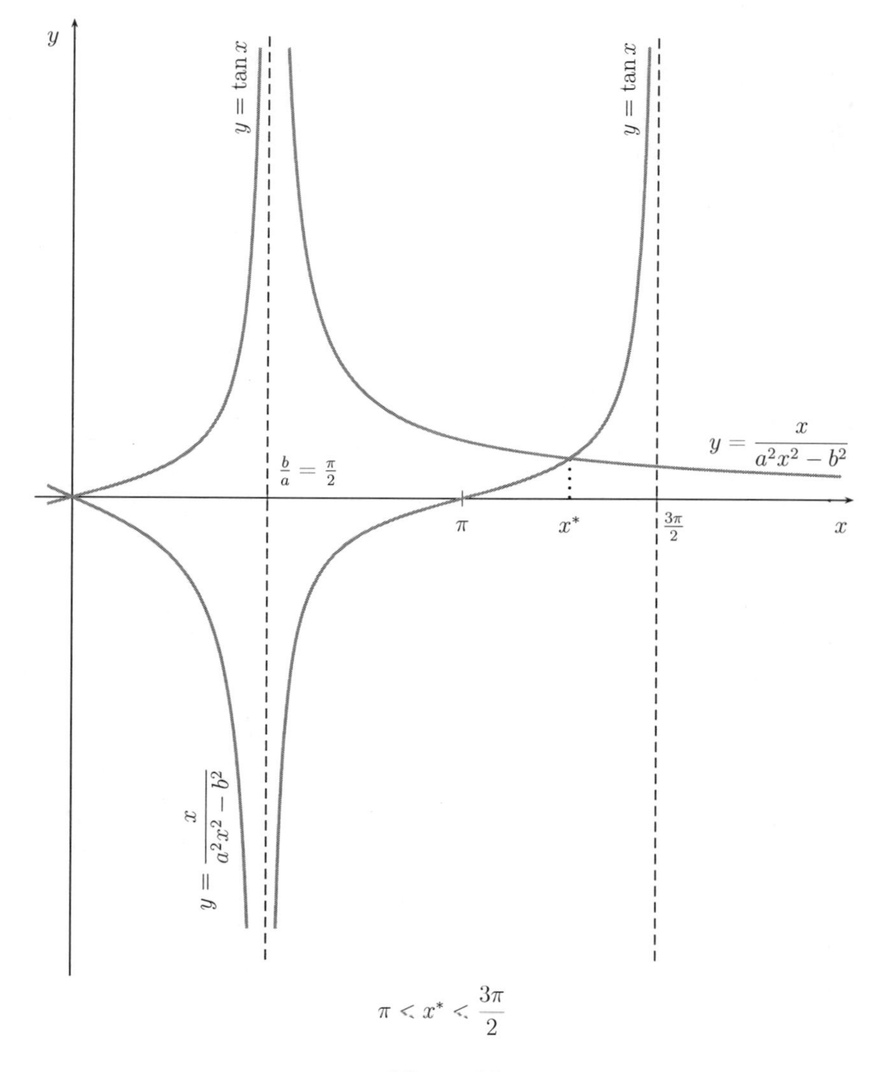

Figure 11

By integrating the equation of our problem we find

$$y_{\pm}(\omega) = \varUpsilon \tan \left\{ \varUpsilon \left(C_{\pm} - (1 + \mu\varsigma)\omega \right) \right\}, \ \forall C_{\pm}. \tag{5.6.4}$$

From the boundary conditions we have

$$C_{\pm} = \pm(1+\mu\varsigma)\frac{\omega_0}{2} \mp \frac{1}{\Upsilon}\arctan\frac{(1+q)\gamma_{\pm}}{\alpha_{\pm}a_{\pm}\Upsilon}. \tag{5.6.5}$$

Finally, by virtue of the conjunction condition, we get the equation for the needed $\varkappa$:

$$\begin{aligned} & a_{+}\cdot\frac{\alpha_{+}a_{+}\Upsilon\tan\left\{(1+\mu\varsigma)\Upsilon\frac{\omega_0}{2}\right\}-(1+q)\gamma_{+}}{\alpha_{+}a_{+}\Upsilon+(1+q)\gamma_{+}\tan\left\{(1+\mu\varsigma)\Upsilon\frac{\omega_0}{2}\right\}} \\ & + a_{-}\cdot\frac{\alpha_{-}a_{-}\Upsilon\tan\left\{(1+\mu\varsigma)\Upsilon\frac{\omega_0}{2}\right\}-(1+q)\gamma_{-}}{\alpha_{-}a_{-}\Upsilon+(1+q)\gamma_{-}\tan\left\{(1+\mu\varsigma)\Upsilon\frac{\omega_0}{2}\right\}} \\ & = \frac{1+q}{\Upsilon}\sigma(0), \quad \text{where } 1+\mu\varsigma = \frac{1+q+\mu}{1+q}. \end{aligned} \tag{5.6.6}$$

Further, from (5.6.4) and (5.6.5) we obtain

$$y_{\pm}(\omega) = \Upsilon\tan\left\{\Upsilon\frac{1+q+\mu}{1+q}\left(\pm\frac{\omega_0}{2}-\omega\right)\mp\arctan\frac{(1+q)\gamma_{\pm}}{\alpha_{\pm}a_{\pm}\Upsilon}\right\} \tag{5.6.7}$$

and, because of $(\ln\psi(\omega))' = y(\omega)$, it follows that

$$\psi_{\pm}(\omega) = \cos^{\frac{1+q}{1+q+\mu}}\left\{\Upsilon\frac{1+q+\mu}{1+q}\left(\pm\frac{\omega_0}{2}-\omega\right)\mp\arctan\frac{(1+q)\gamma_{\pm}}{\alpha_{\pm}a_{\pm}\Upsilon}\right\}. \tag{5.6.8}$$

At last, returning to the function u, by (5.1.1), we establish a solution of (AL)

$$\boxed{u_{\pm}(r,\omega) = r^{\frac{\varkappa}{1+q}}\cos^{\frac{1}{1+q+\mu}}\left\{\Upsilon^*\frac{1+q+\mu}{1+q}\left(\pm\frac{\omega_0}{2}-\omega\right)\mp\arctan\frac{(1+q)\gamma_{\pm}}{\alpha_{\pm}a_{\pm}\Upsilon^*}\right\}}, \tag{5.6.9}$$

where Υ^* is the smallest positive root of the transcendence equation (5.6.6) and $\varkappa$ is defined by (5.6.1).

The Dirichlet problem: $\alpha_{\pm} = 0,\ \gamma_{\pm} \neq 0.$

Direct calculations will give $\Upsilon = \begin{cases} \frac{1+q}{1+q+\mu}\cdot\frac{\pi}{\omega_0}, & \text{if } \sigma(0)=0; \\ \Upsilon^*, & \text{if } \sigma(0)\neq 0, \end{cases}$ where Υ^* is the least positive root of the transcendence equation $\Upsilon\cdot\cot\left(\frac{1+q+\mu}{1+q}\Upsilon\frac{\omega_0}{2}\right) = -\frac{1+q}{a_{+}+a_{-}}\sigma(0)$ and from the graphic solution (see Figure 2 §3.6) we obtain $\frac{1+q}{1+q+\mu}\cdot\frac{\pi}{\omega_0} < \Upsilon^* < \frac{1+q}{1+q+\mu}\cdot\frac{2\pi}{\omega_0}$. The corresponding eigenfunctions are

$$u_{\pm}(r,\omega) = \begin{cases} r^{\widetilde{\lambda}}\cos^{\frac{1}{1+q+\mu}}\left(\frac{\pi\omega}{\omega_0}\right); \quad \widetilde{\lambda} = \frac{\sqrt{(\pi/\omega_0)^2+a_0(1+q+\mu)}}{1+q+\mu}, & \text{if } \sigma(0)=0; \\ r^{\frac{\varkappa}{1+q}}\left|\frac{1+q}{a_{+}+a_{-}}\cdot\frac{\sigma(0)}{\Upsilon^*}\sin\left(\Upsilon^*\frac{1+q+\mu}{1+q}\omega\right)\pm\cos\left(\Upsilon^*\frac{1+q+\mu}{1+q}\omega\right)\right|^{\frac{1}{1+q+\mu}}, & \text{if } \sigma(0)\neq 0, \end{cases}$$

If we consider **the Dirichlet problem without the interface:** $\alpha_\pm = 0$, $a_\pm = 1$, $\sigma(0) = 0$, then we obtain the same result. It is a well-known result (see Example 4.6, p. 374 [7]).

Now we can verify that the derived exact solution satisfies the estimate (5.1.4) of Theorem 5.3. In fact, in our case we have: the value λ for (2.2.1) is equal to $\vartheta = \frac{\pi}{\omega_0}$ and therefore

$$|u(r,\omega)| \le r^{\widetilde{\lambda}} \le r^{\frac{\pi}{\omega_0}\cdot\frac{1}{1+q+\mu}} \le r^{\frac{\pi}{\omega_0}\cdot\frac{1+q-\mu}{(1+q)^2}},$$

since $a_0 \ge 0$ and $\frac{1}{1+q+\mu} \ge \frac{1+q-\mu}{(1+q)^2}$.

The Neumann problem: $\alpha_\pm = 1$, $\gamma_\pm = 0$.

Direct calculations will give $\varUpsilon = \begin{cases} \frac{1+q}{1+q+\mu}\cdot\frac{2\pi}{\omega_0}, & \text{if } \sigma(0)=0; \\ \varUpsilon^*, & \text{if } \sigma(0)\ne 0, \end{cases}$ where $\varUpsilon^*$ is the least positive root of the transcendence equation $\varUpsilon\cdot\tan\left(\frac{1+q+\mu}{1+q}\varUpsilon\frac{\omega_0}{2}\right) = \frac{1+q}{a_+ + a_-}\sigma(0)$ and from the graphic solution (see Figure 3 §3.6) we obtain $0 < \varUpsilon^* < \frac{1+q}{1+q+\mu}\cdot\frac{\pi}{\omega_0}$. The corresponding eigenfunctions are

$$u_\pm(r,\omega) = \begin{cases} r^{\widetilde{\lambda}}\cos^{\frac{1}{1+q+\mu}}\left(\frac{2\pi\omega}{\omega_0}\right); \quad \widetilde{\lambda} = \frac{\sqrt{(2\pi/\omega_0)^2 + a_0(1+q+\mu)}}{1+q+\mu}, & \text{if } \sigma(0)=0; \\ r^{\frac{\varkappa}{1+q}}\left|\frac{1+q}{a_+ + a_-}\cdot\frac{\sigma(0)}{\varUpsilon^*}\sin\left(\varUpsilon^*\frac{1+q+\mu}{1+q}\omega\right) \pm \cos\left(\varUpsilon^*\frac{1+q+\mu}{1+q}\omega\right)\right|^{\frac{1}{1+q+\mu}}, & \text{if } \sigma(0)\ne 0, \end{cases}$$

where $\varkappa$ is defined by (5.6.1).

Mixed problem: $\alpha_+ = 1$, $\alpha_- = 0$; $\gamma_+ = 0$, $\gamma_- = 1$.

Direct calculations will give: $\varUpsilon = \begin{cases} \frac{2}{\omega_0}\cdot\frac{1+q}{1+q+\mu}\arctan\sqrt{\frac{a_-}{a_+}}, & \text{if } \sigma(0)=0; \\ \varUpsilon^*, & \text{if } \sigma(0)\ne 0, \end{cases}$ where $\varUpsilon^*$ is the least positive root of the transcendence equation

$$a_+\tan\left(\frac{1+q+\mu}{1+q}\varUpsilon\frac{\omega_0}{2}\right) - a_-\cot\left(\frac{1+q+\mu}{1+q}\varUpsilon\frac{\omega_0}{2}\right) = \frac{1+q}{\varUpsilon}\sigma(0)$$

and from the graphic solution (see Figure 8) we obtain

$$\frac{1+q}{1+q+\mu}\cdot\frac{2}{\omega_0}\arctan\sqrt{\frac{a_-}{a_+}} < \varUpsilon^* < \frac{1+q}{1+q+\mu}\cdot\frac{\pi}{\omega_0}.$$

The corresponding eigenfunctions are

$$\begin{cases} \psi_\pm(\omega) = \left\{\cos\left(\frac{1+q+\mu}{1+q}\varUpsilon\omega\right) + \sqrt{\left(\frac{a_-}{a_+}\right)^{\pm 1}}\cdot\sin\left(\frac{1+q+\mu}{1+q}\varUpsilon\omega\right)\right\}^{\frac{1+q}{1+q+\mu}}, & \text{if } \sigma(0)=0; \\ \psi_\pm(\omega) = \left\{\cos\left(\frac{1+q+\mu}{1+q}\varUpsilon\omega\right) + T^{\pm 1}\cdot\sin\left(\frac{1+q+\mu}{1+q}\varUpsilon\omega\right)\right\}^{\frac{1+q}{1+q+\mu}}, & \text{if } \sigma(0)\ne 0, \end{cases}$$

where

$$T = \frac{(1+q)\sigma(0) + \sqrt{(1+q)^2\sigma^2(0) + 4(\Upsilon^*)^2 a_+ a_-}}{2\Upsilon^* a_+}.$$

Chapter 6

Transmission problem for strong quasi-linear elliptic equations in a conical domain

6.1 Introduction

In this chapter we study the transmission problem for general elliptic divergence quasi-linear equations in n-dimensional domain with a conical point at the boundary:

$$\begin{cases} -\dfrac{d}{dx_i}a_i(x,u,\nabla u)+b(x,u,\nabla u)=0, & x\in G\setminus\Sigma_0; \\ [u]_{\Sigma_0}=0, & \\ \mathcal{S}[u]\equiv\left[\frac{\partial u}{\partial\nu}\right]_{\Sigma_0}+\frac{1}{|x|^{m-1}}\sigma\left(\frac{x}{|x|}\right)u\cdot|u|^{q+m-2}=h(x,u), & x\in\Sigma_0; \\ \mathcal{B}[u]\equiv\frac{\partial u}{\partial\nu}+\frac{1}{|x|^{m-1}}\gamma\left(\frac{x}{|x|}\right)u\cdot|u|^{q+m-2}=g(x,u), & x\in\partial G\setminus\{\Sigma_0\cup\mathcal{O}\}; \end{cases} \tag{QL}$$

(summation over repeated indices from 1 to n is understood); here:

- $a_i(x,u,\xi)=\begin{cases} a_i^+(x,u_+,\xi_+), & x\in G_+, \\ a_i^-(x,u_-,\xi_-), & x\in G_-, \end{cases}$ $\quad i=1,\ldots,n;\quad$ etc.;
- $\frac{\partial}{\partial\nu}=a_i(x,u,\nabla u)n_i$;
- $\left[\frac{\partial u}{\partial\nu}\right]_{\Sigma_0}$ denotes the saltus of the co-normal derivative of the function $u(x)$ on crossing Σ_0, i.e.,

$$\left[\frac{\partial u}{\partial\nu}\right]_{\Sigma_0}=a_i^+(x,u_+,\nabla u_+)n_i\Big|_{\Sigma_0}-a_i^-(x,u_-,\nabla u_-)n_i\Big|_{\Sigma_0}.$$

M. Borsuk, *Transmission Problems for Elliptic Second-Order Equations in Non-Smooth Domains*, Frontiers in Mathematics, DOI 10.1007/978-3-0346-0477-2_7,

Definition 6.1. Function $u(x)$ is called a *weak* solution of the problem (QL) provided that $u(x) \in \mathbf{C}^0(\overline{G}) \cap \mathbf{V}^1_{m,0}(G)$ and satisfies the integral identity

$$\int_G \{a_i(x,u,u_x)\eta_{x_i} + b(x,u,u_x)\eta(x)\}\,dx + \int_{\Sigma_0} \frac{\sigma(\omega)}{r^{m-1}} u|u|^{q+m-2}\eta(x)ds$$
$$+ \int_{\partial G} \frac{\gamma(\omega)}{r^{m-1}} u|u|^{q+m-2}\eta(x)ds = \int_{\partial G} g(x,u)\eta(x)ds + \int_{\Sigma_0} h(x,u)\eta(x)ds \qquad (II)$$

for all functions $\eta(x) \in \mathbf{C}^0(\overline{G}) \cap \mathbf{V}^1_{m,0}(G)$.

Lemma 6.2. *Let $u(x)$ be a weak solution of (QL). For any function $\eta(x) \in \mathbf{C}^0(\overline{G}) \cap \mathbf{V}^1_{m,0}(G)$ the equality*

$$\int_{G_0^\varrho} \Big\{a_i(x,u,u_x)\eta_{x_i} + b(x,u,u_x)\eta(x)\Big\}dx$$
$$= \int_{\Omega_\varrho} a_i(x,u,u_x)\cos(r,x_i)\eta(x)d\Omega_\varrho$$
$$+ \int_{\Gamma_0^\varrho} \left(g(x,u) - \frac{\gamma(\omega)}{r^{m-1}}u|u|^{q+m-2}\right)\eta(x)ds$$
$$+ \int_{\Sigma_0^\varrho} \left(h(x,u) - \frac{\sigma(\omega)}{r^{m-1}}u|u|^{q+m-2}\right)\eta(x)ds \qquad (II)_{loc}$$

holds for a.e. $\varrho \in (0,d)$.

Proof. The proof is analogous to the proof of Lemma 3.2, Chapter 3. See also the proof of Lemma 5.2 in [14] (pp. 167–170). □

Assumptions. *Let $a = \begin{cases} a_+, & x \in G_+, \\ a_-, & x \in G_-, \end{cases}$ $a_\pm > 0$, $a_* = \min\{a_+, a_-\} > 0$, $a^* = \max\{a_+, a_-\} > 0$; let $1 < m < n$, $mn > p > n > m$, $q \geq 0$, $0 \leq \mu < \frac{q+m-1}{m-1}$ be given numbers; $a_0(x), \alpha(x)$ and $b_0(x)$ be non-negative measurable functions; let $a_i(x,u,\xi)$, $i = 1,\dots,n$; $b(x,u,\xi)$ be Caratheodory functions $G \times \mathbb{R} \times \mathbb{R}^n \to \mathbb{R}$ and continuously differentiable with respect to x_i; $h(x,u)$ be Caratheodory function $\Sigma_0 \times \mathbb{R} \to \mathbb{R}$ and continuously differentiable with respect to variable u, while $g(x,u)$ be Caratheodory function $\partial G \times \mathbb{R} \to \mathbb{R}$ and continuously differentiable with respect to variable u. We assume the following properties:*

1) $a_i(x,u,\xi)\xi_i \geq a|u|^q|\xi|^m - a_0(x)$; $a_0(x) \in \mathbf{L}_{p/m}(G)$;

2) $\sqrt{\sum_{i=1}^{n} a_i^2(x,u,\xi)} + \sqrt{\sum_{i=1}^{n} \left|\frac{\partial a_i(x,u,\xi)}{\partial x_i}\right|^2} \le a|u|^q|\xi|^{m-1} + \alpha(x)$;
$\alpha(x) \in \mathbf{L}_{\frac{p}{m-1}}(G)$;

3a) $|b(x,u,\xi)| \le a\mu|u|^{q-1}|\xi|^m + b_0(x)$; $b_0(x) \in \mathbf{L}_{\frac{p}{m}}(G)$;

3b) $b(x,u,\xi) = \beta(x,u) + \widetilde{b}(x,u,\xi)$, $u \cdot \beta(x,u) \ge a|u|^{q+m}$;
$|\widetilde{b}(x,u,\xi)| \le a\mu|u|^{q-1}|\xi|^m + b_0(x)$, $b_0(x) \in \mathbf{L}_{\frac{p}{m}}(G)$;

4) $\frac{\partial h(x,u)}{\partial u} \le 0$, $\frac{\partial g(x,u)}{\partial u} \le 0$;

5) $\sigma(\omega) \ge \nu_0 \ge 0$ on σ_0; $\gamma(\omega) \ge \nu_0 \ge 0$ on ∂G.

The functions $a_i(x,u,\xi)$ are continuously differentiable with respect to variables u, ξ in $\mathfrak{M}_{d,M_0} = \overline{G_0^d} \times [-M_0, M_0] \times \mathbb{R}^n$ and satisfy, in $\mathfrak{M}_{d,M_0}$, the following conditions:

6) $(m-1)u\frac{\partial a_i(x,u,\xi)}{\partial u} = q\frac{\partial a_i(x,u,\xi)}{\partial \xi_j}\xi_j$; $i = 1, \ldots, n$;

7) $\sqrt{\sum_{i=1}^{n} |a_i(x,u,u_x) - a|u|^q|\nabla u|^{m-2}u_{x_i}|^2} \le a\mathcal{A}(|x|)|u|^q|\nabla u|^{m-1}$, $x \in \overline{G_0^d}$, *where* $\mathcal{A}(r)$ *is a function which is Dini-continuous at zero.*

Let us consider the function change

$$u = v|v|^{\varsigma-1} \text{ with } \quad \varsigma = \frac{m-1}{q+m-1}. \tag{6.1.1}$$

By virtue of the assumption 6), the identity $(II)_{loc}$ takes the form

$$\int_{G_0^\varrho} \Big\langle \mathcal{A}_i(x,v_x)\eta_{x_i} + \mathcal{B}(x,v,v_x)\eta \Big\rangle dx + \int_{\Gamma_0^\varrho} \frac{\gamma(\omega)}{r^{m-1}} v|v|^{m-2}\eta(x)ds$$
$$+ \int_{\Sigma_0^\varrho} \frac{\sigma(\omega)}{r^{m-1}} v|v|^{m-2}\eta(x)ds$$
$$= \int_{\Omega_\varrho} \mathcal{A}_i(x,v_x)\cos(r,x_i)\eta(x)d\Omega_\varrho + \int_{\Gamma_0^\varrho} \mathcal{G}(x,v)\eta(x)ds + \int_{\Sigma_0^\varrho} \mathcal{H}(x,v)\eta(x)ds \tag{6.1.2}$$

for a.e. $\varrho \in (0,d)$, $v(x) \in \mathbf{C}^0(\overline{G}) \cap \mathbf{V}^1_{m,0}(G)$ and any $\eta(x) \in \mathbf{C}^0(\overline{G}) \cap \mathbf{V}^1_{m,0}(G)$, where

$$\mathcal{A}_i(x,v_x) \equiv a_i(x, v|v|^{\varsigma-1}, \varsigma|v|^{\varsigma-1}v_x), \quad \mathcal{B}(x,v,v_x) \equiv b(x, v|v|^{\varsigma-1}, \varsigma|v|^{\varsigma-1}v_x),$$
$$\mathcal{G}(x,v) \equiv g(x, v|v|^{\varsigma-1}), \qquad \mathcal{H}(x,v) \equiv h(x, v|v|^{\varsigma-1}). \tag{6.1.3}$$

We show that coefficients $\mathcal{A}_i$, $i = 1, \ldots, n$ do not depend on v explicitly. In fact, by the change (6.1.1) and the assumption *6)*, we calculate

$$\frac{\partial \mathcal{A}_i}{\partial v} = \frac{\partial a_i(x,u,\xi)}{\partial u} \cdot \frac{\partial}{\partial v}\left(|v^2|^{\frac{\varsigma-1}{2}} \cdot v\right) + \frac{\partial a_i(x,u,\xi)}{\partial \xi_j} \cdot \varsigma v_{x_j} \frac{\partial}{\partial v}\left(|v^2|^{\frac{\varsigma-1}{2}}\right)$$
$$= \varsigma |v|^{\varsigma-1}\frac{\partial a_i}{\partial u} + \varsigma(\varsigma-1)v_{x_j}v|v|^{\varsigma-3} \cdot \frac{\partial a_i}{\partial \xi_j} = \varsigma \cdot \frac{u}{v} \cdot \frac{\partial a_i}{\partial u} + (\varsigma-1) \cdot \frac{\xi_j}{v} \cdot \frac{\partial a_i}{\partial \xi_j}$$
$$= \frac{1}{v}\left(\varsigma u \cdot \frac{\partial a_i}{\partial u} + (\varsigma-1) \cdot \frac{m-1}{q} u \frac{\partial a_i}{\partial u}\right) = \frac{u}{v} \cdot \frac{\partial a_i}{\partial u} \cdot \left(\varsigma + (\varsigma-1) \cdot \frac{m-1}{q}\right) = 0,$$

because of (6.1.1).

Our assumptions can be rewritten as follows:

1)′ $\mathcal{A}_i(x, v_x)v_{x_i} \geq a\varsigma^{m-1}|\nabla v|^m - \frac{1}{\varsigma}|v|^{1-\varsigma}a_0(x)$; $a_0(x) \in \mathbf{L}_{p/m}(G)$;

2)′ $$\sqrt{\sum_{i=1}^{n} \mathcal{A}_i^2(x, v_x)} + \sqrt{\sum_{i=1}^{n} \left|\frac{\partial \mathcal{A}_i(x, v_x)}{\partial x_i}\right|^2} \leq a\varsigma^{m-1}|\nabla v|^{m-1} + \alpha(x);$$
$\alpha(x) \in \mathbf{L}_{\frac{p}{m-1}}(G)$;

3a)′ $|\mathcal{B}(x, v, v_x)| \leq a\mu\varsigma^m |v|^{-1}|\nabla v|^m + b_0(x)$; $b_0(x) \in \mathbf{L}_{\frac{p}{m}}(G)$,

4)′ $\frac{\partial \mathcal{H}(x,v)}{\partial v} \leq 0$, $\frac{\partial \mathcal{G}(x,v)}{\partial v} \leq 0$;

7)′ $$\sqrt{\sum_{i=1}^{n} |\mathcal{A}_i(x, v_x) - a\varsigma^{m-1}|\nabla v|^{m-2}v_{x_i}|^2} \leq a\varsigma^{m-1}\mathcal{A}(|x|)|\nabla v|^{m-1},\ x \in \overline{G_0^d}.$$

The main result in this chapter is the following statement:

Theorem 6.3. *Let u be a weak solution of the problem (QL), assumptions* 1)–7) *be satisfied and ϑ be the smallest positive eigenvalue of the problem $(NEVP)$ (see* §2.1*). Let us assume that $M_0 = \max_{x \in \overline{G}} |u(x)|$ is known. In addition, let $h(x, 0) \in L_\infty(\Sigma_0)$, $g(x, 0) \in L_\infty(\partial G)$ and let there exist real numbers $k_s \geq 0$, $K \geq 0$ such that*

$$k_s =: \sup_{\varrho>0} \varrho^{-ms}\left\{\int_{G_0^\varrho} r^{\frac{q}{q+m-1}}|a_0(x)|^{\frac{m(q+m-1)}{(m-1)(q+m)}}dx + \int_{G_0^\varrho} r^{\frac{1}{m-1}}|b_0(x)|^{\frac{m}{m-1}}dx\right.$$
$$\left. + \int_{\Sigma_0^\varrho} |h(x,0)|^{\frac{m}{m-1}}ds + \int_{\Gamma_0^\varrho} |g(x,0)|^{\frac{m}{m-1}}ds\right\},\quad s > 1; \tag{6.1.4}$$

$$K =: \sup_{\varrho>0} \frac{\varrho^{\frac{n}{m}-1}}{\psi(\varrho)}\left\{\varrho^{m(1-\frac{n}{p})\frac{q+m-1}{(m-1)(q+m)}} \|a_0\|_{\frac{p}{m}, G_0^\varrho}^{\frac{q+m-1}{(m-1)(q+m)}} + \varrho^{1-\frac{n}{p}}\|\alpha(x)\|_{\frac{p}{m-1}, G_0^\varrho}^{\frac{1}{m-1}}\right.$$
$$\left. + \varrho^{(1-\frac{n}{p})\frac{m}{m-1}}\|b_0(x)\|_{\frac{p}{m}, G_0^\varrho}^{\frac{1}{m-1}} + \varrho\left(\|g(x,0)\|_{\infty,\Gamma_0^\varrho}^{\frac{1}{m-1}} + \|h(x,0)\|_{\infty,\Sigma_0^\varrho}^{\frac{1}{m-1}}\right)\right\}, \tag{6.1.5}$$

where

$$\psi(\varrho)=\begin{cases}\varrho^{\frac{\vartheta^{\frac{1}{m}}(m)}{\Xi(m)}\cdot\frac{q+(m-1)(1-\mu)}{q+m-1}}, & s>\frac{\vartheta^{\frac{1}{m}}(m)}{\Xi(m)}\cdot\frac{q+(m-1)(1-\mu)}{q+m-1};\\ \varrho^{\frac{\vartheta^{\frac{1}{m}}(m)}{\Xi(m)}\cdot\frac{q+(m-1)(1-\mu)}{q+m-1}}\ln\frac{d}{\varrho}, & s=\frac{\vartheta^{\frac{1}{m}}(m)}{\Xi(m)}\cdot\frac{q+(m-1)(1-\mu)}{q+m-1};\\ \varrho^{s}, & s<\frac{\vartheta^{\frac{1}{m}}(m)}{\Xi(m)}\cdot\frac{q+(m-1)(1-\mu)}{q+m-1}\end{cases}\tag{6.1.6}$$

and $\Xi(m)$ is determined by (2.4.3). Then there exist $d\in(0,1)$ and a constant $C_0>0$ independent of u such that

$$|u(x)|\le C_0\Big(|x|^{1-\frac{n}{m}}\psi(|x|)\Big)^{\frac{m-1}{q+m-1}},\quad \forall x\in G_0^d.\tag{6.1.7}$$

Furthermore, if coefficients of the problem (QL) satisfy such conditions which guarantee the local a priori estimate $|\nabla u|_{0,G'}\le M_1$ for any smooth $G'\subset\subset\overline{G}\setminus\{\mathcal{O}\}$ (see for example §4 in [6] *or* [64], [47]*), then there is a constant $C_1>0$ independent of u such that*

$$|\nabla u(x)|\le C_1|x|^{-\frac{n(m-1)+qm}{m(q+m-1)}}\psi^{\frac{m-1}{q+m-1}}(|x|),\quad \forall x\in G_0^d.\tag{6.1.8}$$

6.2 Comparison principle

Let us consider the second-order quasi-linear degenerate operator Q of the form

$$\begin{aligned}Q(v,\eta)\equiv&\int_{G_0^d}\Big\langle\mathcal{A}_i(x,v_x)\eta_{x_i}+\mathcal{B}(x,v,v_x)\eta\Big\rangle dx+\int_{\Gamma_0^d}\frac{\gamma(\omega)}{r^{m-1}}v|v|^{m-2}\eta(x)ds\\ &-\int_{\Sigma_0^d}\mathcal{H}(x,v)\eta(x)ds+\int_{\Sigma_0^d}\frac{\sigma(\omega)}{r^{m-1}}v|v|^{m-2}\eta(x)ds\\ &-\int_{\Omega_d}\mathcal{A}_i(x,v_x)\cos(r,x_i)\eta(x)d\Omega_d-\int_{\Gamma_0^d}\mathcal{G}(x,v)\eta(x)ds\end{aligned}\tag{6.2.1}$$

for $v(x)\in\mathbf{C}^0(\overline{G})\cap\mathbf{V}^1_{m,0}(G)$ and for all non-negative η belonging to $\mathbf{C}^0(\overline{G})\cap\mathbf{V}^1_{m,0}(G)$ under the following assumptions:

The functions $\mathcal{A}_i(x,\xi),\mathcal{B}(x,v,\xi),\mathcal{G}(x,v),\mathcal{H}(x,v)$ are Caratheodory, continuously differentiable with respect to the v,ξ variables in $\mathfrak{M}=\overline{\Omega}\times\mathbb{R}\times\mathbb{R}^N$ and satisfy in $\mathfrak{M}$ the following inequalities:

(i) $\dfrac{\partial\mathcal{A}_i(x,\xi)}{\partial\xi_j}p_ip_j\ge a\gamma_m|\xi|^{m-2}p^2,\ \forall p\in\mathbb{R}^n\setminus\{0\};$

(ii) $\sqrt{\sum_{i=1}^N\left|\dfrac{\partial\mathcal{B}(x,v,\xi)}{\partial\xi_i}\right|^2}\le a|v|^{-1}|\xi|^{m-1};\ \dfrac{\partial\mathcal{B}(x,v,\xi)}{\partial v}\ge a|v|^{-2}|\xi|^m;$

(iii) $\dfrac{\partial \mathcal{G}(x,v)}{\partial v} \le 0,\ \dfrac{\partial H(x,v)}{\partial v} \le 0,\ \gamma(\omega) \ge 0, \quad \sigma(\omega) \ge 0.$

Here: $m > 1$, $\gamma_m > 0$ *and* $a > 0$.

Proposition 6.4. *Let operator* Q *satisfy assumptions* (i)–(iii) *and functions* $v, w \in \mathbf{C}^0(\overline{G_0^d}) \cap \mathbf{V}^1_{m,0}(G_0^d)$ $(d \lll 1)$ *satisfy the inequality*

$$Q(v,\eta) \le Q(w,\eta) \tag{6.2.2}$$

for all non-negative $\eta \in \mathbf{C}^0(\overline{G_0^d}) \cap \mathbf{V}^1_{m,0}(G_0^d)$ *and also the inequality*

$$v(x) \le w(x) \ on\ \Omega_d \tag{6.2.3}$$

hold. Then $v(x) \le w(x)$ *in* G_0^d.

Proof. Let us define $z = v - w$ and $v^\tau = \tau v + (1-\tau)w$, $\tau \in [0,1]$. Then we have

$$\begin{aligned}
0 \ge Q(v,\eta) - Q(w,\eta) &= \int\limits_{G_0^d} \Bigg\langle \eta_{x_i} z_{x_j} \int_0^1 \frac{\partial \mathcal{A}_i(x, v^\tau_x)}{\partial v^\tau_{x_j}} d\tau \\
&+ \eta z_{x_i} \int_0^1 \frac{\partial \mathcal{B}(x, v^\tau, v^\tau_x)}{\partial v^\tau_{x_i}} d\tau + \eta z \int_0^1 \frac{\partial \mathcal{B}(x, v^\tau, v^\tau_x)}{\partial v^\tau} d\tau \Bigg\rangle dx \\
&+ \int\limits_{\Gamma_0^d} \frac{\gamma(\omega)}{r^{m-1}} \left(\int_0^1 \frac{\partial (v^\tau |v^\tau|^{m-2})}{\partial v^\tau} d\tau \right) z(x)\eta(x) ds \\
&+ \int\limits_{\Sigma_0^d} \frac{\sigma(\omega)}{r^{m-1}} \left(\int_0^1 \frac{\partial (v^\tau |v^\tau|^{m-2})}{\partial v^\tau} d\tau \right) z(x)\eta(x) ds \\
&- \int\limits_{\Omega_d} \left(\int_0^1 \frac{\partial \mathcal{A}_i(x, v^\tau_x)}{\partial v^\tau_{x_j}} d\tau \right) \cos(r, x_i) \cdot z_{x_j} \eta(x) d\Omega_d \\
&- \int\limits_{\Gamma_0^d} \left(\int_0^1 \frac{\partial \mathcal{G}(x, v^\tau)}{\partial v^\tau} d\tau \right) z(x)\eta(x) ds - \int\limits_{\Sigma_0^d} \left(\int_0^1 \frac{\partial \mathcal{H}(x, v^\tau)}{\partial v^\tau} d\tau \right) z(x)\eta(x) ds
\end{aligned} \tag{6.2.4}$$

for all non-negative $\eta \in \mathbf{C}^0(\overline{G_0^d}) \cap \mathbf{V}^1_{m,0}(G_0^d)$.

Now we introduce the sets

$$\begin{aligned}
(G_0^d)^+ &:= \{x \in G_0^d \mid v(x) > w(x)\} \subset G_0^d, \\
(\Sigma_0^d)^+ &:= \{x \in \Sigma_0^d \mid v(x) > w(x)\} \subset \Sigma_0^d, \\
(\Gamma_0^d)^+ &:= \{x \in \Gamma_0^d \mid v(x) > w(x)\} \subset \Gamma_0^d
\end{aligned}$$

and assume that $(G_0^d)^+ \neq \emptyset$. Let $k \geq 1$ be any odd number. We choose $\eta = \max\{(v-w)^k, 0\}$ as test function in the integral inequality (6.2.4). We have

$$\int_0^1 \frac{\partial(v^\tau |v^\tau|^{m-2})}{\partial v^\tau} d\tau = (m-1)\int_0^1 |v^\tau|^{m-2} d\tau > 0.$$

Then, by assumptions (i)–(iii) and $\eta\big|_{\Omega_d} = 0$, we obtain from (6.2.4)

$$\int\limits_{(G_0^d)^+} \left\{ k\gamma_m a z^{k-1}\Big(\int_0^1 |\nabla v^\tau|^{m-2} d\tau\Big)|\nabla z|^2 dx + a z^{k+1}\Big(\int_0^1 |v^\tau|^{-2}|\nabla v^\tau|^m d\tau\Big) dx \right\}$$

$$\leq \int\limits_{(G_0^d)^+} a z^k \Big(\int_0^1 |v^\tau|^{-1}|\nabla v^\tau|^{m-1} d\tau\Big)|\nabla z| dx. \tag{6.2.5}$$

By the Cauchy inequality

$$z^k|\nabla z||v^\tau|^{-1}|\nabla v^\tau|^{m-1} = \left(|v^\tau|^{-1} z^{\frac{k+1}{2}} |\nabla v^\tau|^{m/2}\right)\cdot\left(z^{\frac{k-1}{2}}|\nabla z||\nabla v^\tau|^{m/2-1}\right)$$

$$\leq \frac{\varepsilon}{2}|v^\tau|^{-2} z^{k+1}|\nabla v^\tau|^m + \frac{1}{2\varepsilon} z^{k-1}|\nabla z|^2|\nabla v^\tau|^{m-2}, \ \forall \varepsilon > 0.$$

Hence, taking $\varepsilon = 2$, we obtain from (6.2.5) the inequality

$$\int\limits_{(G_0^d)^+} a\left(k\gamma_m - \frac{1}{4}\right) z^{k-1}|\nabla z|^2 \left(\int_0^1 |\nabla v^\tau|^{m-2} d\tau\right) dx \leq 0. \tag{6.2.6}$$

Now choosing the odd number $k \geq \max\left(1; \frac{1}{2\gamma_m}\right)$, in view of $z(x) \equiv 0$ on $\partial(G_0^d)^+$, we get from (6.2.6) $z(x) \equiv 0$ in $(G_0^d)^+$. We have arrived at a contradiction to our definition of the set $(G_0^d)^+$. By this fact, the proposition is proved. □

6.3 Maximum principle

In this section we derive an $L_\infty(G)$-*a priori*-estimate of the weak solution to problem (QL).

Theorem 6.5. *Let $u(x)$ be a weak solution of (QL) and assumptions* 1), 3b), 4) *and* 5) *hold. In addition, let*

$$h(x,0) \in L_{\frac{j}{j-1}}(\Sigma_0),\ g(x,0) \in \mathbf{L}_{\frac{j}{j-1}}(\partial G), \quad 1 \leq j < \frac{n-1}{n-m}.$$

Then there exists a constant $M_0 > 0$, depending only on meas G, meas ∂G, meas Σ_0, n, m, μ, q, $\|h(x,0)\|_{L_{\frac{j}{j-1}}(\Sigma_0)}$, $\|g(x,0)\|_{\mathbf{L}_{\frac{j}{j-1}}(\partial G)}$, $\|a_0(x)\|_{\mathbf{L}_{\frac{p}{m}}(G)}$, $\|b_0(x)\|_{\mathbf{L}_{\frac{p}{m}}(G)}$, such that $\|u\|_{L_\infty(G)} \leq M_0$.

Proof. Let us define the set $A(k) = \{x \in \overline{G}, |u(x)| > k\}$ and let $\chi_{A(k)}$ be the characteristic function of the set $A(k)$. We observe that $A(k+d) \subseteq A(k)$ for all $d > 0$.

Putting $\eta((|u|-k)_+)\chi_{A(k)} \cdot \text{sign} u$ as test function in *(II)*, where η is defined by Lemma 1.23 and $k \geq k_0$ (without loss of generality we can assume $k_0 \geq 1$), under assumptions 1), 3b), 5) we obtain the inequality

$$\int_{A(k)} \{a|u|^q|\nabla u|^m \eta'((|u|-k)_+) + a|u|^{q+m-1}\eta((|u|-k)_+)\}\, dx \tag{6.3.1}$$

$$+ \int_{\Sigma_0 \cap A(k)} \frac{\sigma(\omega)}{r^{m-1}}|u|^{q+m-1}\eta((|u|-k)_+)ds + \int_{\partial G \cap A(k)} \frac{\gamma(\omega)}{r^{m-1}}|u|^{q+m-1}\eta((|u|-k)_+)ds$$

$$\leq \int_{A(k)} \{a\mu|u|^{q-1}|\nabla u|^m \eta((|u|-k)_+) + a_0(x)\eta'((|u|-k)_+) + b_0(x)\eta((|u|-k)_+)\} dx$$

$$+ \int_{\Sigma_0 \cap A(k)} h(x,u)\text{sign}\, u \cdot \eta((|u|-k)_+)ds + \int_{\partial G \cap A(k)} g(x,u)\text{sign}\, u \cdot \eta((|u|-k)_+)ds.$$

By virtue of $g(x,u) - g(x,0) = \int_0^1 \frac{d}{d\tau} g(x,\tau u)d\tau = u \cdot \int_0^1 \frac{\partial g(x,\tau u)}{\partial(\tau u)} d\tau$ and the assumption 4),

$$\int_{\partial G \cap A(k)} g(x,u) \cdot \text{sign}\, u \cdot \eta((|u|-k)_+)ds$$

$$= \int_{\partial G \cap A(k)} |u(x)| \left(\int_0^1 \frac{\partial g(x,\tau u)}{\partial(\tau u)} d\tau \right) \eta((|u|-k)_+)ds$$

$$+ \int_{\partial G \cap A(k)} g(x,0) \cdot \text{sign}\, u \cdot \eta((|u|-k)_+)ds$$

$$\leq \int_{\partial G \cap A(k)} |g(x,0)| \cdot \eta((|u|-k)_+)ds,$$

as well

$$\int_{\Sigma_0 \cap A(k)} h(x,u) \cdot \text{sign}\, u \cdot \eta((|u|-k)_+)ds \leq \int_{\Sigma_0 \cap A(k)} |h(x,0)| \cdot \eta((|u|-k)_+)ds.$$

Therefore from (6.3.1), by assumption 5), it follows that

$$\int\limits_{A(k)} \left\{a|u|^q|\nabla u|^m\eta'((|u|-k)_+) + a|u|^{q+m-1}\eta((|u|-k)_+)\right\} dx$$

$$\leq \int\limits_{A(k)} \left\{a\mu k_0^{-1}|u|^q|\nabla u|^m\eta((|u|-k)_+) + a_0(x)\eta'((|u|-k)_+) + b_0(x)\eta((|u|-k)_+)\right\} dx$$

$$+ \int\limits_{\Sigma_0\cap A(k)} |h(x,0)|\eta((|u|-k)_+)ds + \int\limits_{\partial G\cap A(k)} |g(x,0)|\eta((|u|-k)_+)ds. \quad (6.3.2)$$

Now we introduce the function $w_k(x) := \eta\left(\frac{(|u|-k)_+}{m}\right)$. By (1.8.7) from Lemma 1.23, we have

$$\int\limits_{\partial G\cap A(k)} |g(x,0)|\eta((|u|-k)_+)ds \leq M \cdot \int\limits_{\partial G\cap A(k+d)} |g(x,0)||w_k|^m ds$$

$$+ \mathrm{e}^{\varkappa d} \cdot \int\limits_{\partial G\cap\{A(k)\setminus A(k+d)\}} |g(x,0)|ds. \quad (6.3.3)$$

By the Hölder inequality

$$\int\limits_{\partial G\cap A(k+d)} |g(x,0)| \cdot |w_k|^m ds \leq \||w_k|^m\|_{L_j(\partial G\cap A(k))} \cdot \|g(x,0)\|_{L_{\frac{j}{j-1}}(\partial G)}$$

$$= \|w_k\|^m_{L_{mj}(\partial G\cap A(k))} \cdot \|g(x,0)\|_{\mathbf{L}_{\frac{j}{j-1}}(\partial G)}, \ \forall j \geq 1$$

and the Sobolev boundary trace embedding theorem (1.6.8), we derive

$$\int\limits_{\partial G\cap A(k+d)} |g(x,0)| \cdot |w_k|^m ds$$

$$\leq C\|g(x,0)\|_{\mathbf{L}_{\frac{j}{j-1}}(\partial G)} \cdot \int\limits_{A(k)} \left(|\nabla w_k|^m + |w_k|^m\right)dx, \ 1 \leq j < \frac{n-1}{n-m}.$$

In the same way

$$\int\limits_{\Sigma_0\cap A(k+d)} |h(x,0)| \cdot |w_k|^m ds$$

$$\leq C\|h(x,0)\|_{L_{\frac{j}{j-1}}(\Sigma_0)} \cdot \int\limits_{A(k)} \left(|\nabla w_k|^m + |w_k|^m\right)dx, \ 1 \leq j < \frac{n-1}{n-m}.$$

Now from (6.3.2) and (6.3.3) it follows that

$$\int\limits_{A(k)} \{a|u|^q|\nabla u|^m \langle \eta'((|u|-k)_+) - \mu k_0^{-1}\eta((|u|-k)_+)\rangle$$

$$+ a|u|^{q+m-1}\eta((|u|-k)_+)\}dx \tag{6.3.4}$$

$$\leq \int\limits_{A(k)} \{a_0(x)\eta'((|u|-k)_+) + b_0(x)\eta((|u|-k)_+)\}dx$$

$$+ CM\left(\|h(x,0)\|_{L_{\frac{j}{j-1}}(\Sigma_0)} + \|g(x,0)\|_{\mathbf{L}_{\frac{j}{j-1}}(\partial G)}\right) \cdot \int\limits_{A(k)} \left(|\nabla w_k|^m + |w_k|^m\right)dx$$

$$+ \mathrm{e}^{\varkappa d}\Bigg\{ \int\limits_{\Sigma_0\cap\{A(k)\setminus A(k+d)\}} |h(x,0)|ds + \int\limits_{\partial G\cap\{A(k)\setminus A(k+d)\}} |g(x,0)|ds\Bigg\},$$

$1 \leq j < \frac{n-1}{n-m}$. By definition of $\eta(x)$ (see Lemma 1.23) and $w_k(x)$,

$$\mathrm{e}^{\varkappa(|u|-k)_+}|\nabla u|^m = \left(\frac{m}{\varkappa}\right)^m |\nabla w_k|^m, \quad \varkappa > 0. \tag{6.3.5}$$

Therefore, choosing $\varkappa > m + \frac{2\mu}{k_0}$ according to Lemma 1.23 and using (1.8.5)–(1.8.7), from (6.3.4), we obtain

$$k_0^q\left(\frac{m}{\varkappa}\right)^m \int\limits_{A(k)} a|\nabla w_k|^m dx + k_0^{q+m-1}\int\limits_{A(k)} a|w_k|^m dx \leq c_1 M \int\limits_{A(k+d)} B_0(x)|w_k|^m dx$$

$$+ Ma_*^{-1}c_2\left(\|h(x,0)\|_{L_{\frac{j}{j-1}}(\Sigma_0)} + \|g(x,0)\|_{\mathbf{L}_{\frac{j}{j-1}}(\partial G)}\right) \cdot \int\limits_{A(k)} \left(a|\nabla w_k|^m + a|w_k|^m\right)dx$$

$$+ c_3\mathrm{e}^{\varkappa d}\Bigg\{ \int\limits_{\{A(k)\setminus A(k+d)\}} B_0(x)dx + \int\limits_{\Sigma_0\cap\{A(k)\setminus A(k+d)\}} |h(x,0)|ds$$

$$+ \int\limits_{\partial G\cap\{A(k)\setminus A(k+d)\}} |g(x,0)|ds\Bigg\}, \tag{6.3.6}$$

where $1 \leq j < \frac{n-1}{n-m}$ and

$$B_0(x) = a_0(x) + b_0(x). \tag{6.3.7}$$

Under assumptions 1) and 3b) we get that $B_0(x) \in \mathbf{L}_s(G)$, where $s > \frac{n}{m} > 1$. Using the Hölder inequality with exponents s and s', where $\frac{1}{s} + \frac{1}{s'} = 1$, we obtain

$$\int\limits_{A(k+d)} B_0(x)|w_k|^m dx \leq \|B_0(x)\|_{\mathbf{L}_s(G)} \cdot \left(\int\limits_{A(k)} |w_k|^{ms'}dx\right)^{\frac{1}{s'}}. \tag{6.3.8}$$

From the inequality $\frac{1}{s} < \frac{m}{n}$ it follows that $ms' < m^{\#} = \frac{mn}{n-m}$ and then the interpolation inequality (1.5.9) gives

$$\left(\int_{A(k)} |w_k|^{ms'} dx\right)^{\frac{1}{s'}} \leq \left(\int_{A(k)} |w_k|^m dx\right)^{\theta} \cdot \left(\int_{A(k)} |w_k|^{m^{\#}} dx\right)^{\frac{(1-\theta)m}{m^{\#}}}$$

with $\theta \in (0,1)$, which is defined by the equality $\frac{1}{ms'} = \frac{\theta}{m} + \frac{1-\theta}{m^{\#}} \implies \theta = 1 - \frac{n}{ms}$. Now, by using the Young inequality with exponents $\frac{1}{\theta}$ and $\frac{1}{(1-\theta)}$, from (6.3.8) we obtain

$$\int_{A(k+d)} B_0(x)|w_k|^m dx \leq a_*^{-1}\theta\varepsilon^{\frac{\theta-1}{\theta}} \|B_0(x)\|^{\frac{1}{\theta}}_{\mathbf{L}_s(G)} \int_{A(k)} a|w_k|^m dx + (1-\theta)\varepsilon \cdot \left(\int_{A(k)} |w_k|^{m^{\#}} dx\right)^{\frac{m}{m^{\#}}}, \quad \forall \varepsilon > 0. \quad (6.3.9)$$

From (6.3.6), (6.3.9) it follows that

$$\begin{aligned} &\left(k_0^q\left(\frac{m}{\varkappa}\right)^m - c_4\right) \int_{A(k)} a|\nabla w_k|^m dx + \left(k_0^{q+m-1} - c_4 - c_5\varepsilon^{\frac{\theta-1}{\theta}}\right) \int_{A(k)} a|w_k|^m dx \\ &\leq c_6 \left\{ \int_{A(k)} B_0(x)dx + \int_{\Sigma_0 \cap A(k)} |h(x,0)|ds + \int_{\partial G \cap A(k)} |g(x,0)|ds \right\} \\ &+ c_1 M(1-\theta)\varepsilon \cdot \left(\int_{A(k)} |w_k|^{m^{\#}} dx\right)^{\frac{m}{m^{\#}}}, \ \forall \varepsilon > 0, \ \forall k \geq k_0, \end{aligned} \quad (6.3.10)$$

where

$$\begin{aligned} c_4 &= M a_*^{-1} c_2 \left(\|h(x,0)\|_{L_{\frac{j}{j-1}}(\Sigma_0)} + \|g(x,0)\|_{\mathbf{L}_{\frac{j}{j-1}}(\partial G)} \right), \\ c_5 &= M a_*^{-1} c_1 \theta \|B_0(x)\|^{\frac{1}{\theta}}_{\mathbf{L}_s(G)}, \quad c_6 = c_3 \mathrm{e}^{\varkappa d}. \end{aligned}$$

Now we apply the Sobolev embedding Theorem 1.15; as a result we get from (6.3.10) the inequality

$$\left(k_0^q\left(\frac{m}{\varkappa}\right)^m - c_4 - c_7\varepsilon\right)\int\limits_{A(k)} a|\nabla w_k|^m dx$$

$$+\left(k_0^{q+m-1} - c_4 - c_5\varepsilon^{\frac{\theta-1}{\theta}} - c_7\varepsilon\right)\int\limits_{A(k)} a|w_k|^m dx \tag{6.3.11}$$

$$\leq c_6\Bigg\{\int\limits_{A(k)} B_0(x)dx + \int\limits_{\Sigma_0\cap A(k)} |h(x,0)|ds + \int\limits_{\partial G\cap A(k)} |g(x,0)|ds\Bigg\},\ \forall\varepsilon>0,\ \forall k\geq k_0,$$

where $c_7 = \widetilde{c}^m a_*^{-1} c_1 M(1-\theta)$. Let us choose

$$c_5\varepsilon^{\frac{\theta-1}{\theta}} = c_7\varepsilon \quad\Rightarrow\quad \varepsilon = \left(\frac{c_5}{c_7}\right)^\theta \tag{6.3.12}$$

and

$$\begin{cases} k_0^q\left(\frac{m}{\varkappa}\right)^m \geq 2(c_4+c_7\varepsilon) = 2(c_4+c_7^{1-\theta}c_5^\theta);\\ k_0^{q+m-1} \geq 2(c_4+2c_7\varepsilon) = 2c_4+4c_7^{1-\theta}c_5^\theta \end{cases} \Longrightarrow$$

$$k_0 \geq \max\left\{1;\ 2^{\frac{1}{q}}\left(\frac{\varkappa}{m}\right)^{\frac{m}{q}}\cdot\left(c_4+c_7^{1-\theta}c_5^\theta\right)^{\frac{1}{q}};\ \left(2c_4+4c_7^{1-\theta}c_5^\theta\right)^{\frac{1}{q+m-1}}\right\}. \tag{6.3.13}$$

Thus, from (6.3.10)–(6.3.13) we obtain the inequality

$$\int\limits_{A(k)} \left(a|\nabla w_k|^m + a|w_k|^m\right)dx$$

$$\leq \frac{c_6}{c_4+c_7^{1-\theta}c_5^\theta}\Bigg\{\int\limits_{A(k)} B_0(x)dx + \int\limits_{\Sigma_0\cap A(k)} |h(x,0)|ds + \int\limits_{\partial G\cap A(k)} |g(x,0)|ds\Bigg\}$$

for all $k\geq k_0$. Applying the Sobolev embedding Theorems 1.15 and 1.20, we have

$$\left(\int\limits_{A(k)} |w_k|^{m^\#}dx\right)^{\frac{m}{m^\#}} + \left(\int\limits_{\Sigma_0\cap A(k)} |w_k|^{j^*}ds\right)^{\frac{m}{j^*}} + \left(\int\limits_{\partial G\cap A(k)} |w_k|^{j^*}ds\right)^{\frac{m}{j^*}}$$

$$\leq \widetilde{c}\int\limits_{A(k)} \left(a|\nabla w_k|^m + a|w_k|^m\right)dx \leq \frac{\widetilde{c}c_6}{c_4+c_7^{1-\theta}c_5^\theta}\Bigg\{\int\limits_{A(k)} B_0(x)dx \tag{6.3.14}$$

$$+\int\limits_{\Sigma_0\cap A(k)} |h(x,0)|ds + \int\limits_{\partial G\cap A(k)} |g(x,0)|ds\Bigg\},\quad j^* = \frac{m(n-1)}{n-m},\ \forall k\geq k_0.$$

At last, by the Hölder inequality, we get

$$\int\limits_{A(k)} B_0(x)dx \le \|B_0(x)\|_{\mathbf{L}_s(G)} \text{ meas }^{1-\frac{1}{s}} A(k);$$

$$\int\limits_{\Sigma_0 \cap A(k)} |h(x,0)|ds \le \|h(x,0)\|_{L_{\frac{j}{j-1}}(\Sigma_0)} \big[\text{meas}(\Sigma_0 \cap A(k))\big]^{\frac{1}{j}};$$

$$\int\limits_{\partial G \cap A(k)} |g(x,0)|ds \le \|g(x,0)\|_{\mathbf{L}_{\frac{j}{j-1}}(\partial G)} \cdot \big[\text{meas}(\partial G \cap A(k))\big]^{\frac{1}{j}}, \quad \frac{n-m}{n-1} < \frac{1}{j} \le 1.$$

Next from (6.3.14) it follows that

$$\begin{aligned} &\Bigg(\int\limits_{A(k)} |w_k|^{m^\#} dx\Bigg)^{\frac{m}{m^\#}} + \Bigg(\int\limits_{\Sigma_0 \cap A(k)} |w_k|^{j^*} ds\Bigg)^{\frac{m}{j^*}} + \Bigg(\int\limits_{\partial G \cap A(k)} |w_k|^{j^*} ds\Bigg)^{\frac{m}{j^*}} \\ &\le \frac{\widetilde{c}c_6}{c_4 + c_7^{1-\theta} c_5^{\theta}} \Big\{ \|B_0(x)\|_{\mathbf{L}_s(G)} \text{meas}^{1-\frac{1}{s}} A(k) \qquad (6.3.15) \\ &\quad + \|h(x,0)\|_{L_{\frac{j}{j-1}}(\Sigma_0)} \cdot \big[\text{meas}(\Sigma_0 \cap A(k))\big]^{\frac{1}{j}} \\ &\quad + \|g(x,0)\|_{\mathbf{L}_{\frac{j}{j-1}}(\partial G)} \cdot \big[\text{meas}(\partial G \cap A(k))\big]^{\frac{1}{j}} \Big\}, \end{aligned}$$

$\frac{n-m}{n-1} < \frac{1}{j} \le 1$, $j^* = \frac{m(n-1)}{n-m}$, for all $k \ge k_0$. Let now $l > k > k_0$. By (1.8.8) and the definition of the function $w_k(x)$, we have $|w_k| \ge \frac{1}{m}(|u|-k)_+$ and therefore

$$\int\limits_{A(l)} |w_k|^{m^\#} dx \ge \left(\frac{l-k}{m}\right)^{m^\#} \cdot \text{meas} A(l);$$

$$\int\limits_{\Sigma_0 \cap A(l)} |w_k|^{j^*} ds \ge \left(\frac{l-k}{m}\right)^{j^*} \cdot \text{meas } (\Sigma_0 \cap A(l));$$

$$\int\limits_{\partial G \cap A(l)} |w_k|^{j^*} ds \ge \left(\frac{l-k}{m}\right)^{j^*} \cdot \text{meas } (\partial G \cap A(l)).$$

From (6.3.15) it follows:

$$\begin{aligned} &\text{meas } A(l) + \big[\text{meas}(\Sigma_0 \cap A(l))\big]^{\frac{m^\#}{j^*}} + \big[\text{meas}(\partial G \cap A(l))\big]^{\frac{m^\#}{j^*}} \\ &\le \left(\frac{m}{l-k}\right)^{m^\#} \Bigg\{ \int\limits_{A(k)} |w_k|^{m^\#} dx + \Bigg(\int\limits_{\Sigma_0 \cap A(k)} |w_k|^{j^*} ds\Bigg)^{\frac{m^\#}{j^*}} + \Bigg(\int\limits_{\partial G \cap A(k)} |w_k|^{j^*} ds\Bigg)^{\frac{m^\#}{j^*}} \Bigg\} \end{aligned}$$

$$\leq c_8 \left(\frac{m}{l-k}\right)^{m^\#} \cdot \left(\|B_0(x)\|_{\mathbf{L}_s(G)} + \|h(x,0)\|_{L_{\frac{j}{j-1}}(\Sigma_0)} + |g(x,0)\|_{\mathbf{L}_{\frac{j}{j-1}}(\partial G)}\right)^{\frac{m^\#}{m}}$$
$$\times \left\{\text{meas}^{\frac{m^\#}{m}(1-\frac{1}{s})} A(k) + \left[\text{meas}(\Sigma_0 \cap A(k))\right]^{\frac{m^\#}{m}\cdot\frac{1}{j}} + \left[\text{meas}(\partial G \cap A(k))\right]^{\frac{m^\#}{m}\cdot\frac{1}{j}}\right\},$$
$$\frac{n-m}{n-1} < \frac{1}{j} \leq 1, \quad j^* = \frac{m(n-1)}{n-m}, \; \forall l > k \geq k_0. \tag{6.3.16}$$

Let us introduce

$$\psi(k) = \text{meas}\, A(k) + \left[\text{meas}(\Sigma_0 \cap A(k))\right]^{\frac{m^\#}{j^*}} + \left[\text{meas}(\partial G \cap A(k))\right]^{\frac{m^\#}{j^*}}.$$

Then from (6.3.16) it follows that

$$\psi(l) \leq c_9 \left(\frac{m}{l-k}\right)^{m^\#} \cdot \left\langle [\psi(k)]^{\frac{m^\#}{m}(1-\frac{1}{s})} + [\psi(k)]^{\frac{j^*}{m}\cdot\frac{1}{j}} \right\rangle, \; \forall l > k \geq k_0; \tag{6.3.17}$$
$$m^\# = \frac{mn}{n-m}, \quad s > \frac{n}{m} > 1, \quad \frac{n-m}{n-1} < \frac{1}{j} \leq 1, \quad j^* = \frac{m(n-1)}{n-m}. \tag{6.3.18}$$

By (6.3.18), we observe that $\gamma = \min\left\{\frac{m^\#}{m}\left(1-\frac{1}{s}\right), \frac{j^*}{m}\cdot\frac{1}{j}\right\} > 1$. Then from (6.3.17)–(6.3.18) we get $\psi(l) \leq \frac{c_{10}}{(l-k)^{m^\#}} \psi^\gamma(k)$, $\gamma > 1$; $\forall l > k \geq k_0$ and therefore, because of the Stampacchia Lemma, we have that $\psi(k_0 + \delta) = 0$ with δ depending only on quantities given in the formulation of Theorem 6.5. This fact means that $|u(x)| < k_0 + \delta$ for almost all $x \in G$. Theorem 6.5 is proved. □

6.4 Local estimate at the boundary

In this section we derive the local boundedness (near the conical point) of the weak solution of problem (QL).

Theorem 6.6. *Let $u(x)$ be a weak solution of the problem (QL) and assumptions* 1), 2), 3a), 4), 5), 6), (6.1.5) *be satisfied. In addition, let $h(x,0) \in L_\infty(\Sigma_0)$, $g(x,0) \in \mathbf{L}_\infty(\partial G)$. Then there exists a constant $C > 0$ depending only on $n, m, p, t, q, \varkappa, a_*, a^*, m_*, m^*, d, \|a_0(x)\|_{\frac{p}{m},G}, \|\alpha(x)\|_{\frac{p}{m-1},G}, \|b_0(x)\|_{\frac{p}{m},G}$ such that the inequality*

$$\sup_{x \in G_0^{\varkappa\varrho}} |u(x)| \leq C\Big\{\varrho^{-n\varsigma/t}\|u\|_{\frac{t}{\varsigma},G_0^\varrho} + \varrho^{\frac{m\varsigma(p-n)}{p(m-1+\varsigma)}} \cdot \|a_0(x)\|_{\frac{p}{m},G_0^\varrho}^{\frac{\varsigma}{m-1+\varsigma}}$$
$$+ \varrho^{\varsigma(1-\frac{n}{p})}\|\alpha(x)\|_{\frac{p}{m-1},G_0^\varrho}^{\frac{\varsigma}{m-1}} + \varrho^{\varsigma(1-\frac{n}{p})\frac{m}{m-1}}\|b_0(x)\|_{\frac{p}{m},G_0^\varrho}^{\frac{\varsigma}{m-1}} \tag{6.4.1}$$
$$+ \varrho^\varsigma\left(\|g(x,0)\|_{\infty,\Gamma_0^\varrho}^{\frac{\varsigma}{m-1}} + \|h(x,0)\|_{\infty,\Sigma_0^\varrho}^{\frac{\varsigma}{m-1}}\right)\Big\}, \; p > n > m$$

holds for any $t>0$, $\varkappa\in(0,1)$, $\varrho\in(0,d)$ *and* $\varsigma=\frac{m-1}{q+m-1}$.

Proof. We apply the Moser iteration method. Let us introduce the change of function (6.1.1). By the assumption 6), the identity (II) takes the form (see §1):

$$\int\limits_G \Big\langle \mathcal{A}_i(x,v_x)\eta_{x_i}+\mathcal{B}(x,v,v_x)\eta\Big\rangle dx+\int\limits_{\partial G}\frac{\gamma(\omega)}{r^{m-1}}v|v|^{m-2}\eta(x)ds$$
$$+\int\limits_{\Sigma_0}\frac{\sigma(\omega)}{r^{m-1}}v|v|^{m-2}\eta(x)ds=\int\limits_{\partial G}\mathcal{G}(x,v)\eta(x)ds+\int\limits_{\Sigma_0}\mathcal{H}(x,v)\eta(x)ds \qquad (\widetilde{II})$$

with coefficients that are determined by (6.1.3).

At first, let $t\ge m$. We consider the integral identity $(\widetilde{II})$ and make the coordinate transformation $x=\varrho x'$. Let G' be the image of G, $\partial G'$ be the image of ∂G, Σ_0' be the image of Σ_0, and $z(x')=v(\varrho x')$. We have $dx=\varrho^n dx'$, $ds=\varrho^{n-1}ds'$. Then $(\widetilde{II})$ means

$$\int\limits_{G'}\left\{\mathcal{A}_i(\varrho x',\varrho^{-1}z_{x'})\eta_{x_i'}+\varrho\mathcal{B}(\varrho x',z,\varrho^{-1}z_{x'})\eta(x')\right\}dx'$$
$$+\frac{1}{\varrho^{m-1}}\int\limits_{\Sigma_0'}\frac{\sigma(\omega)}{|x'|^{m-1}}z|z|^{m-2}\eta(x')ds'+\frac{1}{\varrho^{m-1}}\int\limits_{\partial G'}\frac{\gamma(\omega)}{|x'|^{m-1}}z|z|^{m-2}\eta(x')ds'$$
$$=\int\limits_{\partial G'}\mathcal{G}(\varrho x',z)\eta(x')ds'+\int\limits_{\Sigma_0'}\mathcal{H}(\varrho x',z)\eta(x')ds' \qquad (II)'$$

for all $\eta(x')\in\mathbf{C}^0(\overline{G'})\cap\mathbf{V}^1_{m,0}(G')$. Let us define the quantity k by

$$\begin{aligned}k=k(\varrho)=&\left(\frac{\varrho}{\varsigma}\right)^{\frac{m}{m-1+\varsigma}}\cdot\|a_0(\varrho x')\|^{\frac{1}{m-1+\varsigma}}_{\frac{p}{m},G_0^1}\\&+\left(\frac{\varrho}{\varsigma}\right)\cdot\Big\{\|\alpha(\varrho x')\|^{\frac{1}{m-1}}_{\frac{p}{m-1},G_0^1}+\left(\varrho\|b_0(\varrho x')\|_{\frac{p}{m},G_0^1}\right)^{\frac{1}{m-1}}\\&+\|\mathcal{G}(\varrho x',0)\|^{\frac{1}{m-1}}_{\infty,\Gamma_0^1}+\|\mathcal{H}(\varrho x',0)\|^{\frac{1}{m-1}}_{\infty,\Sigma_0^1}\Big\}\end{aligned} \qquad (6.4.2)$$

and set

$$\overline{z}(x')=|z(x')|+k. \qquad (6.4.3)$$

We choose $\eta(x')=\left(\frac{\varrho}{\varsigma}\right)^{m-1}\cdot z(x')\overline{z}^{t-m}(x')\zeta^m(|x'|)$, where $\zeta(|x'|)\in\mathbf{C}_0^\infty([0,1])$ is a non-negative function to be further specified, as the test function in the integral identity $(II)'$. By the chain and product rules, η is a valid test function in $(II)'$ and also

$$\eta_{x_i'} = \left(\frac{\varrho}{\varsigma}\right)^{m-1} \cdot \left(1 + (t-m)\frac{|z|}{\overline{z}}\right) \overline{z}^{t-m} z_{x_i'} \zeta^m(|x'|)$$
$$+ m\left(\frac{\varrho}{\varsigma}\right)^{m-1} \cdot z(x')\overline{z}^{t-m}(x')\zeta^{m-1}\zeta_{x_i'}.$$

Then taking into account that $0 \le |z| \le \overline{z}$ and $t \ge m$, by virtue of assumptions 1)′, 2)′, 3a)′, 4)′ and 5) we obtain

$$\int\limits_{G_0^1} a\overline{z}^{t-m}|\nabla' z|^m \zeta^m(|x'|)dx' \le \int\limits_{G_0^1} \Big\{ am\overline{z}^{t-m+1}|\nabla' z|^{m-1}\zeta^{m-1}(|x'|)|\nabla'\zeta|$$
$$+ a\mu\varsigma\overline{z}^{t-m}|\nabla' z|^m\zeta^m + (t-m+1)\left(\frac{\varrho}{\varsigma}\right)^m \cdot a_0(\varrho x')\overline{z}^{t-m+1-\varsigma}(x')\zeta^m(|x'|)$$
$$+ m\left(\frac{\varrho}{\varsigma}\right)^{m-1} \cdot |\alpha(\varrho x')| \cdot |\nabla'\zeta|\overline{z}^{t-m+1}\zeta^{m-1}(|x'|) + \frac{\varrho^m}{\varsigma^{m-1}} b_0(\varrho x')\overline{z}^{t-m+1}\zeta^m(|x'|)\Big\} dx'$$
$$+ \left(\frac{\varrho}{\varsigma}\right)^{m-1} \cdot \int\limits_{\Gamma_0^1} z(x')\mathcal{G}(\varrho x', z)\overline{z}^{t-m}(x')\zeta^m(|x'|)ds'$$
$$+ \left(\frac{\varrho}{\varsigma}\right)^{m-1} \cdot \int\limits_{\Sigma_0^1} z(x')\mathcal{H}(\varrho x', z)\overline{z}^{t-m}(x')\zeta^m(|x'|)ds'. \tag{6.4.4}$$

We estimate every term by the Young inequality with regard to $\overline{z} \ge k$:

$$am\overline{z}^{t-m+1}|\nabla' z|^{m-1}\zeta^{m-1}|\nabla'\zeta| = am\left(\overline{z}^{(t-m)\frac{m-1}{m}}|\nabla' z|^{m-1}\zeta^{m-1}\right) \cdot \left(\overline{z}^{\frac{t}{m}}|\nabla'\zeta|\right)$$
$$\le \varepsilon(m-1)a\overline{z}^{t-m}|\nabla' z|^m\zeta^m + \varepsilon^{1-m}a\overline{z}^t|\nabla'\zeta|^m, \ \forall\varepsilon > 0;$$
$$\left(\frac{\varrho}{\varsigma}\right)^m \cdot a_0(\varrho x')\overline{z}^{t-m+1-\varsigma}\zeta^m \le \frac{1}{k^{m-1+\varsigma}}\left(\frac{\varrho}{\varsigma}\right)^m \cdot a_0(\varrho x')\overline{z}^t\zeta^m;$$
$$m\left(\frac{\varrho}{\varsigma}\right)^{m-1} \cdot |\alpha(\varrho x')| \cdot |\nabla'\zeta|\overline{z}^{t-m+1}\zeta^{m-1}$$
$$= m \cdot \left(\overline{z}^{\frac{t}{m}}|\nabla'\zeta|\right) \times \left(\left(\frac{\varrho}{\varsigma}\right)^{m-1} \cdot |\alpha(\varrho x')|\overline{z}^{(t-m)\frac{m-1}{m}}\zeta^{m-1}\right)$$
$$\le (m-1)\left(\frac{\varrho}{\varsigma}\right)^m \cdot |\alpha(\varrho x')|^{\frac{m}{m-1}} \cdot \overline{z}^{t-m}\zeta^m$$
$$+ \overline{z}^t|\nabla'\zeta|^m \le \overline{z}^t|\nabla'\zeta|^m + (m-1)\left(\frac{\varrho}{k\varsigma}\right)^m \cdot |\alpha(\varrho x')|^{\frac{m}{m-1}} \cdot \overline{z}^t\zeta^m;$$
$$\frac{\varrho^m}{\varsigma^{m-1}} b_0(\varrho x')\overline{z}^{t-m+1}\zeta^m = \frac{\varrho^m}{\varsigma^{m-1}}\left(\overline{z}^t\zeta^m\right)\frac{b_0(\varrho x')}{\overline{z}^{m-1}} \le \frac{\varrho^m}{(k\varsigma)^{m-1}} b_0(\varrho x') \cdot \overline{z}^t\zeta^m.$$

Further,

$$z \cdot \mathcal{G}(\varrho x', z) = z \cdot \mathcal{G}(\varrho x', 0) + z \cdot \int_0^1 \frac{d}{d\tau} \mathcal{G}(\varrho x', \tau z) d\tau$$
$$= z \cdot \mathcal{G}(\varrho x', 0) + z^2 \cdot \int_0^1 \frac{\partial \mathcal{G}(\varrho x', \tau z)}{\partial(\tau z)} d\tau \leq \overline{z} \cdot |\mathcal{G}(\varrho x', 0)|,$$

because of $\frac{\partial \mathcal{G}(\varrho x', \tau z)}{\partial(\tau z)} \leq 0$. Therefore just as above

$$\int_{\Gamma_0^1} \left(\frac{\varrho}{\varsigma}\right)^{m-1} z \cdot \mathcal{G}(\varrho x', z) \overline{z}^{t-m} \zeta^m ds'$$
$$\leq \int_{\Gamma_0^1} \zeta^m \cdot \overline{z}^{\frac{t}{m}} \cdot \left\{ |\mathcal{G}(\varrho x', 0)| \left(\frac{\varrho}{\varsigma}\right)^{m-1} \cdot \overline{z}^{(t-m)\frac{m-1}{m}} \right\} ds'$$
$$\leq \int_{\Gamma_0^1} \left\{ \overline{z}^t + |\mathcal{G}(\varrho x', 0)|^{\frac{m}{m-1}} \left(\frac{\varrho}{\varsigma}\right)^m \overline{z}^{t-m} \right\} \zeta^m ds'$$
$$\leq \int_{\Gamma_0^1} \left\{ 1 + \left(\frac{\varrho \cdot \|\mathcal{G}(\varrho x', 0)\|_\infty^{\frac{1}{m-1}}}{k\varsigma} \right)^m \right\} \overline{z}^t \zeta^m ds'.$$

In the same way

$$\int_{\Sigma_0^1} \left(\frac{\varrho}{\varsigma}\right)^{m-1} z \cdot \mathcal{H}(\varrho x', z) \overline{z}^{t-m}(x') \zeta^m(|x'|) ds'$$
$$\leq \int_{\Sigma_0^1} \left\{ 1 + \left(\frac{\varrho \cdot \|\mathcal{H}(\varrho x', 0)\|_\infty^{\frac{1}{m-1}}}{k\varsigma} \right)^m \right\} \overline{z}^t \zeta^m ds'.$$

Now from (6.4.4) it follows that

$$(1 - \varsigma\mu) \int_{G_0^1} a \overline{z}^{t-m} |\nabla' z|^m \zeta^m dx'$$
$$\leq \int_{G_0^1} \left\{ \varepsilon(m-1) a \overline{z}^{t-m} |\nabla' z|^m \zeta^m + \varepsilon^{1-m} a \overline{z}^t |\nabla' \zeta|^m \right.$$
$$+ \overline{z}^t |\nabla' \zeta|^m + \frac{t}{k^{m-1+\varsigma}} \left(\frac{\varrho}{\varsigma}\right)^m \cdot a_0(\varrho x') \overline{z}^t \zeta^m + \varrho^m \overline{z}^t \zeta^m + \frac{\varrho^m}{(k\varsigma)^{m-1}} b_0(\varrho x') \cdot \overline{z}^t \zeta^m$$
$$\left. + (m-1) \left(\frac{\varrho}{k\varsigma}\right)^m \cdot |\alpha(\varrho x')|^{\frac{m}{m-1}} \cdot \overline{z}^t \zeta^m \right\} dx'$$

$$+\left\{1+\left(\frac{\varrho}{k\varsigma}\right)^m\cdot\left(\|\mathcal{G}(\varrho x',0)\|_{\infty,\Gamma_0^1}^{\frac{m}{m-1}}+\|\mathcal{H}(\varrho x',0)\|_{\infty,\Sigma_0^1}^{\frac{m}{m-1}}\right)\right\}\int\limits_{\Gamma_0^1\cup\Sigma_0^1}\overline{z}^t\zeta^m ds' \quad (6.4.5)$$

for all $\varepsilon>0$. By inequality (1.5.11)

$$\int\limits_{\Gamma_0^1\cup\Sigma_0^1}\overline{z}^t\zeta^m ds'\le\widetilde{c}\int\limits_{G_0^1}\left(\overline{z}^t\zeta^m+|\nabla'(\overline{z}^t\zeta^m)|\right)dx'.$$

Again, by the Young inequality, we have

$$\begin{aligned}&|\nabla'(\overline{z}^t\zeta^m)|\le t\overline{z}^{t-1}|\nabla'z|\zeta^m+m\overline{z}^t\zeta^{m-1}|\nabla'\zeta|\\&=t\left(\overline{z}^{\frac{t-m}{m}}|\nabla'z|\right)\cdot\left(\overline{z}^{t\frac{m-1}{m}}\right)\zeta^m+m\overline{z}^t\zeta^{m-1}|\nabla'\zeta|\\&\le\frac{t\delta}{m}\overline{z}^{t-m}|\nabla'z|^m\zeta^m+t\frac{m-1}{m}\delta^{\frac{1}{1-m}}\overline{z}^t\zeta^m+\left(|\nabla'\zeta|^m+(m-1)\zeta^m\right)\overline{z}^t,\ \forall\delta>0.\end{aligned}$$

From above we get

$$\int\limits_{\Gamma_0^1\cup\Sigma_0^1}\overline{z}^t\zeta^m ds' \quad (6.4.6)$$

$$\le\int\limits_{G_0^1}\left\{\frac{t\delta}{m}\overline{z}^{t-m}|\nabla'z|^m\zeta^m+t\frac{m-1}{m}\delta^{\frac{1}{1-m}}\overline{z}^t\zeta^m+\left(|\nabla'\zeta|^m+m\zeta^m\right)\overline{z}^t\right\}dx'$$

for any $\delta>0$. Thus, from (6.4.5)–(6.4.6) according to the definition (6.4.2) of the number k and choosing $\varepsilon=\frac{1-\varsigma\mu}{4(m-1)}$, $\delta=\frac{ma_*(1-\varsigma\mu)}{8\widetilde{c}t}$, we obtain

$$\frac{1-\varsigma\mu}{2}\int\limits_{G_0^1}a\overline{z}^{t-m}|\nabla'z|^m\zeta^m dx' \quad (6.4.7)$$

$$\le c_1\left(1+t^{\frac{m}{m-1}}\right)\int\limits_{G_0^1}\left(\zeta^m+|\nabla'\zeta|^m\right)\overline{z}^t dx'+c_2t\int\limits_{G_0^1}F(x')\overline{z}^t\zeta^m dx',$$

where $c_1=const\,(m,\mu,\varsigma,a_*,\widetilde{c})$, $c_2=const\,(m)$ and

$$F(x')=\frac{a_0(\varrho x')}{k^{m-1+\varsigma}}\left(\frac{\varrho}{\varsigma}\right)^m+|\alpha(\varrho x')|^{\frac{m}{m-1}}\cdot\left(\frac{\varrho}{k\varsigma}\right)^m+b_0(\varrho x')\cdot\frac{\varrho^m}{(k\varsigma)^{m-1}}. \quad (6.4.8)$$

Let us define the function

$$w(x')=a^{\frac{1}{m}}\left(\frac{m}{t}\right)\overline{z}^{\frac{t}{m}}\implies\overline{z}^t=a^{-1}\left(\frac{t}{m}\right)^m w^m,\ |\nabla'w|^m=a\overline{z}^{t-m}|\nabla'z|^m. \quad (6.4.9)$$

Then, by virtue of $1 - \varsigma\mu > 0$, (6.4.7) takes the form

$$\int_{G_0^1} |\nabla' w|^m \zeta^m dx' \tag{6.4.10}$$
$$\leq C_1 t^m \left(1 + t^{\frac{m}{m-1}}\right) \int_{G_0^1} (\zeta^m + |\nabla' \zeta|^m) w^m dx' + C_2 t^{m+1} \int_{G_0^1} F(x') w^m \zeta^m dx'.$$

The required iteration process can now be developed from (6.4.10). For this we apply the Hölder inequality

$$\int_{G_0^1} |F(x')| \cdot w^m(x') \zeta^m(x') dx' \leq \|F\|_{p/m, G_0^1} \cdot \|w\zeta\|^m_{\frac{mp}{p-m}, G_0^1}, \quad p > m, \tag{6.4.11}$$

the interpolation inequality for L_p-norms

$$\|\zeta w\|_{\frac{mp}{p-m}, G_0^1} \leq \varepsilon \|\zeta w\|_{\frac{mn}{n-m}, G_0^1} + \varepsilon^{\frac{n}{n-p}} \|\zeta w\|_{m, G_0^1}, \quad p > n > m, \ \forall \varepsilon > 0, \tag{6.4.12}$$

and the Sobolev embedding inequality

$$\|\zeta w\|^m_{\frac{mn}{n-m}, G_0^1} \leq C^* \int_{G_0^1} \left\{ (|\nabla' \zeta|^m + \zeta^m) |w|^m + \zeta^m |\nabla' w|^m \right\} dx', \quad n > m, \tag{6.4.13}$$

where C^* depends only on n, m and the domain G. From (6.4.10)–(6.4.13), we derive

$$\|\zeta w\|_{\frac{mn}{n-m}, G_0^1} \leq c_3 t (1 + t^{\frac{1}{m-1}}) \cdot \|(\zeta + |\nabla' \zeta|) w\|_{m, G_0^1}$$
$$+ c_4 t^{\frac{m+1}{m}} \|F\|^{1/m}_{p/m, G_0^1} \left(\varepsilon \|w\zeta\|_{\frac{mn}{n-m}, G_0^1} + \varepsilon^{\frac{n}{n-p}} \|\zeta w\|_{m, G_0^1} \right), \ p > n > m, \ \forall \varepsilon > 0.$$

By (6.4.2) and (6.4.8), $\|F\|^{1/m}_{p/m, G_0^1} \leq c(p, m)$. If we choose $\varepsilon = \frac{1}{2c_4} t^{-\frac{m+1}{m}} \|F\|^{-\frac{1}{m}}_{p/m, G_0^1}$, we obtain

$$\|\zeta w\|_{\frac{mn}{n-m}, G_0^1} \leq C (1 + t)^{\frac{m+1}{m} \cdot \frac{p}{p-n}} \|(\zeta + |\nabla' \zeta|) w\|_{m, G_0^1}, \quad mn > p > n > m, \tag{6.4.14}$$

where C depends only on $m, \mu, \varsigma, a_*, \widetilde{c}, n, p, \|F\|^{1/m}_{p/m, G_0^1}$ and it is independent of t. Let us recall the definition of w by (6.4.9) and $t \geq m > 1$ by (6.4.14). Finally, we establish the inequality

$$\|\zeta \cdot \overline{z}^{t/m}\|_{\frac{mn}{n-m}, G_0^1} \leq C t^{\frac{m+1}{m} \cdot \frac{p}{p-n}} \|(\zeta + |\nabla' \zeta|) \cdot \overline{z}^{t/m}\|_{m, G_0^1}, \quad mn > p > n > m. \tag{6.4.15}$$

This inequality can now be iterated to yield the desired estimate.

Let us define sets $G'_{(j)} \equiv G_0^{\varkappa+(1-\varkappa)2^{-j}}$, $j = 0, 1, 2, \dots$ for all $\varkappa \in (0,1)$. It is easy to see that $G_0^{\varkappa} \equiv G'_{(\infty)} \subset \dots \subset G'_{(j+1)} \subset G'_{(j)} \subset \dots \subset G'_{(0)} \equiv G_0^1$. Let us introduce also the sequence of cut-off functions $\zeta_j(x') \in \mathbf{C}^\infty(G'_{(j)})$ such that

$$0 \le \zeta_j(x') \le 1 \text{ in } G'_{(j)} \quad \text{and } \zeta_j(x') \equiv 1 \text{ in } G'_{(j+1)}, \quad \zeta_j(x') \equiv 0$$

for $|x'| > \varkappa + 2^{-j}(1-\varkappa)$;

$$|\nabla'\zeta_j| \le \frac{2^{j+1}}{1-\varkappa} \quad \text{for} \quad \varkappa + 2^{-j-1}(1-\varkappa) < |x'| < \varkappa + 2^{-j}(1-\varkappa)$$

and the number sequence $t_j = t\left(\frac{n}{n-m}\right)^j$, $j = 0, 1, 2, \dots$. Now we can rewrite the inequality (6.4.15) replacing $\zeta(|x'|)$ by $\zeta_j(x')$ and t by t_j. Then taking the t_j-th root, we obtain

$$\|\overline{z}\|_{t_{j+1}, G'_{(j+1)}} \le \left(\frac{C}{1-\varkappa}\right)^{m/t_j} \cdot 2^{\frac{mj}{t_j}} \cdot (t_j)^{\frac{(m+1)p}{p-n} \cdot \frac{1}{t_j}} \|\overline{z}\|_{t_j, G'_{(j)}}.$$

After iteration, we find

$$\|\overline{z}\|_{t_{j+1}, G'_{(j+1)}} \le \left(\frac{C}{1-\varkappa}\right)^{m\sum\limits_{j=0}^{\infty}\frac{1}{t_j}} \cdot 2^{m\sum\limits_{j=0}^{\infty}\frac{j}{t_j}} \cdot \left(\frac{nt}{n-m}\right)^{\frac{(m+1)p}{p-n}\sum\limits_{j=0}^{\infty}\frac{1}{t_j}} \cdot \|\overline{z}\|_{t, G_0^1}.$$

The series $\sum\limits_{j=0}^{\infty}\frac{j}{t_j}$ is convergent according to the d'Alembert ratio test, while the series $\sum\limits_{j=0}^{\infty}\frac{1}{t_j} = \frac{1}{t}\sum\limits_{j=0}^{\infty}\left(\frac{n-m}{n}\right)^j = \frac{n}{mt}$ as a geometric series. Hence we get

$$\|\overline{z}\|_{t_{j+1}, G'_{(j+1)}} \le \frac{C}{(1-\varkappa)^{n/t}}\|\overline{z}\|_{t, G_0^1}.$$

Consequently, letting $j \to \infty$, we have $\sup\limits_{x' \in G_0^{\varkappa}} \overline{z}(x') \le \frac{C}{(1-\varkappa)^{n/t}}\|\overline{z}\|_{t, G_0^1}$. Thus, by (6.4.3) and (6.4.2), we obtain

$$\begin{aligned}
&\sup_{x' \in G_0^{\varkappa}} |z(x')| \\
&\le C\Bigg\{ \|z\|_{t, G_0^1} + \left(\frac{\varrho}{\varsigma}\right)^{\frac{m}{m-1+\varsigma}} \cdot \|a_0(\varrho x')\|_{\frac{p}{m}, G_0^1}^{\frac{1}{m-1+\varsigma}} + \left(\frac{\varrho}{\varsigma}\right) \cdot \Big\langle \|\alpha(\varrho x')\|_{\frac{p}{m-1}, G_0^1}^{\frac{1}{m-1}} \\
&\quad + \left(\varrho \|b_0(\varrho x')\|_{\frac{p}{m}, G_0^1}\right)^{\frac{1}{m-1}} + \|\mathcal{G}(\varrho x', 0)\|_{\infty, \Gamma_0^1}^{\frac{1}{m-1}} + \|\mathcal{H}(\varrho x', 0)\|_{\infty, \Sigma_0^1}^{\frac{1}{m-1}} \Big\rangle \Bigg\}.
\end{aligned}$$

Returning to the variables x, v we obtain the estimate

$$\sup_{x \in G_0^{\varkappa\varrho}} |v(x)| \le C\left\{\varrho^{-n/t}\|v\|_{t, G_0^{\varrho}} + K(\varrho)\right\}, \ t \ge m, \tag{6.4.16}$$

where

$$K(\varrho) = \varrho^{\frac{m(p-n)}{p(m-1+\varsigma)}} \cdot \|a_0(x)\|^{\frac{1}{m-1+\varsigma}}_{\frac{p}{m},G_0^\varrho} + \varrho^{(1-\frac{n}{p})\frac{m}{m-1}} \|b_0(x)\|^{\frac{1}{m-1}}_{\frac{p}{m},G_0^\varrho} \tag{6.4.17}$$
$$+ \varrho^{1-\frac{n}{p}} \|\alpha(x)\|^{\frac{1}{m-1}}_{\frac{p}{m-1},G_0^\varrho} + \varrho \left(\|g(x,0)\|^{\frac{1}{m-1}}_{\infty,\Gamma_0^\varrho} + \|h(x,0)\|^{\frac{1}{m-1}}_{\infty,\Sigma_0^\varrho} \right),$$

$p > n > m$ and the constant C depends on $m, q, \varkappa$.

Let now $0 < t < m$. We put in (6.4.16) $t = m$:

$$\sup_{x \in G_0^{\varkappa\varrho}} |v(x)| \leq C \left\{ \varrho^{-n/m} \|v\|_{m,G_0^\varrho} + K(\varrho) \right\}. \tag{6.4.18}$$

Using the Young inequality with $s = \frac{m}{t}$ and $s' = \frac{m}{m-t}$ we can write

$$C\varrho^{-\frac{n}{m}} \|v\|_{m,G_0^\varrho} = C\varrho^{-\frac{n}{m}} \left(\int_{G_0^\varrho} |v|^t \cdot |v|^{m-t} \right)^{1/m} \tag{6.4.19}$$
$$\leq \left(\sup_{G_0^\varrho} |v(x)| \right)^{1-t/m} \cdot C\varrho^{-\frac{n}{m}} \|v\|^{t/m}_{t,G_0^\varrho} \leq \frac{m-t}{m} \sup_{G_0^\varrho} |v(x)| + C_1 \varrho^{-\frac{n}{t}} \|v\|_{t,G_0^\varrho}.$$

Let us define the function $\psi(s) = \sup_{x \in G_0^s} |v(x)|$. Then from (6.4.18)–(6.4.19) it follows that

$$\psi(\varkappa\varrho) \leq \frac{m-t}{m} \psi(\varrho) + C_1 \varrho^{-\frac{n}{t}} \|v\|_{t,G_0^\varrho} + CK(\varrho), \quad \varkappa \in (0,1). \tag{6.4.20}$$

Further, we apply Lemma 1.24: setting $r = \varkappa\varrho$, $R = \varrho$, $\delta = 1 - \frac{t}{m}$, $\alpha = \frac{n}{t}$, $A = C_1 \|v\|_{t,G_0^\varrho}$, $B = CK(\varrho)$ from (6.4.20) we obtain the validity of estimate (6.4.16) in the case $0 < t < m$.

Returning to the variable u by (6.1.1), from (6.4.16)–(6.4.17) we get the estimate (6.4.1). Thus, the proof of Theorem 6.6 is complete. □

6.5 Integral estimates

Now we will derive a global estimate for the Dirichlet integral.

Theorem 6.7. *Let $u(x)$ be a weak solution of the problem (QL) and assumptions 1), 3a), 4) and 5) with $\nu_0 > 0$ be satisfied. In addition, we assume that $M_0 = \max_{x \in \overline{G}} |u(x)|$ is known. If $a_0(x) \in \mathbf{L}_1(G)$, $b_0(x) \in \mathbf{L}_1(G)$, $h(x,0) \in L_1(\Sigma_0)$, $g(x,0) \in L_1(\partial G)$, then the inequality*

$$\int_G a|u|^{\frac{qm}{m-1}} |\nabla u|^m dx + \int_{\Sigma_0} \frac{\sigma(\omega)}{r^{m-1}} |u|^{\frac{m}{m-1}(q+m-1)} ds + \int_{\partial G} \frac{\gamma(\omega)}{r^{m-1}} |u|^{\frac{m}{m-1}(q+m-1)} ds$$

$$\leq c(M_0, a_*, \nu_0, q, m, \mu, n, meas\, G) \times \Big(\int\limits_G (a_0(x) + b_0(x))\, dx + \int\limits_{\Sigma_0} |h(x,0)| ds + \int\limits_{\partial G} |g(x,0)| ds \Big) \tag{6.5.1}$$

holds.

Proof. Using the function change (6.1.1) and putting $\eta(x) = v(x)$ in the integral identity $(\widetilde{II})$ for $v(x)$ (see §6.4), we have

$$\int\limits_G \Big\langle \mathcal{A}_i(x, v_x) v_{x_i} + \mathcal{B}(x, v, v_x) v \Big\rangle dx + \int\limits_{\partial G} \frac{\gamma(\omega)}{r^{m-1}} |v|^m ds + \int\limits_{\Sigma_0} \frac{\sigma(\omega)}{r^{m-1}} |v|^m ds$$

$$= \int\limits_{\partial G} \mathcal{G}(x, v) v(x) ds + \int\limits_{\Sigma_0} \mathcal{H}(x, v) v(x) ds.$$

According to assumptions 1)′, 3a)′, 4)′, since $\varsigma^{m-1}(1 - \varsigma\mu) < 1$ by (6.1.1), we obtain

$$\varsigma^{m-1}(1 - \varsigma\mu) \Big\{ \int\limits_G a |\nabla v|^m dx + \int\limits_{\Sigma_0} \frac{\sigma(\omega)}{r^{m-1}} |v|^m ds + \int\limits_{\partial G} \frac{\gamma(\omega)}{r^{m-1}} |v|^m ds \Big\}$$

$$\leq \int\limits_G |v| b_0(x) dx + \frac{1}{\varsigma} \int\limits_G |v|^{1-\varsigma} a_0(x) dx + \int\limits_{\Sigma_0} |h(x,0)| \cdot |v| ds + \int\limits_{\partial G} |g(x,0)| \cdot |v| ds.$$

From $M_0 = \sup\limits_G |u(x)|$, by the change (6.1.1), it follows that $|v(x)| \leq M_0^{\frac{1}{\varsigma}}$. Therefore we get

$$\int\limits_G a |\nabla v|^m dx + \int\limits_{\Sigma_0} \frac{\sigma(\omega)}{r^{m-1}} |v|^m ds + \int\limits_{\partial G} \frac{\gamma(\omega)}{r^{m-1}} |v|^m ds$$

$$\leq c(M_0, m, q, \mu, \text{meas}\, G) \times \Big(\int\limits_G (a_0(x) + b_0(x))\, dx + \int\limits_{\Sigma_0} |h(x,0)| ds + \int\limits_{\partial G} |g(x,0)| ds \Big). \tag{6.5.2}$$

Returning to the function $u(x)$, by (6.1.1), we obtain the desired estimate (6.5.1). □

Further, we establish a local integral weighted estimate.

Theorem 6.8. *Let $u(x)$ be a weak solution of the problem (QL) and $\vartheta(m)$ be the smallest positive eigenvalue of (NEVP). Let us assume that $M_0 = \max\limits_{x \in \overline{G}} |u(x)|$ is known and assumptions of Theorem 6.7 and assumption 7) are satisfied. In* *addition, assume that there exists a real number $k_s \geq 0$ defined by* (6.1.4)*. Then there exist $d \in (0,1)$ and a constant $c > 0$ independent of u and depending only on $m, n, s, q, d, \vartheta(m), k_1, k_s, \operatorname{meas}\Omega$ and M_0 such that, for any $\varrho \in (0,d)$,*

$$\int\limits_{G_0^\varrho} a|u|^{\frac{qm}{m-1}}|\nabla u|^m dx + \int\limits_{\Sigma_0^\varrho} \frac{\sigma(\omega)}{r^{m-1}}|u|^{\frac{m}{m-1}(q+m-1)}ds + \int\limits_{\Gamma_0^\varrho} \frac{\gamma(\omega)}{r^{m-1}}|u|^{\frac{m}{m-1}(q+m-1)}ds \leq c\psi^m(\varrho), \tag{6.5.3}$$

where $\psi(\varrho)$ is defined by (6.1.6) *with* (2.4.3)*.*

Proof. We perform the change (6.1.1). By virtue of Theorem 6.7, we have that

$$V(\varrho) = \int\limits_{G_0^\varrho} a|\nabla v|^m dx + \int\limits_{\Sigma_0^\varrho} \frac{\sigma(\omega)}{r^{m-1}}|v|^m ds + \int\limits_{\Gamma_0^\varrho} \frac{\gamma(\omega)}{r^{m-1}}|v|^m ds < \infty, \ \varrho \in (0,d). \tag{6.5.4}$$

Therefore we can set $\eta(x) = v(x)$ in the identity (6.1.2):

$$\int\limits_{G_0^\varrho} \Big\langle \mathcal{A}_i(x, v_x)v_{x_i} + \mathcal{B}(x, v, v_x)v(x)\Big\rangle dx + \int\limits_{\Sigma_0^\varrho} \frac{\sigma(\omega)}{r^{m-1}}|v|^m ds + \int\limits_{\Gamma_0^\varrho} \frac{\gamma(\omega)}{r^{m-1}}|v|^m ds$$
$$= \int\limits_{\Omega_\varrho} \mathcal{A}_i(x, v_x)\cos(r, x_i)\cdot v(x)d\Omega_\varrho + \int\limits_{\Gamma_0^\varrho} \mathcal{G}(x, v)\cdot v(x)ds + \int\limits_{\Sigma_0^\varrho} \mathcal{H}(x, v)\cdot v(x)ds. \tag{6.5.5}$$

By assumptions 1)′, 3a)′, 4)′, 5), 7)′ and $\mu\varsigma < 1$, we get

$$(1 - \varsigma\mu)\varsigma^{m-1}V(\varrho) \leq \int\limits_{G_0^\varrho} |v|b_0(x)dx + \frac{1}{\varsigma}\int\limits_{G_0^\varrho} |v|^{1-\varsigma}a_0(x)dx$$
$$+ \varsigma^{m-1}\mathcal{A}(\varrho)\int\limits_{\Omega_\varrho} a|v|\cdot|\nabla v|^{m-1}d\Omega_\varrho + \varsigma^{m-1}\int\limits_{\Omega_\varrho} a|\nabla v|^{m-2}\cdot v\frac{\partial v}{\partial r}d\Omega_\varrho$$
$$+ \int\limits_{\Sigma_0^\varrho} |h(x,0)|\cdot|v|ds + \int\limits_{\Gamma_0^\varrho} |g(x,0)|\cdot|v|ds. \tag{6.5.6}$$

By Lemma 2.11, from (6.5.6) it follows that

$$(1 - \varsigma\mu)V(\varrho) \leq \frac{\Xi(m)}{m\vartheta^{\frac{1}{m}}(m)}\varrho V'(\varrho) + \mathcal{A}(\varrho)\int\limits_{\Omega_\varrho} a|v|\cdot|\nabla v|^{m-1}d\Omega_\varrho$$
$$+ \frac{1}{\varsigma^m}\int\limits_{G_0^\varrho} |v|^{1-\varsigma}a_0(x)dx + \varsigma^{1-m}\int\limits_{G_0^\varrho} |v|b_0(x)dx + \varsigma^{1-m}\int\limits_{\Sigma_0^\varrho} |h(x,0)|\cdot|v|ds$$

$$+\varsigma^{1-m}\int\limits_{\Gamma_0^\varrho}|g(x,0)|\cdot|v|ds. \tag{6.5.7}$$

We estimate every term on the right-hand side in (6.5.7). By the Hölder inequality for integrals,

$$\int\limits_{\Omega_\varrho} a|v|\cdot|\nabla v|^{m-1}d\Omega_\varrho \le \left(\int\limits_{\Omega_\varrho} a|v|^m d\Omega_\varrho\right)^{\frac{1}{m}}\cdot\left(\int\limits_{\Omega_\varrho} a|\nabla v|^m d\Omega_\varrho\right)^{\frac{m-1}{m}}. \tag{6.5.8}$$

Because of the inequality $|\nabla_\omega v|\le\varrho|\nabla v|$, the inequality $(W)_m$ and formula (2.4.4), we have

$$\begin{aligned}\int\limits_{\Omega_\varrho} a|v|^m d\Omega_\varrho &\le \frac{\varrho^{n-1}}{\vartheta}\Big\{\int\limits_{\Omega} a|\nabla_\omega v|^m d\Omega+\int\limits_{\sigma_0}\sigma(\omega)|v|^m d\sigma+\int\limits_{\partial\Omega}\gamma(\omega)|v|^m d\sigma\Big\}\\ &\le \frac{\varrho^m}{\vartheta}V'(\varrho).\end{aligned} \tag{6.5.9}$$

From (6.5.8)–(6.5.9) it follows that

$$\int\limits_{\Omega_\varrho} a|v|\cdot|\nabla v|^{m-1}d\Omega_\varrho \le \frac{\varrho}{\vartheta^{\frac{1}{m}}}V'(\varrho). \tag{6.5.10}$$

By the Young inequality and (2.2.2),

$$\begin{aligned}\frac{1}{\varsigma}\int\limits_{G_0^\varrho}|v|^{1-\varsigma}a_0(x)dx &= \frac{1}{\varsigma}\int\limits_{G_0^\varrho}\left(r^{\frac{(1-\varsigma)(1-\varsigma-m)}{m}}|v|^{1-\varsigma}\right)\cdot\left(r^{\frac{(1-\varsigma)(m+\varsigma-1)}{m}}a_0(x)\right)dx\\ &\le \int\limits_{G_0^\varrho}\left(\frac{1-\varsigma}{m\varsigma}r^{1-\varsigma-m}|v|^m+\frac{m+\varsigma-1}{m\varsigma}r^{1-\varsigma}|a_0(x)|^{\frac{m}{m+\varsigma-1}}\right)dx\\ &\le \frac{1-\varsigma}{a_*m\varsigma\vartheta(m)}\varrho^{1-\varsigma}V(\varrho)+\frac{m+\varsigma-1}{m\varsigma}\int\limits_{G_0^\varrho}r^{1-\varsigma}|a_0(x)|^{\frac{m}{m+\varsigma-1}}dx;\\ \int\limits_{G_0^\varrho}|v|b_0(x)dx &= \int\limits_{G_0^\varrho}\left(r^{-\frac{1}{m}}|v|\right)\left(r^{\frac{1}{m}}b_0(x)\right)dx\\ &\le \int\limits_{G_0^\varrho}\left(\frac{1}{m}r^{-1}|v|^m+\frac{m-1}{m}r^{\frac{1}{m-1}}|b_0(x)|^{\frac{m}{m-1}}\right)dx\\ &\le \frac{1}{a_*m\vartheta(m)}\varrho^{m-1}V(\varrho)+\frac{m}{m-1}\int\limits_{G_0^\varrho}r^{\frac{1}{m-1}}|b_0(x)|^{\frac{m}{m-1}}dx\end{aligned} \tag{6.5.11}$$

$$\int_{\Sigma_0^\varrho} |h(x,0)| \cdot |v| ds + \int_{\Gamma_0^\varrho} |g(x,0)| \cdot |v| ds \le \frac{1}{m}\Big(\int_{\Sigma_0^\varrho} |v|^m ds + \int_{\Gamma_0^\varrho} |v|^m ds\Big)$$
$$+ \frac{m-1}{m}\Big(\int_{\Sigma_0^\varrho} |h(x,0)|^{\frac{m}{m-1}} ds + \int_{\Gamma_0^\varrho} |g(x,0)|^{\frac{m}{m-1}} ds\Big) \le \frac{\varrho^{m-1}}{m\nu_0}\Big(\int_{\Sigma_0^\varrho} \frac{\sigma(\omega)}{r^{m-1}} |v|^m ds$$
$$+ \int_{\Gamma_0^\varrho} \frac{\gamma(\omega)}{r^{m-1}} |v|^m ds\Big) + \frac{m-1}{m}\Big(\int_{\Sigma_0^\varrho} |h(x,0)|^{\frac{m}{m-1}} ds + \int_{\Gamma_0^\varrho} |g(x,0)|^{\frac{m}{m-1}} ds\Big)$$
$$\le \frac{\varrho^{m-1}}{m\nu_0} V(\varrho) + \frac{m-1}{m}\Big(\int_{\Sigma_0^\varrho} |h(x,0)|^{\frac{m}{m-1}} ds + \int_{\Gamma_0^\varrho} |g(x,0)|^{\frac{m}{m-1}} ds\Big), \tag{6.5.12}$$

by assumption 5) and (6.5.4).

Thus, from (6.5.7)–(6.5.12) it follows that

$$(1-\varsigma\mu)\,(1-\delta(\varrho))\,V(\varrho) \le \frac{\Xi(m)}{m\vartheta^{\frac{1}{m}}(m)}\Big(1+\widetilde{\mathcal{A}}(\varrho)\Big)\,\varrho V'(\varrho)$$
$$+ \frac{m\varsigma^{1-m}}{m-1}\int_{G_0^\varrho} r^{\frac{1}{m-1}} |b_0(x)|^{\frac{m}{m-1}} dx + \frac{m+\varsigma-1}{m\varsigma^m}\int_{G_0^\varrho} r^{1-\varsigma} |a_0(x)|^{\frac{m}{m+\varsigma-1}} dx$$
$$+ \frac{m-1}{m\varsigma^{m-1}}\Big(\int_{\Sigma_0^\varrho} |h(x,0)|^{\frac{m}{m-1}} ds + \int_{\Gamma_0^\varrho} |g(x,0)|^{\frac{m}{m-1}} ds\Big), \tag{6.5.13}$$

where $\delta(\varrho) = \text{const}\,(m,\mu,a_*,q,\nu_0,\vartheta(m)) \cdot \Big(\varrho^{m-1} + \frac{q}{q+m-1}\varrho^{\frac{q}{q+m-1}}\Big)$ and $\widetilde{\mathcal{A}}(\varrho) = \frac{m}{\Xi(m)}\mathcal{A}(\varrho)$. We observe that

$$\int_0 \frac{\delta(\varrho)}{\varrho} d\varrho < \infty, \qquad \int_0 \frac{\widetilde{\mathcal{A}}(\varrho)}{\varrho} d\varrho < \infty. \tag{6.5.14}$$

Thus, from (6.5.13), by virtue of assumption (6.1.4), we derive the Cauchy problem (CP) from §1.7 with

$$\mathcal{P}(\varrho) = \frac{1}{\varrho} \cdot \frac{1-\delta(\varrho)}{1+\widetilde{\mathcal{A}}(\varrho)} \cdot \frac{m\vartheta^{\frac{1}{m}}(m)}{\Xi(m)}(1-\varsigma\mu), \quad \mathcal{Q}(\varrho) = k\varrho^{ms-1}, \quad \mathcal{N}(\varrho) \equiv 0;$$
$$k = \text{const}\,(k_s, m, q, \vartheta(m)). \tag{6.5.15}$$

Now we have:

$$V(d) = \int_{G_0^d} a|\nabla v|^m dx + \int_{\Sigma_0^d} \frac{\sigma(\omega)}{r^{m-1}} |v|^m ds + \int_{\Gamma_0^d} \frac{\gamma(\omega)}{r^{m-1}} |v|^m ds$$

$$\leq \int\limits_G a|\nabla v|^m dx + \int\limits_{\Sigma_0} \frac{\sigma(\omega)}{r^{m-1}}|v|^m ds + \int\limits_{\partial G} \frac{\gamma(\omega)}{r^{m-1}}|v|^m ds$$
$$\leq c(M_0, m, q, \mu, \operatorname{meas} G) \cdot V_0,$$
$$V_0 \equiv \int\limits_G (a_0(x) + b_0(x))\, dx + \int\limits_{\Sigma_0} |h(x,0)| ds + \int\limits_{\partial G} |g(x,0)| ds,$$

by virtue of (6.5.2). The solution of problem (CP) is determined by (1.7.1) from Theorem 1.21. Direct calculations give:

$$-\int\limits_{\varrho}^{\tau} \mathcal{P}(\xi) d\xi = -\frac{m\vartheta^{\frac{1}{m}}(m)(1-\varsigma\mu)}{\Xi(m)} \int\limits_{\varrho}^{\tau} \frac{1}{\xi} \cdot \frac{1-\delta(\xi)}{1+\widetilde{\mathcal{A}}(\xi)} d\xi$$
$$= -\frac{m\vartheta^{\frac{1}{m}}(m)(1-\varsigma\mu)}{\Xi(m)} \int\limits_{\varrho}^{\tau} \frac{1}{\xi} \cdot \left(1 - \frac{\delta(\xi)+\widetilde{\mathcal{A}}(\xi)}{1+\widetilde{\mathcal{A}}(\xi)}\right) d\xi$$
$$\leq \frac{m\vartheta^{\frac{1}{m}}(m)(1-\varsigma\mu)}{\Xi(m)} \left(\ln\frac{\varrho}{\tau} + \int\limits_0^d \frac{\delta(\xi)+\widetilde{\mathcal{A}}(\xi)}{\xi} d\xi\right) \quad \Longrightarrow$$
$$\exp\left(-\int\limits_{\varrho}^{\tau} \mathcal{P}(\xi) d\xi\right)$$
$$\leq \left(\frac{\varrho}{\tau}\right)^{\frac{m\vartheta^{\frac{1}{m}}(m)(1-\varsigma\mu)}{\Xi(m)}} \cdot \exp\left(\frac{m\vartheta^{\frac{1}{m}}(m)(1-\varsigma\mu)}{\Xi(m)} \int\limits_0^d \frac{\delta(\xi)+\widetilde{\mathcal{A}}(\xi)}{\xi} d\xi\right);$$
$$\int\limits_{\varrho}^{d} \mathcal{Q}(\tau) \exp\left(-\int\limits_{\varrho}^{\tau} \mathcal{P}(\xi) d\xi\right) d\tau \leq kC_1 \varrho^{\frac{m\vartheta^{\frac{1}{m}}(m)(1-\varsigma\mu)}{\Xi(m)}} \int\limits_{\varrho}^{d} \tau^{ms-1} \tau^{-\frac{m\vartheta^{\frac{1}{m}}(m)(1-\varsigma\mu)}{\Xi(m)}} d\tau$$
$$= kC_1 \varrho^{\frac{m\vartheta^{\frac{1}{m}}(m)(1-\varsigma\mu)}{\Xi(m)}}$$
$$\times \begin{cases} \dfrac{d^{m\left(s-\frac{\vartheta^{\frac{1}{m}}(m)(1-\varsigma\mu)}{\Xi(m)}\right)} - \varrho^{m\left(s-\frac{\vartheta^{\frac{1}{m}}(m)(1-\varsigma\mu)}{\Xi(m)}\right)}}{m\left(s-\frac{\vartheta^{\frac{1}{m}}(m)(1-\varsigma\mu)}{\Xi(m)}\right)}, & s \neq \dfrac{\vartheta^{\frac{1}{m}}(m)(1-\varsigma\mu)}{\Xi(m)}; \\ \ln\frac{d}{\varrho}, & s = \dfrac{\vartheta^{\frac{1}{m}}(m)(1-\varsigma\mu)}{\Xi(m)}, \end{cases}$$

where $C_1 = \exp\left(\frac{m\vartheta^{\frac{1}{m}}(m)(1-\varsigma\mu)}{\Xi(m)} \int\limits_0^d \frac{\delta(\xi)+\widetilde{\mathcal{A}}(\xi)}{\xi} d\xi\right)$.

$$V_0 \cdot \exp\left(-\int\limits_{\varrho}^{d} \mathcal{P}(\xi) d\xi\right) \leq V_0 C_1 \left(\frac{\varrho}{d}\right)^{\frac{m\vartheta^{\frac{1}{m}}(m)(1-\varsigma\mu)}{\Xi(m)}} \quad \Longrightarrow$$

$$V(\varrho) \le cC_1(V_0+k)\cdot \begin{cases} \varrho^{\frac{m\vartheta^{\frac{1}{m}}(m)(1-\varsigma\mu)}{\Xi(m)}}, & s > \frac{\vartheta^{\frac{1}{m}}(m)(1-\varsigma\mu)}{\Xi(m)}; \\ \varrho^{\frac{m\vartheta^{\frac{1}{m}}(m)(1-\varsigma\mu)}{\Xi(m)}} \ln\frac{d}{\varrho}, & s = \frac{\vartheta^{\frac{1}{m}}(m)(1-\varsigma\mu)}{\Xi(m)}; \\ \varrho^{ms}, & s < \frac{\vartheta^{\frac{1}{m}}(m)(1-\varsigma\mu)}{\Xi(m)}, \end{cases} \tag{6.5.16}$$

where $c = const(m,s,\vartheta(m))$. Thus we have proved the statement of our theorem. □

6.6 The power modulus of continuity at the conical point for weak solutions

Proof of Theorem 6.3. We consider the function $\psi(\varrho)$, $0<\varrho<d$ that is determined by (6.1.6). By Theorem 6.6 about the local bound of the weak solution modulus we have (see (6.4.17)–(6.4.18))

$$\sup_{x\in G_0^{\varrho/2}} |v(x)|^m \le C\left\{\varrho^{-n}\|v\|^m_{m,G_0^\varrho} + K^m(\varrho)\right\}. \tag{6.6.1}$$

Because of inequality $(H-W)_m$ according to notation (2.4.1), we get

$$\varrho^{-n}\int_{G_0^\varrho} a|v(x)|^m dx \le \frac{\varrho^{m-n}}{\vartheta(m)}V(\varrho) \le C\varrho^{m-n}\psi^m(\varrho), \tag{6.6.2}$$

by inequality (6.5.16). From (6.6.1)–(6.6.2) it follows that

$$\sup_{x\in G_0^{\varrho/2}} |v(x)| \le C\left\{\varrho^{1-\frac{n}{m}}\psi(\varrho) + K(\varrho)\right\}. \tag{6.6.3}$$

Now, from (6.4.17), by virtue of assumption (6.1.4), it follows that $K(\varrho) \le K\varrho^{1-\frac{n}{m}}\psi(\varrho)$. Therefore, hence and from (6.6.3) we get

$$|v(x)| \le \widetilde{C}_0\varrho^{1-\frac{n}{m}}\psi(\varrho),\ x\in G_{\varrho/4}^{\varrho/2}.$$

Setting $|x| = \frac{\varrho}{3}$ we obtain

$$|v(x)| \le C_0|x|^{1-\frac{n}{m}}\psi(|x|),\ x\in G_0^d. \tag{6.6.4}$$

By (6.1.1), from (6.6.4) we derive the first required estimate (6.1.7).

Repeating verbatim the second part of the proof of Theorem 6.9 (see Appendix 6.7 below) we obtain the inequality

$$|\nabla v(x)| \le C_1|x|^{-\frac{n}{m}}\psi(|x|),\ x\in G_0^d. \tag{6.6.5}$$

Further, since, due to (6.1.1),

$$|\nabla u(x)| \le |v|^{\varsigma-1}\cdot|\nabla v(x)|,$$

from (6.6.4)–(6.6.5) we derive the second required estimate (6.1.8). □

6.7 Appendix

If a neighborhood of the conical point is a **convex** set, then we can construct the barrier function and apply the comparison principle for the estimating of weak solutions of the transmission problem.

Assumptions.

1)–7) *from Section* 6.1 *with*

$$\nu_0 > \frac{a^* 2^m}{\left(\cos\frac{\omega_0}{2}\right)^{m-1}} \max\left(1; 2^{m-2}\right). \tag{6.7.1}$$

Functions $a_i(x,u,\xi)$ and $b(x,u,\xi)$ are continuously differentiable with respect to the x,u,ξ variables in $\mathfrak{M}_{d,M_0} = \overline{G_0^d} \times [-M_0, M_0] \times \mathbb{R}^n$ and satisfy in $\mathfrak{M}_{d,M_0}$ the following inequalities:

8) $\frac{\partial a_i(x,u,\xi)}{\partial \xi_j} p_i p_j \geq a|u|^q|\xi|^{m-2}p^2,\ \forall p \in \mathbb{R}^n \setminus \{0\}$;

9) $\left|\frac{\partial a_i(x,u,\xi)}{\partial x_i} - b(x,u,\xi)\right| \leq \mathcal{B}(r)|u|^{q-1}|\xi|^m + b_1 r^{s-2},\ s > 1$;

10) $\sqrt{\sum\limits_{i=1}^{n}\left|\frac{\partial b(x,u,\xi)}{\partial \xi_i}\right|^2} \leq a\mu|u|^{q-1}|\xi|^{m-1} + b_0(x)$;

11) $\frac{\partial b(x,u,\xi)}{\partial u} \geq a|u|^{q-2}|\xi|^m$;

12) $\alpha(x) + b_0(x) + |h(x,0)| + |g(x,0)| \leq k_1|x|^{s-1}$,

where $\mathcal{B}(r)$ is a non-negative monotonically increasing function that is continuous at zero with $\mathcal{B}(0) = 0$.

After the function change (6.1.1) these assumptions take the following form:

8)′ $\frac{\partial \mathcal{A}_i(x,v_x)}{\partial v_{x_j}} p_i p_j \geq a\varsigma^{m-1}|\nabla v|^{m-2}p^2,\ \forall p \in \mathbb{R}^n \setminus \{0\}$;

9)′ $\left|\frac{\partial \mathcal{A}_i(x,\eta)}{\partial x_i} - \mathcal{B}(x,v,\eta)\right| \leq \mathcal{B}(r)v^{-1}|\eta|^m + b_1 r^{s-2}$;

10)′ $\sqrt{\sum\limits_{i=1}^{n}\left|\frac{\partial \mathcal{B}(x,v,\eta)}{\partial \eta_i}\right|^2} \leq a\mu\varsigma^{m-1}|v|^{-1}\cdot|\eta|^{m-1} + b_0(x)$;

11)′ $\frac{\partial \mathcal{B}(x,v,\eta)}{\partial v} \geq a\varsigma^{m-1}|v|^{-2}|\eta|^m$.

6.7.1 The barrier function. The preliminary estimate of the solution modulus

Now we can estimate $|u(x)|$ and $|\nabla u(x)|$ for (QL) in a neighborhood of the conical point.

Theorem 6.9. *Let G_0^d be a convex rotational cone. Let u be a weak solution of the problem (QL), assumptions 1)–12) be satisfied and $M_0 = \max_{x\in\overline{G}} |u(x)|$ be known. Then there exist $\varkappa \in (0,1)$, $d \in (0,1)$ and a constant $C_0 > 0$ independent of u such that*

$$|u(x)| \le C_0 |x|^{\frac{(m-1)(1+\varkappa)}{q+m-1}}, \quad \forall x \in G_0^d. \tag{6.7.2}$$

In addition, if coefficients of the problem (QL) satisfy such conditions which guarantee the local a priori estimate $|\nabla u|_{0,G'} \le M_1$ for any smooth $G' \subset\subset \overline{G} \setminus \{\mathcal{O}\}$ (see for example §4 in [6] or [64]), then there exists a constant $C_1 > 0$ independent of u such that

$$|\nabla u(x)| \le C_1 |x|^{\frac{(m-1)(1+\varkappa)}{q+m-1}-1}, \quad \forall x \in G_0^d. \tag{6.7.3}$$

Proof. We perform the function change (6.1.1) and consider the function $v(x)$. For the proof we construct the barrier function $w(x)$ and apply the comparison principle to $v(x)$ and $w(x)$. We shall show that $Q(Aw,\eta) \ge 0$ for all non-negative $\eta \in \mathbf{C}^0(\overline{G_0^d}) \cap \mathbf{V}^1_{m,0}(G_0^d)$ and some $A > 0$. We consider operator Q, which is defined by (6.2.1). Integrating by parts in the first integral, we have:

$$\begin{aligned}
Q(Aw,\eta) \equiv \int\limits_{G_0^d} \Big\langle -\frac{d\mathcal{A}_i(x, A\nabla w)}{dx_i} + \mathcal{B}(x, Aw, A\nabla w)\Big\rangle \eta(x)dx \\
+ \int\limits_{\Gamma_0^d} \Big\langle \mathcal{A}_i(x, A\nabla w)\cos(\overrightarrow{n}, x_i) + A^{m-1}\frac{\gamma(\omega)}{r^{m-1}} w|w|^{m-2} - \mathcal{G}(x, Aw)\Big\rangle \eta(x)ds \\
+ \int\limits_{\Sigma_0^d} \Big\langle [\mathcal{A}_i(x, A\nabla w)\cos(\overrightarrow{n}, x_i)] \\
+ A^{m-1}\frac{\sigma(\omega)}{r^{m-1}} w|w|^{m-2} - \mathcal{H}(x, Aw)\Big\rangle \eta(x)ds.
\end{aligned} \tag{6.7.4}$$

Let us recall functions $\mathcal{A}_i$, $\mathcal{B}$, $\mathcal{G}$, $\mathcal{H}$, which were determined by (6.1.3) with ς from (6.1.1).

Let $(x, y, x') \in \mathbb{R}^n$, where $x = x_1, y = x_2, x' = (x_3, \dots, x_n)$. In half-plane $\{x_1 \ge 0\}$ we consider the cone K with the vertex in $\mathcal{O}$ such that $K \supset G_0^d$ (we recall that $G_0^d \subset \{x_1 \ge 0\}$). Let ∂K be the lateral surface of K and let $x = \pm hy$, where $h = \cot\frac{\omega_0}{2}$, $0 < \omega_0 < \pi$, be the equation of $\partial K \cap yOx = \Gamma_\pm$ such that in the interior of K the inequality $x > h|y|$ holds. We introduce the **barrier** function:

$$w(x; y, x') \equiv x^{\varkappa-1}(x^2 - h^2y^2) + Bx^{\varkappa+1} \ \text{ with some } \varkappa \in (0;1),\ B \ge 1. \tag{6.7.5}$$

Step 1. Firstly, we show that in G_0^d,

$$-\frac{d\mathcal{A}_i(x, A\nabla w)}{dx_i} + \mathcal{B}(x, Aw, A\nabla w) \ge 0 \tag{6.7.6}$$

for some $A > 0$, $B > 1, 0 < d \ll 1$.

By direct calculations:

$$\begin{aligned} w_x &= (1+\varkappa)(1+B)x^{\varkappa} - (\varkappa-1)h^2y^2x^{\varkappa-2}, \quad w_y = -2h^2yx^{\varkappa-1}, \\ w_{xx} &= \varkappa(1+\varkappa)(1+B)x^{\varkappa-1} - (1-\varkappa)(2-\varkappa)h^2y^2x^{\varkappa-3}, \\ w_{xy} &= 2h^2(1-\varkappa)yx^{\varkappa-2}, \quad w_{yy} = -2h^2x^{\varkappa-1} \end{aligned} \tag{6.7.7}$$

we get for $(x, y) \in G_0^d$,

$$B\left(\frac{h}{\sqrt{1+h^2}}\right)^{\varkappa+1} \cdot r^{\varkappa+1} \le w(x,y) \le (1+B)r^{\varkappa+1}; \tag{6.7.8}$$

$$\left(\frac{h}{\sqrt{1+h^2}}\right)^{\varkappa} \cdot r^{\varkappa} \le |\nabla w| \le 2(1+h+B)r^{\varkappa}, \tag{6.7.9}$$

as well as

$$\begin{aligned} -\frac{\partial \mathcal{A}_i(x,\nabla w)}{\partial w_{x_j}} w_{x_ix_j} &= 2h^2\frac{\partial A_2}{\partial w_y}x^{\varkappa-1} - 2h^2(1-\varkappa)\left(\frac{\partial A_1}{\partial w_y} + \frac{\partial A_2}{\partial w_x}\right)yx^{\varkappa-2} \\ &\quad + \left\langle (1-\varkappa)(2-\varkappa)h^2y^2x^{\varkappa-3} - \varkappa(1+\varkappa)(1+B)x^{\varkappa-1}\right\rangle \frac{\partial A_1}{\partial w_x} \\ &\equiv \phi(\varkappa)\cdot x^{\varkappa-1}. \end{aligned} \tag{6.7.10}$$

Because of the ellipticity condition 8)′,

$$\begin{aligned} \phi(0) &= 2h^2\left\langle \frac{y^2}{x^2}\cdot\frac{\partial A_1}{\partial w_x} - \frac{y}{x}\cdot\left(\frac{\partial A_1}{\partial w_y} + \frac{\partial A_2}{\partial w_x}\right) + \frac{\partial A_2}{\partial w_y}\right\rangle \\ &\ge 2ah^2\varsigma^{m-1}|\nabla w|^{m-2}\left(1+\frac{y^2}{x^2}\right). \end{aligned} \tag{6.7.11}$$

$\phi(\varkappa)$ is a square function with negative leading coefficient $\frac{y^2h^2}{x^2} - 1 - B < -B$ and according to (6.7.11) there exists a number $\varkappa_0 > 0$ such that

$$\phi(\varkappa) \ge ah^2\varsigma^{m-1}|\nabla w|^{m-2}\left(1+\frac{y^2}{x^2}\right)$$

for $\varkappa \in [0; \varkappa_0]$. Therefore, from (6.7.10) it follows that

$$\begin{aligned} -\frac{\partial \mathcal{A}_i(x, A\nabla w)}{\partial (Aw_{x_j})}Aw_{x_ix_j} &\ge a_*h^2(A\varsigma)^{m-1}|\nabla w|^{m-2}r^{\varkappa-1}(\cos\omega)^{\varkappa-3} \\ &\ge a_*h^2(A\varsigma)^{m-1}|\nabla w|^{m-2}r^{\varkappa-1}, \end{aligned} \tag{6.7.12}$$

by virtue of $\varkappa < 1$ and $\cos\omega \le 1$.

Hence, by the assumption 9)′, we obtain in G_0^d,

$$
\begin{aligned}
&-\frac{d\mathcal{A}_i(x, A\nabla w)}{dx_i} + \mathcal{B}(x, Aw, A\nabla w) \\
&\geq a_* h^2 (A\varsigma)^{m-1} |\nabla w|^{m-2} r^{\varkappa-1} - A^{m-1} w^{-1} |\nabla w|^m \mathcal{B}(r) - b_1 r^{s-2} \\
&= A^{m-1} |\nabla w|^{m-2} \left\langle a_* h^2 \varsigma^{m-1} r^{\varkappa-1} - \mathcal{B}(r) w^{-1} |\nabla w|^2 \right\rangle - b_1 r^{s-2} \\
&\geq A^{m-1} |\nabla w|^{m-2} r^{\varkappa-1} \left\langle a_* h^2 \varsigma^{m-1} - \mathcal{B}(d) \frac{4(1+h+B)^2}{B\left(\frac{h}{\sqrt{1+h^2}}\right)^{\varkappa+1}} \right\rangle - b_1 r^{s-2}.
\end{aligned}
$$

By virtue of the continuity at zero of function $\mathcal{B}(r)$ and $\mathcal{B}(0) = 0$, we can choose $d > 0$ so small that

$$
\mathcal{B}(d) \leq \frac{1}{2} a_* h^2 \varsigma^{m-1} \frac{B\left(\frac{h}{\sqrt{1+h^2}}\right)^{\varkappa+1}}{4(1+h+B)^2}. \tag{6.7.13}
$$

Further, from (6.7.9) we obtain

$$
|\nabla w|^{m-2} \geq m_1 r^{\varkappa(m-2)}, \quad m_1 = \begin{cases} \left(\frac{h}{\sqrt{1+h^2}}\right)^{\varkappa(m-2)}, & \text{if } m \geq 2; \\ \langle 2(1+h+B) \rangle^{m-2}, & \text{if } 1 < m < 2. \end{cases} \tag{6.7.14}
$$

By (6.7.13), we get

$$
-\frac{d\mathcal{A}_i(x, A\nabla w)}{dx_i} + \mathcal{B}(x, Aw, A\nabla w) \geq \frac{1}{2} a_* h^2 (A\varsigma)^{m-1} m_1 r^{\varkappa(m-1)-1} - b_1 r^{s-2}.
$$

Hence the desired inequality (6.7.6) follows if we choose

$$
\varkappa \leq \frac{s-1}{m-1} \quad \text{and} \quad A \geq \frac{1}{\varsigma} \left(\frac{2b_1}{a_* h^2 m_1} \right)^{\frac{1}{m-1}}. \tag{6.7.15}
$$

Step 2. Secondly, we shall estimate integrals over Γ_0^d and show that

$$
\mathcal{A}_i(x, A\nabla w) \cos(\overrightarrow{n}, x_i) + A^{m-1} \frac{\gamma(\omega)}{r^{m-1}} w|w|^{m-2} - \mathcal{G}(x, Aw) \Big|_{\Gamma_0^d} \geq 0. \tag{6.7.16}
$$

By (6.7.5) and $x = \pm hy$ on $\Gamma_\pm$, we have

$$
\begin{aligned}
|\nabla w|\Big|_{\Gamma_\pm} &= r^{\varkappa} \left(\frac{h}{\sqrt{1+h^2}} \right)^{\varkappa} \sqrt{\{(1+\varkappa)B + 2\}^2 + 4h^2} \\
&\leq 2(B+1+h) r^{\varkappa} \left(\frac{h}{\sqrt{1+h^2}} \right)^{\varkappa},
\end{aligned} \tag{6.7.17}
$$

since $\varkappa \leq 1$ and $\sqrt{a^2+b^2} \leq |a| + |b|$. Next, in virtue of assumption 4)′ and (6.1.3),

$$
\mathcal{G}(x, w) = \mathcal{G}(x, 0) + w \cdot \int\limits_0^1 \frac{\partial \mathcal{G}(x, \tau w)}{\partial(\tau w)} d\tau \leq |g(x, 0)|.
$$

Further, since $\varsigma \le 1, a \le a^*$, by assumptions 5), 2)$'$, 12) and (6.7.17), on Γ_0^d:

$$\begin{aligned}
&\mathcal{A}_i(x, A\nabla w)\cos(\overrightarrow{n}, x_i) + A^{m-1}\frac{\gamma(\omega)}{r^{m-1}}w|w|^{m-2} - \mathcal{G}(x, Aw)\\
&\ge \nu_0(AB)^{m-1}r^{\varkappa(m-1)}\left(\frac{h}{\sqrt{1+h^2}}\right)^{(\varkappa+1)(m-1)}\\
&\quad - a^*A^{m-1}|\nabla w|^{m-1} - \alpha(x) - |g(x,0)|\\
&\ge \nu_0(AB)^{m-1}r^{\varkappa(m-1)}\left(\frac{h}{\sqrt{1+h^2}}\right)^{(\varkappa+1)(m-1)}\\
&\quad - a^*(2A)^{m-1}\left(\frac{hr}{\sqrt{1+h^2}}\right)^{\varkappa(m-1)}(B+1+h)^{m-1} - k_1r^{s-1}\\
&\ge A^{m-1}\left(\frac{hr}{\sqrt{1+h^2}}\right)^{\varkappa(m-1)}\left\{\left(\nu_0\left(\frac{h}{\sqrt{1+h^2}}\right)^{(m-1)} - a^*2^{m-1}\max\{1, 2^{m-2}\}\right)B^{m-1}\right.\\
&\quad \left. - a^*2^{m-1}\max\{1,2^{m-2}\}(1+h)^{m-1}\right\} - k_1r^{s-1}.
\end{aligned} \tag{6.7.18}$$

Now, by virtue of assumption (6.7.1), we derive from (6.7.18)

$$\begin{aligned}
&\mathcal{A}_i(x, A\nabla w)\cos(\overrightarrow{n}, x_i) + A^{m-1}\frac{\gamma(\omega)}{r^{m-1}}w|w|^{m-2} - \mathcal{G}(x, Aw)\\
&\ge A^{m-1}\left(\frac{hr}{\sqrt{1+h^2}}\right)^{\varkappa(m-1)}a^*2^{m-1}\max\{1,2^{m-2}\}(B^{m-1} - (1+h)^{m-1}) - k_1r^{s-1}.
\end{aligned}$$

If we choose

$$B \ge 2^{\frac{1}{m-1}}(1+h), \quad A \ge \left(\frac{k_1}{a^*}\right)^{\frac{1}{m-1}} \cdot \left(\frac{\sqrt{1+h^2}}{h}\right)^{\varkappa} \cdot \frac{1}{2(1+h)\max\{1, 2^{\frac{m-2}{m-1}}\}} \tag{6.7.19}$$

and $\varkappa > 0$ as in (6.7.15), we get the required (6.7.16).

Step 3. Analogously as above we shall derive that

$$\left[\mathcal{A}_i(x, A\nabla w)\cos(\overrightarrow{n}, x_i)\right]_{\Sigma_0^d} + \left\{A^{m-1}\frac{\sigma(\omega)}{r^{m-1}}w|w|^{m-2} - \mathcal{H}(x, Aw)\right\}\Big|_{\Sigma_0^d} \ge 0. \tag{6.7.20}$$

On Σ_0 we have:

$$w\Big|_{\Sigma_0} = (1+B)r^{1+\varkappa}; \quad \cos(\vec{n}, x) = 0, \ \cos(\vec{n}, y) = \pm 1;$$

$$w_x\Big|_{\Sigma_0} = (1+\varkappa)(1+B)r^{\varkappa}; \ w_y\Big|_{\Sigma_0} = 0 \Longrightarrow |\nabla w|\Big|_{\Sigma_0} = (1+\varkappa)(1+B)r^{\varkappa}.$$

Next, by assumptions 4)′, 12), and (6.1.3),

$$\mathcal{H}(x,w)=\mathcal{H}(x,0)+w\cdot\int\limits_0^1\frac{\partial\mathcal{H}(x,\tau w)}{\partial(\tau w)}d\tau\le|h(x,0)|.$$

Therefore, since $\varsigma\le 1, a\le a^*$, because of assumptions 5), 7)′, on Σ_0^d we obtain

$$\begin{aligned}&\Big[\mathcal{A}_i(x,A\nabla w)\cos(\overrightarrow{n},x_i)\Big]+A^{m-1}\frac{\sigma(\omega)}{r^{m-1}}w|w|^{m-2}-\mathcal{H}(x,Aw)\\ &\ge\nu_0A^{m-1}\frac{1}{r^{m-1}}w|w|^{m-2}-a^*A^{m-1}|\nabla w|^{m-1}-\alpha(x)-|h(x,0)|\\ &\ge\nu_0A^{m-1}(1+B)^{m-1}r^{\varkappa(m-1)}-a^*A^{m-1}(1+\varkappa)^{m-1}(1+B)^{m-1}r^{\varkappa(m-1)}\\ &\quad-k_1r^{s-1}\\ &=A^{m-1}(1+B)^{m-1}r^{\varkappa(m-1)}\left(\nu_0-a^*(1+\varkappa)^{m-1}\right)-k_1r^{s-1}\\ &\ge A^{m-1}r^{\varkappa(m-1)}\left(\nu_0-a^*(1+\varkappa)^{m-1}\right)-k_1r^{s-1}\ge 0,\end{aligned}$$

by virtue of (6.7.1), if we choose $\varkappa>0$ as in (6.7.15) as well as

$$0<\varkappa\le\left(\frac{\nu_0}{2a^*}\right)^{\frac{1}{m-1}}-1 \tag{6.7.21}$$

and

$$A\ge\left(\frac{2k_1}{\nu_0}\right)^{\frac{1}{m-1}}. \tag{6.7.22}$$

Thus (6.7.20) is proved.

Step 4. Let us consider barrier function $w(x)$, which is defined by (6.7.5), and the function $v(x)$ from (6.1.1) that satisfies the integral identity (6.1.2). For them we shall verify Proposition 6.4. From the above proved estimates we obtain

$$Q(w,\eta)\ge 0=Q(v,\eta) \tag{6.7.23}$$

for all non-negative $\eta\in\mathbf{C}^0(\overline{G_0^d})\cap\mathbf{V}^1_{m,0}(G_0^d)$.

Further, we compare $v(x)$ and $w(x)$ on Ω_d. Since $x^2\ge h^2y^2$ in $\overline{K}$ from (6.7.5) we have

$$Aw(x)\Big|_{r=d}\ge Ad^{1+\varkappa}\left(\frac{h}{\sqrt{1+h^2}}\right)^{\varkappa+1}.$$

On the other hand

$$v(x)\Big|_{\Omega_d}\le|u(x)|^{\frac{1}{\varsigma}}\Big|_{\Omega_d}\le M_0^{\frac{q+m-1}{m-1}}$$

and therefore it follows that

$$Aw(x)\Big|_{\Omega_d}\ge Ad^{\varkappa+1}\left(\frac{h}{\sqrt{1+h^2}}\right)^{\varkappa+1}\ge M_0^{\frac{q+m-1}{m-1}}\ge v\Big|_{\Omega_d},$$

if we choose

$$A \geq \frac{M_0^{\frac{q+m-1}{m-1}} \cdot \left(\frac{\sqrt{1+h^2}}{h}\right)^{\varkappa+1}}{d^{\varkappa+1}}. \qquad (6.7.24)$$

Thus, if we choose at first a large $B \geq 1$ according to (6.7.19), then a small $d > 0$ according to (6.7.13), a value $m_1 > 0$ according to (6.7.14), $\varkappa \in (0, \varkappa_0]$ according to (6.7.15), (6.7.21) and finally a large $A > 0$ according to (6.7.15), (6.7.19), (6.7.22) and (6.7.24), we assure the validity of Proposition 6.4.

Therefore, by the comparison principle (Proposition 6.4), we have

$$v(x) \leq Aw(x), \; x \in \overline{G_0^d}. \qquad (6.7.25)$$

Analogously, we derive the estimate $v(x) \geq -Aw(x)$, if we replace $v(x)$ with $-v(x)$. Returning to the function $u(x)$ by (6.1.1) we establish the first required estimate (6.7.2).

Now we want to estimate the gradient modulus of the problem (QL) solution near a conical point. Let us consider problem (QL) for the function $v(x)$ after the change (6.1.1) in the set $G^{\varrho}_{\varrho/2} \subset G$, $0 < \rho < d$. We perform the transformation $x = \varrho x'$; $z(x') = \varrho^{-\varkappa-1} v(\varrho x')$. Function $z(x')$ satisfies the problem

$$\begin{cases} -\frac{d\mathcal{A}_i(\varrho x', \varrho^{\varkappa} z_{x'})}{dx'_i} + \varrho\mathcal{B}(\varrho x', \varrho^{\varkappa+1} z, \varrho^{\varkappa} z_{x'}) = 0, \; x' \in G^1_{1/2}, \\ [z(x')]_{\Sigma^1_{1/2}} = 0, \\ [\mathcal{A}_i(\varrho x', \varrho^{\varkappa} z_{x'}) \cos(\overrightarrow{n}, x'_i)]_{\Sigma^1_{1/2}} + \frac{\sigma(\omega)}{|x'|^{m-1}} \varrho^{\varkappa(m-1)} z|z|^{m-2} = \mathcal{H}(\varrho x', \varrho^{\varkappa+1} z), \\ \hfill x' \in \Sigma^1_{1/2}, \\ \mathcal{A}_i(\varrho x', \varrho^{\varkappa} z_{x'}) \cos(\overrightarrow{n}, x'_i) + \frac{\gamma(\omega)}{|x'|^{m-1}} \varrho^{\varkappa(m-1)} z|z|^{m-2} = \mathcal{G}(\varrho x', \varrho^{\varkappa+1} z), \; x' \in \Gamma^1_{1/2}. \end{cases} \qquad (QL)'$$

We apply our assumption about a priori estimate of the gradient modulus of the problem $(QL)'$ solution

$$\max_{x' \in G^1_{1/2}} |\nabla' z(x')| \leq M'_1 \qquad (6.7.26)$$

(see §4 Theorem 4.4 in [6] or §§3, 5 in [64]). Returning to variable x and function $v(x)$ we obtain from (6.7.26)

$$|\nabla v(x)| \leq M'_1 \varrho^{\varkappa}, \; x \in G^{\varrho}_{\varrho/2}, \; 0 < \varrho < d.$$

Setting now $|x| = \frac{2}{3}\varrho$ we obtain the estimate

$$|\nabla v(x)| \leq M'_1 |x|^{\varkappa}, \; x \in G^d_0.$$

Returning to function $u(x)$ by (6.1.1), we obtain the required estimate (6.7.3). □

Corollary 6.10. *Let $u(x)$ be a weak solution of the problem (QL) and suppose that assumptions of Theorem 6.9 are satisfied. Then $u(0) = 0$.*

Proof. From the problem boundary condition it follows that

$$\gamma(\omega)u(x)|u(x)|^{q+m-2} = |x|^{m-1} \cdot \{g(x, u(x)) - a_i(x, u(x), u_x(x)) \cdot \cos(\overrightarrow{n}, x_i)\},$$

$x \in \partial G \backslash \mathcal{O}$. By assumptions 2), 4), 5), 12) and estimates (6.7.2)–(6.7.3), we obtain

$$\begin{aligned}\nu_0|u(x)|^{q+m-1} &\leq \gamma(\omega)|u(x)|^{q+m-1} \leq |x|^{m-1} \cdot \left\{|g(x,0)| + a|u(x)|^q |\nabla u(x)|^{m-1}\right\} \\ &\leq k_1|x|^{s+m-2} + C_0^q C_1^{m-1} a^* |x|^{\varkappa(m-1)}.\end{aligned}$$

By letting $|x|$ tend to 0 we get, because of the continuity of $u(x)$, that $\nu_0|u(0)| \leq 0$ and taking into account $\nu_0 > 0$, we get that $u(0) = 0$. □

Chapter 7

Best possible estimates of solutions to the transmission problem for a quasi-linear elliptic divergence second-order equation in a domain with a boundary edge

7.1 Introduction. Assumptions

This chapter is devoted to the estimate of weak solutions to the transmission problem for elliptic quasi-linear *degenerate* second-order equations. We investigate the behavior of weak solutions of the transmission problem with Robin boundary condition for quasi-linear second-order elliptic equations with triple degeneracy and singularity in the coefficients in a neighborhood of the boundary edge.

Let $G \subset \mathbb{R}^n$, $n \geq 3$ be a domain bounded by an $(n-1)$-dimensional manifold ∂G and let Γ_-, Γ_+ be open nonempty disjoint submanifolds such that $\partial G = \overline{\Gamma}_- \cup \overline{\Gamma}_+$, where $\overline{\Gamma}_- \cap \overline{\Gamma}_+$ is a smooth $(n-2)$-dimensional submanifold that contains an edge $\Gamma_0 \subseteq \overline{\Gamma}_- \cap \overline{\Gamma}_+$. For $x = (x_1, \ldots, x_n)$ let us introduce the cylindrical coordinates (r, ω, x'), where $x' = (x_3, \ldots, x_n)$, $r = \sqrt{x_1^2 + x_2^2}$, $\omega = \arctan \frac{x_2}{x_1}$. We assume that G is divided into two open subdomains G_- and G_+ by an $(n-1)$-dimensional hyperplane $\{(r, 0, x') \big| \, x' \in \mathbb{R}^{n-2}, \, r > 0\}$, thus $G = G_+ \cup G_-$. We set $\Sigma_0 = G \cap \{(r, 0, x') \big| \, x' \in \mathbb{R}^{n-2}, \, r > 0\}$ and assume that $\Gamma_0 \subset \overline{\Sigma_0}$.

M. Borsuk, *Transmission Problems for Elliptic Second-Order Equations in Non-Smooth Domains*, 171
Frontiers in Mathematics, DOI 10.1007/978-3-0346-0477-2_8, © Springer Basel AG 2010

We derive *the exact estimate* of the weak solution in a neighborhood of the boundary edge for the problem

$$\begin{cases} -\frac{d}{dx_i}a^i(x,u,\nabla u) + b(x,u,\nabla u) + a_0 c(x,u) = f(x), & x \in G \setminus \Sigma_0; \\ [u]_{\Sigma_0} = 0, & x \in \Sigma_0; \\ \mathcal{S}[u] \equiv \left[aa^i(x,u,\nabla u)n_i\right]_{\Sigma_0} + \beta_0 r^{\tau-m+1}u|u|^{q+m-2} & \\ \qquad = h(x,u), & x \in \Sigma_0; \\ \mathcal{B}[u] \equiv aa^i(x,u,\nabla u)n_i + \gamma(\omega) r^{\tau-m+1}u|u|^{q+m-2} & \\ \qquad = g(x,u), & x \in \partial G \setminus \{\Sigma_0 \cup \Gamma_0\}, \end{cases} \tag{TDQL}$$

where $q \geq 0$, $m > 1$, $a_\pm > 0$, $a_0 \geq 0$, $\beta_0 > 0$, $\tau \geq m-2$ are given numbers.

For a sufficiently small number $d > 0$ we also define the sets

$$G_0^d = G \cap \{(r,\omega,x')\big|\ 0 < r < d,\ \omega \in (-\omega_0/2, \omega_0/2),\ x' \in \mathbb{R}^{n-2}\};$$
$$\Gamma^d = \Gamma_+^d \cup \Gamma_-^d, \text{ where } \Gamma_\pm^d = \Gamma_\pm \cap \overline{G_0^d} \subset \partial G_0^d;$$
$$\Omega_d = G \cap \{(r,\omega,x')\big|\ r = d,\ \omega \in (-\omega_0/2, \omega_0/2),\ x' \in \mathbb{R}^{n-2}\}.$$

We will assume the following:

- $\partial G \setminus \Gamma_0$ is a smooth submanifold in $\mathbb{R}^n$;
- there exists a number $d > 0$ such that $\Gamma_0^d = \{(0,0,x')|\,|x'| < d\} \subset \Gamma_0$ is the straight edge with the center in the origin;
- G_0^d is locally diffeomorphic to the dihedral cone

  $$\mathbb{D}_d = \{(r,\omega)\big|\ 0 < r < d,\ \omega \in (-\omega_0/2, \omega_0/2)\} \times \mathbb{R}^{n-2}; \quad 0 < \omega_0 < 2\pi;$$

  thus we assume that $G_0^d \subset G$ and, consequently, the domain G is a *"wedge"* in some vicinity of the edge;
- $\omega\ |_{\Gamma_\pm^d} = \pm\omega_0/2$.

We denote by $\mathfrak{N}^1_{m,q,\tau}(G)$ the set of functions $u(x) \in \mathbf{C}^0(\overline{G})$ having first weak derivatives with the finite integral

$$\int_G \left(ar^\tau |u|^q |\nabla u|^m + a_0 r^{\tau-m}|u|^{q+m}\right) dx < \infty,$$
$$q \geq 0,\ m > 1,\ a > 0,\ a_0 \geq 0,\ \tau \geq m-2. \tag{7.1.1}$$

Regarding the equation we assume that the following **conditions** are satisfied:

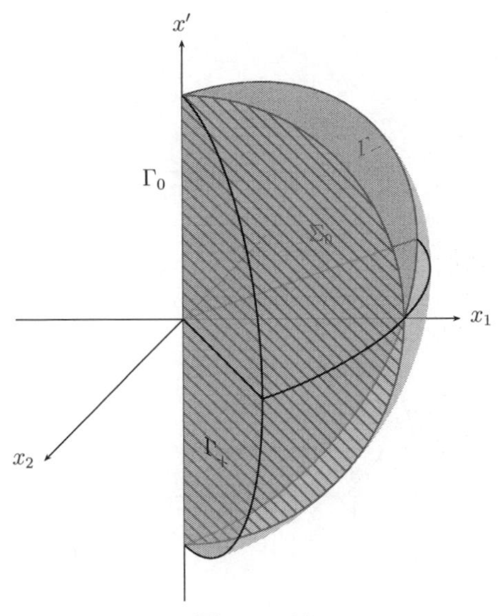

Figure 12

$a^i(x,u,\xi)$, $i = 1,\dots,n$; $b(x,u,\xi)$ are the Caratheodory functions $G \times \mathbb{R} \times \mathbb{R}^n \to \mathbb{R}$ and continuously differentiable with respect to x_i, u, ξ_i; $c(x,u)$ is the Caratheodory function $G \times \mathbb{R} \to \mathbb{R}$ and continuously differentiable with respect to the variable u, $h(x,u)$ is a function $\Sigma_0 \times \mathbb{R} \to \mathbb{R}$ continuously differentiable with respect to the variable u, while $g(x,u)$ is a function $\partial G \times \mathbb{R} \to \mathbb{R}$ continuously differentiable with respect to variable u, as well as $f(x) \in \mathbf{L}_1(G)$ and $h(x,u(x)) \in L_1(\Sigma_0)$, $g(x,u(x)) \in \mathbf{L}_1(\partial G)$ $\forall u(x) \in \mathbf{C}^0(\overline{G})$, possessing the properties:

1) $\dfrac{\partial a^i(x,u,\xi)}{\partial \xi_j} p_i p_j \ge r^\tau |u|^q |\xi|^{m-2} p^2$, $\forall p \in \mathbb{R}^n \setminus \{0\}$;

2) $(m-1)u\dfrac{\partial a^i(x,u,\xi)}{\partial u} = q\dfrac{\partial a^i(x,u,\xi)}{\partial \xi_j}\xi_j$; $i = 1,\dots,n$;

3) $\left|\dfrac{\partial a^i(x,u,\xi)}{\partial \xi_j} - r^\tau |u|^q |\xi|^{m-4}\big(\delta_i^j |\xi|^2 + (m-2)\xi_i\xi_j\big)\right| \le \mathcal{A}(r) r^\tau |u|^q |\xi|^{m-2}$;

4) $\left|\dfrac{\partial a^i(x,u,\xi)}{\partial x_i} - \tau r^{\tau-2} |u|^q |\xi|^{m-2} x_i \xi_i\right| \le \mathcal{A}(r) r^{\tau-1} |u|^q |\xi|^{m-1}$;

5) $\sqrt{\displaystyle\sum_{i=1}^n |a^i(x,u,\xi) - r^\tau |u|^q |\xi|^{m-2}\xi_i|^2} \le \mathcal{A}(r) r^\tau |u|^q |\xi|^{m-1}$;

6) $|b(x,u,\xi)| \leq \mu_0 r^\tau |u|^{q-1}|\xi|^m$, $\sqrt{\sum_{i=1}^{n}\left|\frac{\partial b(x,u,\xi)}{\partial \xi_i}\right|^2} \leq \mu_0 r^\tau |u|^{q-1}|\xi|^{m-1}$, $0 \leq \mu_0 < \frac{q+m-1}{m-1}$;

7) $\frac{\partial b(x,u,\xi)}{\partial u} \geq \gamma_m r^\tau |u|^{q-2}|\xi|^m$; $uc(x,u) \geq r^{\tau-m}|u|^{q+m}$;
$\frac{\partial c(x,u)}{\partial u} \geq \gamma_m r^{\tau-m}|u|^{q+m-2}$;

8) $\frac{\partial h(x,u)}{\partial u} \leq 0$, $\frac{\partial g(x,u)}{\partial u} \leq 0$;

9) $\gamma(\pm\omega_0/2) > 0$;

10) $|f(x)| \leq f_1 r^{\tau-m+(q+m-1)\lambda}$, $|h(x,0)| \leq h_1 r^{\tau-m+1+(q+m-1)\lambda}$,
$|g(x,0)| \leq g_1 r^{\tau-m+1+(q+m-1)\lambda}$,

where $\gamma_m > 0$, $f_1 \geq 0$, $h_1 \geq 0$, $g_1 \geq 0$ and $\mathcal{A}(r)$ is a nonnegative non-decreasing function continuous at zero with $\mathcal{A}(0) = 0$.

Definition 7.1. The function $u(x)$ is called a *weak* solution of the problem $(TDQL)$ provided that $u(x) \in \mathbf{C}^0(\overline{G}) \cap \mathfrak{N}^1_{m,q,\tau}(G)$ and satisfies the integral identity

$$\int_G \{aa^i(x,u,\nabla u)\eta_{x_i} + ab(x,u,u_x)\eta(x) + aa_0 c(x,u)\eta(x)\}\, dx$$
$$+ \beta_0 \int_{\Sigma_0} r^{\tau-m+1} u|u|^{q+m-2}\eta(x)ds + \int_{\partial G} \gamma(\omega) r^{\tau-m+1} u|u|^{q+m-2}\eta(x)ds$$
$$= \int_G af(x)\eta(x)dx + \int_{\partial G} g(x,u)\eta(x)ds + \int_{\Sigma_0} h(x,u)\eta(x)ds \tag{II}$$

for all functions $\eta(x) \in \mathbf{C}^0(\overline{G}) \cap \mathfrak{N}^1_{m,q,\tau}(G)$.

Lemma 7.2. *Let $u(x)$ be a weak solution of $(TDQL)$. For any function $\eta(x) \in \mathbf{C}^0(\overline{G}) \cap \mathfrak{N}^1_{m,q,\tau}(G)$ the following equality holds for a.e. $\varrho \in (0,d)$:*

$$\int_{G_0^\varrho} \Big\{aa^i(x,u,\nabla u)\eta_{x_i} + ab(x,u,u_x)\eta(x) + aa_0 c(x,u)\eta(x)\Big\} dx$$
$$= \int_{G_0^\varrho} af(x)\eta(x)dx + \int_{\Omega_\varrho} aa^i(x,u,\nabla u)\cos(r,x_i)\eta(x)d\Omega_\varrho$$
$$+ \int_{\Gamma^\varrho} \left(g(x,u) - \gamma(\omega) r^{\tau-m+1} u|u|^{q+m-2}\right)\eta(x)ds$$

$$+ \int\limits_{\Sigma_0^\varrho} \left(h(x,u) - \beta_0 r^{\tau-m+1} u|u|^{q+m-2}\right) \eta(x) ds. \qquad (II)_{loc}$$

Proof. The proof is similar to the proof of Lemma 3.2 Chapter 3; see also the proof of Lemma 5.2 in [14] (pp. 167–170). □

We make the function change $u = v|v|^{\varsigma-1}$ with $\varsigma = \frac{m-1}{q+m-1}$. Then the identity $(II)_{loc}$ can be presented in the following form:

$$\int\limits_{G_0^\varrho} \left\langle a\mathcal{A}^i(x,v_x)\eta_{x_i} + a\mathcal{B}(x,v,v_x)\eta + aa_0\mathcal{C}(x,v)\eta \right\rangle dx$$

$$+ \int\limits_{\Gamma^\varrho} \gamma(\omega) r^{\tau-m+1} v|v|^{m-2}\eta(x)ds + \beta_0 \int\limits_{\Sigma_0^\varrho} r^{\tau-m+1} v|v|^{m-2}\eta(x)ds$$

$$= \int\limits_{\Omega_\varrho} a\mathcal{A}^i(x,v_x)\cos(r,x_i)\eta(x)d\Omega_\varrho + \int\limits_{\Gamma^\varrho} \mathcal{G}(x,v)\eta(x)ds$$

$$+ \int\limits_{\Sigma_0^\varrho} \mathcal{H}(x,v)\eta(x)ds + \int\limits_{G_0^\varrho} af(x)\eta(x)dx \qquad (\widehat{II})_{loc}$$

for a.e. $\varrho \in (0,d)$, $v(x) \in \mathbf{C}^0(\overline{G}) \cap \mathfrak{N}^1_{m,0,\tau}(G)$ and arbitrary $\eta(x) \in \mathbf{C}^0(\overline{G}) \cap \mathfrak{N}^1_{m,0,\tau}(G)$, with

$$\mathcal{A}^i(x,v_x) \equiv a^i(x, v|v|^{\varsigma-1}, \varsigma|v|^{\varsigma-1}v_x), \quad \mathcal{B}(x,v,v_x) \equiv b(x, v|v|^{\varsigma-1}, \varsigma|v|^{\varsigma-1}v_x),$$
$$\mathcal{C}(x,v) \equiv c(x, v|v|^{\varsigma-1}), \quad \mathcal{G}(x,v) \equiv g(x, v|v|^{\varsigma-1}), \quad \mathcal{H}(x,v) \equiv h(x, v|v|^{\varsigma-1}).$$

The explicit independence of $\mathcal{A}^i$ from v is guaranteed by the assumption *2)* (see in detail §6.1).

Now our assumptions can be rewritten as follows:

1)′ $\dfrac{\partial \mathcal{A}^i(x,\eta)}{\partial \eta_j} p_i p_j \geq \varsigma^{m-1} r^\tau |\eta|^{m-2} p^2$, $\forall p \in \mathbb{R}^n \setminus \{0\}$;

3)′ $\left| \dfrac{\partial \mathcal{A}^i(x,\eta)}{\partial \eta_j} - \varsigma^{m-1} r^\tau |\eta|^{m-4} \left(\delta_i^j |\eta|^2 + (m-2)\eta_i\eta_j\right) \right| \leq \varsigma^{m-1} \mathcal{A}(r) r^\tau |\eta|^{m-2}$;

4)′ $\left| \dfrac{\partial \mathcal{A}^i(x,\eta)}{\partial x_i} - \tau\varsigma^{m-1} r^{\tau-2} |\eta|^{m-2} x_i\eta_i \right| \leq \varsigma^{m-1} \mathcal{A}(r) r^{\tau-1} |\eta|^{m-1}$;

5)′ $\sqrt{\sum_{i=1}^n |\mathcal{A}^i(x,\eta) - \varsigma^{m-1} r^\tau |\eta|^{m-2}\eta_i|^2} \leq \varsigma^{m-1} \mathcal{A}(r) r^\tau |\eta|^{m-1}$;

6)′ $|\mathcal{B}(x,v,\eta)| \le \mu_0 \varsigma^m r^\tau v^{-1}|\eta|^m$, $\sqrt{\sum_{i=1}^{n}\left|\frac{\partial\mathcal{B}(x,v,\eta)}{\partial\eta_i}\right|^2} \le \mu_0\varsigma^m r^\tau|v|^{-1}|\eta|^{m-1}$,

$0 \le \mu_0 < \frac{q+m-1}{m-1}$;

7)′ $\dfrac{\partial\mathcal{B}(x,v,\eta)}{\partial v} \ge \gamma_m\varsigma^{m+1}r^\tau|v|^{-2}|\eta|^m$; $v\mathcal{C}(x,v) \ge r^{\tau-m}|v|^m$;

$\dfrac{\partial\mathcal{C}(x,v)}{\partial v} \ge \varsigma\gamma_m r^{\tau-m}|v|^{m-2}$;

8)′ $\dfrac{\partial\mathcal{H}(x,v)}{\partial v} \le 0$, $\dfrac{\partial\mathcal{G}(x,v)}{\partial v} \le 0$,

where $\gamma_m > 0$ and $\mathcal{A}(r)$ is a non-negative continuous at zero function with $\mathcal{A}(0) =$ 0.

7.2 The comparison principle

We consider the second-order quasi-linear degenerate operator Q defined as

$$\begin{aligned} Q(v,\eta) \equiv &\int_{G_0^d}\Big\langle \mathcal{A}^i(x,v_x)\eta_{x_i} + \mathcal{B}(x,v,v_x)\eta + a_0\mathcal{C}(x,v)\eta\Big\rangle dx \\ &+ \int_{\Gamma^d}\gamma(\omega)r^{\tau-m+1}v|v|^{m-2}\eta(x)ds - \int_{\Sigma_0^d}\mathcal{H}(x,v)\eta(x)ds \\ &+ \beta_0\int_{\Sigma_0^d} r^{\tau-m+1}v|v|^{m-2}\eta(x)ds - \int_{\Omega_d}\mathcal{A}^i(x,v_x)\cos(r,x_i)\eta(x)d\Omega_d \\ &- \int_{\Gamma^d}\mathcal{G}(x,v)\eta(x)ds - \int_{G_0^d} af(x)\eta(x)dx \end{aligned} \tag{7.2.1}$$

for $v(x) \in \mathbf{C}^0(\overline{G}) \cap \mathfrak{N}^1_{m,0,\tau}(G)$ and for all non-negative η belonging to $\mathbf{C}^0(\overline{G}) \cap \mathfrak{N}^1_{m,0,\tau}(G)$ under the following assumptions:

$\mathcal{A}^i(x,\xi)$, $\mathcal{B}(x,v,\xi)$, $\mathcal{C}(x,v)$, $\mathcal{G}(x,v)$, $\mathcal{H}(x,v)$ *are Caratheodory functions in* $\mathfrak{M} = \overline{G_0^d} \times \mathbb{R} \times \mathbb{R}^n$, *continuously differentiable with respect to the variables* v, ξ, $f(x) \in \mathbf{L}_1(G_0^d)$ *and satisfy in* $\mathfrak{M}$ *the following inequalities:*

(i) $\dfrac{\partial\mathcal{A}^i(x,\xi)}{\partial\xi_j}p_ip_j \ge a\gamma_m r^\tau|\xi|^{m-2}|p|^2$, $\forall p \in \mathbb{R}^n \setminus \{0\}$;

(ii) $\sqrt{\sum_{i=1}^{N}\left|\dfrac{\partial\mathcal{B}(x,v,\xi)}{\partial\xi_i}\right|^2} \le ar^\tau|v|^{-1}|\xi|^{m-1}$; $\dfrac{\partial\mathcal{B}(x,v,\xi)}{\partial v} \ge ar^\tau|v|^{-2}|\xi|^m$;

(iii) $\dfrac{\partial \mathcal{C}(x,v)}{\partial v} \geq a\gamma_m r^{\tau-m}|v|^{m-2}$;

(iv) $\dfrac{\partial \mathcal{G}(x,v)}{\partial v} \leq 0, \ \dfrac{\partial H(x,v)}{\partial v} \leq 0, \ \gamma(\omega) \geq 0, \quad \beta_0 \geq 0, \ a_0 \geq 0.$

(here: $\tau \geq m-2$, $m > 1$, $\gamma_m > 0$ and $a > 0$).

Proposition 7.3. *Let operator Q satisfy assumptions* (i)–(iv) *and $d \lll 1$. We also assume that functions $v, w \in \mathbf{C}^0(\overline{G_0^d}) \cap \mathfrak{N}^1_{m,0,\tau}(G_0^d)$ and satisfy the inequality*

$$Q(v,\eta) \leq Q(w,\eta) \tag{7.2.2}$$

for all non-negative $\eta \in \mathbf{C}^0(\overline{G_0^d}) \cap \mathfrak{N}^1_{m,0,\tau}(G_0^d)$ and also the inequality

$$v(x) \leq w(x) \text{ on } \Omega_d \tag{7.2.3}$$

holds. Then $v(x) \leq w(x)$ in G_0^d.

Proof. For the proof see word for word Proposition 6.4. □

7.3 Construction of the barrier function

In this section, for an n-dimensional *infinite* dihedral cone

$$G_0^\infty = \left\{x = (r,\omega,x') \big| \ x' \in \mathbb{R}^{n-2}, \ 0 < r < \infty, \ -\frac{\omega_0}{2} < \omega < \frac{\omega_0}{2}, \ \omega_0 \in (0,2\pi)\right\}$$

with the edge $\Gamma_0 = \{(0,0,x') | \ x' \in \mathbb{R}^{n-2}\}$, that contains the origin, and with lateral faces

$$\Gamma_-^\infty = \left\{(r, -\frac{\omega_0}{2}, x') \big| \ x' \in \mathbb{R}^{n-2}, \ 0 < r < \infty\right\},$$
$$\Gamma_+^\infty = \left\{(r, +\frac{\omega_0}{2}, x') \big| \ x' \in \mathbb{R}^{n-2}, \ 0 < r < \infty\right\},$$

we consider the following elliptic transmission problem for the model equation:

$$\begin{cases} -\frac{d}{dx_i}(r^\tau |w|^q |\nabla w|^{m-2} w_{x_i}) \\ +a_0 r^{\tau-m} w|w|^{q+m-2} - \mu r^\tau w|w|^{q-2}|\nabla w|^m = 0, & x \in G_0^\infty \setminus \Sigma_0, \\ [w]_{\Sigma_0} = 0, \\ \left[a|w|^q|\nabla w|^{m-2}\frac{\partial w}{\partial \vec{n}}\right]_{\Sigma_0} + \beta r^{1-m} w|w|^{q+m-2} = 0, & x \in \Sigma_0, \\ a|w|^q|\nabla w|^{m-2}\frac{\partial w}{\partial \vec{n}} + \gamma r^{1-m} w|w|^{q+m-2} = 0, & x \in \partial G_0^\infty \setminus (\Gamma_0 \cup \Sigma_0), \end{cases} \tag{MVP}$$

where

$$a = \begin{cases} a_+, & x \in G_+, \\ a_-, & x \in G_-, \end{cases} \quad \gamma = \begin{cases} \gamma_+, & x \in \Gamma_+^\infty, \\ \gamma_-, & x \in \Gamma_-^\infty, \end{cases}$$

$$a_\pm > 0,\ m > 1,\ q \geq 0,\ \tau \geq m-2,\ a_0 \geq 0,\ \mu > \mu_0,\ 0 < \beta < \beta_0, \qquad (7.3.1)$$
$$0 < \gamma_\pm < \gamma\left(\pm\frac{\omega_0}{2}\right)$$

are constants.

We shall seek a solution of the problem (MVP) of the form

$$w(x) = r^\lambda \Phi(\omega), \quad \omega \in [-\frac{\omega_0}{2}, \frac{\omega_0}{2}], \quad \lambda > 0. \qquad (7.3.2)$$

By substituting the function (7.3.2) in (MVP) and calculating in the cylindrical coordinates, we get the following Sturm-Liouville boundary problem for the function $\Phi(\omega)$:

$$\begin{cases} \frac{d}{d\omega}\left[\left(\lambda^2\Phi^2 + \Phi'^2\right)^{\frac{m-2}{2}} |\Phi|^q \Phi'\right] \\ +\lambda[\lambda(q+m-1) - m + 2 + \tau]\Phi|\Phi|^q \left(\lambda^2\Phi^2 + \Phi'^2\right)^{\frac{m-2}{2}} \\ = a_0\Phi|\Phi|^{q+m-2} - \mu\Phi|\Phi|^{q-2}\left(\lambda^2\Phi^2 + \Phi'^2\right)^{\frac{m}{2}}, \ \omega \in (-\omega_0/2, 0) \cup (0, \omega_0/2), \\ [\Phi]_{\omega=0} = 0, \quad \left[a|\Phi|^q\left(\lambda^2\Phi^2 + \Phi'^2\right)^{\frac{m-2}{2}} \Phi'\right]_{\omega=0} = \beta\Phi(0)|\Phi(0)|^{q+m-2}, \\ \pm a_\pm|\Phi_\pm|^q\left(\lambda^2\Phi_\pm^2 + {\Phi'_\pm}^2\right)^{\frac{m-2}{2}} \Phi'_\pm\Big|_{\omega=\pm\omega_0/2} + \gamma_\pm\Phi_\pm|\Phi_\pm|^{q+m-2}\Big|_{\omega=\pm\omega_0/2} = 0. \end{cases} \qquad (StL)$$

Performing the function change $w = z|z|^{\varsigma-1}$ with $\varsigma = \frac{m-1}{q+m-1}$ we get the problem for *the barrier function*, z:

$$\begin{cases} \mathcal{L}_0 z \equiv -\overline{a}\frac{d}{dx_i}(r^\tau|\nabla z|^{m-2}z_{x_i}) + \overline{a}\overline{a}_0 r^{\tau-m}z|z|^{m-2} \\ \qquad\qquad -\overline{a}\overline{\mu}r^\tau z^{-1}|\nabla z|^m = 0, \quad x \in G_0^\infty \setminus \Sigma_0, \\ [z]_{\Sigma_0} = 0, \ \left[\overline{a}|\nabla z|^{m-2}\frac{\partial z}{\partial \vec{n}}\right]_{\Sigma_0} + \beta r^{1-m}z|z|^{m-2} = 0, \quad x \in \Sigma_0, \\ \overline{a}|\nabla z|^{m-2}\frac{\partial z}{\partial \vec{n}} + \gamma r^{1-m}z|z|^{m-2} = 0, \qquad x \in \partial G_0^\infty \setminus (\Gamma_0 \cup \Sigma_0), \end{cases} \qquad \overline{(MVP)}$$

where $\overline{a} = a\varsigma^{m-1}$, $\overline{a}_0 = a_0\varsigma^{1-m}$, $\overline{\mu} = \mu\varsigma$. For the solution z of the problem $\overline{(MVP)}$ we have the presentation

$$z(x) = r^{\overline{\lambda}}\psi(\omega), \quad \psi(\omega) = \Phi(\omega)|\Phi(\omega)|^{\frac{1-\varsigma}{\varsigma}}, \quad \omega \in [-\frac{\omega_0}{2}, \frac{\omega_0}{2}];$$
$$\overline{\lambda} = \frac{\lambda}{\varsigma}, \quad \frac{\psi'(\omega)}{\psi(\omega)} = \frac{1}{\varsigma}\frac{\Phi'(\omega)}{\Phi(\omega)}. \qquad (7.3.3)$$

By substituting the function (7.3.3) in $\overline{(MVP)}$ and calculating in the polar coordinates, we get the following Sturm-Liouville boundary problem for the function $\psi(\omega)$:

$$\begin{cases} \frac{d}{d\omega}\left\{\left(\overline{\lambda}^2\psi^2+\psi'^2\right)^{\frac{m-2}{2}}\psi'\right\}+\overline{\lambda}\left\langle\overline{\lambda}(m-1)-m+2+\tau\right\rangle\psi\left(\overline{\lambda}^2\psi^2+\psi'^2\right)^{\frac{m-2}{2}} \\ \qquad =\overline{a}_0\psi|\psi|^{m-2}-\overline{\mu}\psi^{-1}\left(\overline{\lambda}^2\psi^2+\psi'^2\right)^{\frac{m}{2}}, \quad \omega\in(-\omega_0/2,0)\cup(0,\omega_0/2), \\ [\psi]_{\omega=0}=0, \quad \left[\overline{a}\left(\overline{\lambda}^2\psi^2+\psi'^2\right)^{\frac{m-2}{2}}\psi'\right]_{\omega=0}=\beta\psi(0)|\psi(0)|^{m-2}, \\ \pm\overline{a}_\pm\left(\overline{\lambda}^2\psi_\pm^2+{\psi'_\pm}^2\right)^{\frac{m-2}{2}}\psi'_\pm\Big|_{\omega=\pm\omega_0/2}+\gamma_\pm\psi_\pm|\psi_\pm|^{m-2}\Big|_{\omega=\pm\omega_0/2}=0. \end{cases} \quad \overline{(StL)}$$

This problem is **a nonlinear eigenvalue problem**. Now we want to study the properties of eigenfunctions

$$\psi(\omega)=\begin{cases}\psi_+(\omega), & \omega\in[0,\frac{\omega_0}{2}],\\ \psi_-(\omega), & \omega\in[-\frac{\omega_0}{2},0].\end{cases}$$

Lemma 7.4. *Let $\lambda>0$ be an eigenvalue and $\psi(\omega)$ be the associated eigenfunction of $\overline{(StL)}$. Suppose that following inequalities are satisfied:*

$$\lambda\geq\left(\frac{a_0}{\mu+\frac{m-2}{4\varsigma}}\right)^{\frac{1}{m}}; \quad \mu>\frac{2-m}{4\varsigma}; \quad \varsigma=\frac{m-1}{q+m-1}, \quad \text{if} \quad 1<m<2; \tag{7.3.4}$$

$$\begin{cases}\lambda\geq\left(\frac{a_0}{\mu}\right)^{\frac{1}{m}} \quad \text{for } \mu>0,\\ (q+m-1)\lambda^m+(2-m+\tau)\lambda^{m-1}\geq a_0 \quad \text{for } \mu=0,\\ \text{(see inequality (9.1.4) in §9.1 [14])}\end{cases} \quad \text{if} \quad m\geq 2. \tag{7.3.5}$$

Then $\psi\psi''\leq 0$.

Proof. We rewrite the $\overline{(StL)}$ equation in the form

$$\begin{aligned} &-\psi\psi''\left\langle\overline{\lambda}^2\psi^2+(m-1)\psi'^2\right\rangle(\overline{\lambda}^2\psi^2+\psi'^2)^{\frac{m-4}{2}}\\ &=\overline{\mu}(\overline{\lambda}^2\psi^2+\psi'^2)^{\frac{m}{2}}-\overline{a}_0|\psi|^m+(m-2)\overline{\lambda}^2\psi^2\psi'^2(\overline{\lambda}^2\psi^2+\psi'^2)^{\frac{m-4}{2}}\\ &\quad+\overline{\lambda}\left\langle\overline{\lambda}(m-1)+2-m+\tau\right\rangle(\overline{\lambda}^2\psi^2+\psi'^2)^{\frac{m-2}{2}}\psi^2\equiv j(\psi). \end{aligned} \tag{7.3.6}$$

At first we consider the case when $1<m<2$. By the Cauchy inequality

$$\overline{\lambda}|\psi||\psi'|\leq\frac{1}{2}(\overline{\lambda}^2\psi^2+\psi'^2) \implies (m-2)\overline{\lambda}^2\psi^2\psi'^2\geq\frac{m-2}{4}(\overline{\lambda}^2\psi^2+\psi'^2)^2.$$

Since $\overline{\lambda}^2\psi^2 \le \overline{\lambda}^2\psi^2 + \psi'^2$,

$$\overline{a}_0|\psi|^m = \frac{\overline{a}_0}{\overline{\lambda}^m}(\overline{\lambda}^2\psi^2)^{\frac{m}{2}} \le \frac{\overline{a}_0}{\overline{\lambda}^m}(\overline{\lambda}^2\psi^2 + \psi'^2)^{\frac{m}{2}}. \tag{7.3.7}$$

Hence it follows that

$$j(\psi) \ge \left(\overline{\mu} + \frac{m-2}{4} - \frac{\overline{a}_0}{\overline{\lambda}^m}\right)(\overline{\lambda}^2\psi^2 + \psi'^2)^{\frac{m}{2}}, \tag{7.3.8}$$

by virtue of $2 - m + \tau \ge 0$. Now from (7.3.4) the required statement follows.

Let us consider the case when $m \ge 2$. If $\mu > 0$, then from (7.3.6)–(7.3.7) we obtain

$$-\psi\psi''\left\langle \overline{\lambda}^2\psi^2 + (m-1)\psi'^2\right\rangle \ge \left(\overline{\mu} - \frac{\overline{a}_0}{\overline{\lambda}^m}\right)(\overline{\lambda}^2\psi^2 + \psi'^2)^2 + (m-2)\overline{\lambda}^2\psi^2\psi'^2$$
$$+ \overline{\lambda}\langle\overline{\lambda}(m-1) + 2 - m + \tau\rangle(\overline{\lambda}^2\psi^2 + \psi'^2)\psi^2 \ge 0,$$

by (7.3.5). Finally, if $\mu = 0$, then

$$j(\psi) \ge \left\langle (m-1)\overline{\lambda}^m + (2-m+\tau)\overline{\lambda}^{m-1} - \overline{a}_0\right\rangle |\psi|^m \ge 0,$$

by (7.3.5). □

7.4 The case $\frac{\gamma_+}{a_+} = \frac{\gamma_-}{a_-}$

7.4.1 The barrier function

We observe that the solutions of $\overline{(StL)}$ are determined uniquely up to a scalar multiple. Therefore we can assume that $\psi_\pm(0) \ge 0$. Let

$$\psi_\pm(0) = 0.$$

Then it is possible that $\psi_-(\omega) < 0$ on $[-\frac{\omega_0}{2}, 0)$ and $\psi_+(\omega) > 0$ on $(0, \frac{\omega_0}{2}]$; we will show below in Lemma 7.8 under what assumptions it is possible.

Lemma 7.5. *Suppose that $\frac{\gamma_+}{a_+} = \frac{\gamma_-}{a_-}$, $\psi_+(0) = 0$ and the function $\psi_+(\omega)$ satisfies the boundary condition of $\overline{(StL)}$ at the point $\omega = \frac{\omega_0}{2}$. Then the function*

$$\psi_-(\omega) = -\left(\frac{a_+}{a_-}\right)^{\frac{1}{m-1}} \cdot \psi_+(-\omega), \quad \omega \in [-\frac{\omega_0}{2}, 0]$$

satisfies the boundary condition of $\overline{(StL)}$ at the point $\omega = -\frac{\omega_0}{2}$ and conjunction conditions of $\overline{(StL)}$ are valid.

Proof. In fact, we have

$$\psi_-(-\omega) = -\left(\frac{a_+}{a_-}\right)^{\frac{1}{m-1}} \cdot \psi_+(\omega), \quad \omega \in [0, \frac{\omega_0}{2}], \tag{7.4.1}$$

$$\frac{d\psi_-(-\omega)}{d\omega} = \left(\frac{a_+}{a_-}\right)^{\frac{1}{m-1}} \cdot \psi_+'(\omega); \quad \frac{d^2\psi_-(-\omega)}{d\omega^2} = -\left(\frac{a_+}{a_-}\right)^{\frac{1}{m-1}} \cdot \psi_+''(\omega). \tag{7.4.2}$$

Now from the boundary condition at the point $\omega = \frac{\omega_0}{2}$ we obtain

$$\begin{aligned} 0 &= \left(\overline{\lambda}^2\psi_+^2 + {\psi_+'}^2\right)^{\frac{m-2}{2}} \psi_+'\Big|_{\omega=\omega_0/2} + \frac{\gamma_+}{\overline{a}_+}\psi_+|\psi_+|^{m-2}\Big|_{\omega=\omega_0/2} \\ &= \frac{\overline{a}_-}{\overline{a}_+} \cdot \left\langle \left(\overline{\lambda}^2\psi_-^2 + {\psi_-'}^2\right)^{\frac{m-2}{2}} \psi_-'\Big|_{\omega=-\omega_0/2} - \frac{\gamma_-}{\overline{a}_-}\psi_-|\psi_-|^{m-2}\Big|_{\omega=-\omega_0/2} \right\rangle, \end{aligned}$$

by (7.4.1)–(7.4.2). Thus the boundary condition at the point $\omega = -\frac{\omega_0}{2}$ holds. Further, because $\psi_+(0) = 0$, we have $\psi_-(0) = 0$, by (7.4.1) and (7.4.2),

$$\overline{a}_+ \left(\overline{\lambda}^2\psi_+^2 + {\psi_+'}^2\right)^{\frac{m-2}{2}} \psi_+'\Big|_{\omega=0} = \overline{a}_+\psi_+'(0)|\psi_+'(0)|^{m-2} = \overline{a}_-\psi_-'(0)|\psi_-'(0)|^{m-2}.$$

Thus the conjunction conditions hold. □

Lemma 7.6. *Let $z_+(x) - r^{\overline{\lambda}}\psi_+(\omega)$, $\omega \in [0, \frac{\omega_0}{2}]$ and*

$$z_-(r, -\omega) = -\left(\frac{a_+}{a_-}\right)^{\frac{1}{m-1}} \cdot z_+(r, \omega), \quad \omega \in \left[0, \frac{\omega_0}{2}\right]. \tag{7.4.3}$$

Then the equality

$$\begin{aligned} a_-\Bigg\langle -\sum_{i=1}^{2} \frac{d}{dx_i}\left(r^\tau|\nabla z_-|^{m-2}\frac{\partial z_-}{\partial x_i}\right)\Bigg|_{(r,-\omega)} &+ \overline{a}_0 r^{\tau-m} z_-(r,-\omega)|z_-(r,-\omega)|^{m-2} \\ &- \overline{\mu} r^\tau z_-^{-1}(r,-\omega)|\nabla z_-(r,-\omega)|^m\Bigg\rangle \\ = -a_+\Bigg\langle -\sum_{i=1}^{2} \frac{d}{dx_i}\left(r^\tau|\nabla z_+|^{m-2}\frac{\partial z_+}{\partial x_i}\right)\Bigg|_{(r,\omega)} &+ \overline{a}_0 r^{\tau-m} z_+(r,\omega)|z_+(r,\omega)|^{m-2} \\ &- \overline{\mu} r^\tau z_+^{-1}(r,\omega)|\nabla z_+(r,\omega)|^m\Bigg\rangle \end{aligned} \tag{7.4.4}$$

holds for all $\omega \in \left(0, \frac{\omega_0}{2}\right)$.

Proof. By virtue of $x_1 = r\cos\omega$, $x_2 = r\sin\omega$, we calculate

$$
\begin{aligned}
\frac{\partial}{\partial x_1} &= \cos\omega \cdot \frac{\partial}{\partial r} - \frac{\sin\omega}{r} \cdot \frac{\partial}{\partial \omega}; \qquad \frac{\partial}{\partial x_2} = \sin\omega \cdot \frac{\partial}{\partial r} + \frac{\cos\omega}{r} \cdot \frac{\partial}{\partial \omega}; \\
\frac{\partial^2}{\partial x_1^2} &= \cos^2\omega \cdot \frac{\partial^2}{\partial r^2} - \frac{\sin 2\omega}{r} \cdot \frac{\partial^2}{\partial r \partial \omega} + \frac{\sin^2\omega}{r^2} \cdot \frac{\partial^2}{\partial \omega^2} + \frac{\sin^2\omega}{r} \cdot \frac{\partial}{\partial r} \\
&\quad + \frac{\sin 2\omega}{r^2} \cdot \frac{\partial}{\partial \omega}; \\
\frac{\partial^2}{\partial x_2^2} &= \sin^2\omega \cdot \frac{\partial^2}{\partial r^2} + \frac{\sin 2\omega}{r} \cdot \frac{\partial^2}{\partial r \partial \omega} + \frac{\cos^2\omega}{r^2} \cdot \frac{\partial^2}{\partial \omega^2} + \frac{\cos^2\omega}{r} \cdot \frac{\partial}{\partial r} \\
&\quad - \frac{\sin 2\omega}{r^2} \cdot \frac{\partial}{\partial \omega}; \\
\frac{\partial^2}{\partial x_1 \partial x_2} &= \frac{\sin 2\omega}{2} \cdot \frac{\partial^2}{\partial r^2} + \frac{\cos 2\omega}{r} \cdot \frac{\partial^2}{\partial r \partial \omega} - \frac{\sin 2\omega}{2r^2} \cdot \frac{\partial^2}{\partial \omega^2} - \frac{\sin 2\omega}{2r} \cdot \frac{\partial}{\partial r} \\
&\quad - \frac{\cos 2\omega}{r^2} \cdot \frac{\partial}{\partial \omega}.
\end{aligned} \tag{7.4.5}
$$

Therefore, by (7.4.1)–(7.4.2), we get:

$$
\begin{aligned}
\frac{\partial z_+(r,\omega)}{\partial x_1} &= r^{\overline{\lambda}-1}\left(\overline{\lambda}\cos\omega \cdot \psi_+(\omega) - \sin\omega \cdot \psi_+'(\omega)\right); \\
\frac{\partial z_+(r,\omega)}{\partial x_2} &= r^{\overline{\lambda}-1}\left(\overline{\lambda}\sin\omega \cdot \psi_+(\omega) + \cos\omega \cdot \psi_+'(\omega)\right); \\
\frac{\partial z_-(r,-\omega)}{\partial x_1} &= r^{\overline{\lambda}-1}\left(-\overline{\lambda}\cos\omega \cdot \psi_+(\omega) + \sin\omega \cdot \psi_+'(\omega)\right)\left(\frac{a_+}{a_-}\right)^{\frac{1}{m-1}} \\
&= -\left(\frac{a_+}{a_-}\right)^{\frac{1}{m-1}} \frac{\partial z_+(r,\omega)}{\partial x_1}; \\
\frac{\partial z_-(r,-\omega)}{\partial x_2} &= r^{\overline{\lambda}-1}\left(\overline{\lambda}\sin\omega \cdot \psi_+(\omega) + \cos\omega \cdot \psi_+'(\omega)\right)\left(\frac{a_+}{a_-}\right)^{\frac{1}{m-1}} \\
&= \left(\frac{a_+}{a_-}\right)^{\frac{1}{m-1}} \frac{\partial z_+(r,\omega)}{\partial x_2}; \\
\frac{\partial^2 z_+(r,\omega)}{\partial x_1^2} &= r^{\overline{\lambda}-2}\Big\langle \sin^2\omega \cdot \psi_+''(\omega) + (1-\overline{\lambda})\sin 2\omega \cdot \psi_+'(\omega) \\
&\qquad + (\overline{\lambda}^2\cos^2\omega - \overline{\lambda}\cos 2\omega) \cdot \psi_+(\omega)\Big\rangle; \\
2\frac{\partial^2 z_+(r,\omega)}{\partial x_1 \partial x_2} &= r^{\overline{\lambda}-2}\Big\langle -\sin 2\omega \cdot \psi_+''(\omega) + 2(\overline{\lambda}-1)\cos 2\omega \cdot \psi_+'(\omega) \\
&\qquad + (\overline{\lambda}^2 - 2\overline{\lambda})\sin 2\omega \cdot \psi_+(\omega)\Big\rangle;
\end{aligned}
$$

$$\frac{\partial^2 z_+(r,\omega)}{\partial x_2^2} = r^{\overline{\lambda}-2}\Big\langle \cos^2\omega \cdot \psi_+''(\omega) - (1-\overline{\lambda})\sin 2\omega \cdot \psi_+'(\omega) + (\overline{\lambda}^2 \sin^2\omega + \overline{\lambda}\cos 2\omega)\cdot \psi_+(\omega)\Big\rangle. \tag{7.4.6}$$

Hence it follows that

$$\begin{aligned}
|\nabla z_-|^{m-2}\frac{\partial z_-}{\partial x_1}\Big|_{(r,-\omega)} &= -\frac{a_+}{a_-}|\nabla z_+|^{m-2}\frac{\partial z_+}{\partial x_1}\Big|_{(r,\omega)};\\
|\nabla z_-|^{m-2}\frac{\partial z_-}{\partial x_2}\Big|_{(r,-\omega)} &= \frac{a_+}{a_-}|\nabla z_+|^{m-2}\frac{\partial z_+}{\partial x_2}\Big|_{(r,\omega)};\\
\frac{\partial^2 z_-}{\partial x_1^2}\Big|_{(r,-\omega)} &= -\left(\frac{a_+}{a_-}\right)^{\frac{1}{m-1}}\cdot \frac{\partial^2 z_+}{\partial x_1^2}\Big|_{(r,\omega)};\\
\frac{\partial^2 z_-}{\partial x_1\partial x_2}\Big|_{(r,-\omega)} &= \left(\frac{a_+}{a_-}\right)^{\frac{1}{m-1}}\cdot \frac{\partial^2 z_+}{\partial x_1\partial x_2}\Big|_{(r,\omega)};\\
\frac{\partial^2 z_-}{\partial x_2^2}\Big|_{(r,-\omega)} &= -\left(\frac{a_+}{a_-}\right)^{\frac{1}{m-1}}\cdot \frac{\partial^2 z_+}{\partial x_2^2}\Big|_{(r,\omega)}.
\end{aligned} \tag{7.4.7}$$

Further, by virtue of

$$\begin{aligned}
\sum_{i=1}^{2} x_i \frac{\partial z_+(r,\omega)}{\partial x_i} &= r^{\overline{\lambda}}\cos\omega\cdot\big(\overline{\lambda}\cos\omega\cdot\psi_+(\omega) - \sin\omega\cdot\psi_+'(\omega)\big)\\
&\quad + r^{\overline{\lambda}}\sin\omega\cdot\big(\overline{\lambda}\sin\omega\cdot\psi_+(\omega) + \cos\omega\cdot\psi_+'(\omega)\big)\\
&= \overline{\lambda} r^{\overline{\lambda}}\psi_+(\omega) = \overline{\lambda} z_+(r,\omega)
\end{aligned} \tag{7.4.8}$$

as well as

$$\begin{aligned}
\sum_{i=1}^{2}\frac{d}{dx_i}\left(r^{\tau}|\nabla z|^{m-2}z_{x_i}\right) &= r^{\tau}|\nabla z|^{m-4}\sum_{i,j=1}^{2}\Big\{\delta_i^j|\nabla z|^2 + (m-2)z_{x_i}z_{x_j}\Big\}z_{x_ix_j}\\
&\quad + \overline{\lambda}\tau r^{\tau-2}z|\nabla z|^{m-2},
\end{aligned} \tag{7.4.9}$$

we obtain

$$a_-\sum_{i=1}^{2}\frac{d}{dx_i}\left(r^{\tau}|\nabla z_-|^{m-2}\frac{\partial z_-}{\partial x_i}\right)\Big|_{(r,-\omega)} = -a_+\sum_{i=1}^{2}\frac{d}{dx_i}\left(r^{\tau}|\nabla z_+|^{m-2}\frac{\partial z_+}{\partial x_i}\right)\Big|_{(r,\omega)} \tag{7.4.10}$$

and

$$\begin{aligned}
&\overline{a}_0 r^{\tau-m} z_-(r,-\omega)|z_-(r,-\omega)|^{m-2} - \overline{\mu} r^{\tau} z_-^{-1}(r,-\omega)|\nabla z_-(r,-\omega)|^m\\
&= -\frac{a_+}{a_-}\Big\langle \overline{a}_0 r^{\tau-m} z_+(r,\omega)|z_+(r,\omega)|^{m-2} - \overline{\mu} r^{\tau} z_+^{-1}(r,\omega)|\nabla z_+(r,\omega)|^m\Big\rangle.
\end{aligned} \tag{7.4.11}$$

Hence it follows the required (7.4.4). □

Lemmas 7.5 and 7.6 give the following

Corollary 7.7. *Let* $\frac{\gamma_+}{a_+} = \frac{\gamma_-}{a_-}$ *and* $\psi_+(\omega) \not\equiv 0$ *be a solution of the problem*

$$\begin{cases} \frac{d}{d\omega}\left\{\left(\overline{\lambda}^2\psi_+^2 + {\psi'_+}^2\right)^{\frac{m-2}{2}}\psi'_+\right\} \\ +\overline{\lambda}\left\langle\overline{\lambda}(m-1) - m + 2 + \tau\right\rangle\psi_+\left(\overline{\lambda}^2\psi_+^2 + {\psi'_+}^2\right)^{\frac{m-2}{2}} \\ = \overline{a}_0\psi_+|\psi_+|^{m-2} - \overline{\mu}\psi_+^{-1}\left(\overline{\lambda}^2\psi_+^2 + {\psi'_+}^2\right)^{\frac{m}{2}}, \qquad \omega \in (0, \omega_0/2), \\ \\ \psi_+(0) = 0, \quad \overline{a}_+\left(\overline{\lambda}^2\psi_+^2 + {\psi'_+}^2\right)^{\frac{m-2}{2}}\psi'_+\Big|_{\omega=\omega_0/2} + \gamma_+\psi_+|\psi_+|^{m-2}\Big|_{\omega=\omega_0/2} = 0. \end{cases} \tag{MiP}$$

Then the function

$$z(r,\omega) = r^{\overline{\lambda}}\begin{cases} a_-^{\frac{1}{m-1}}\psi_+(\omega), & \omega \in [0, \omega_0/2], \\ -a_+^{\frac{1}{m-1}}\psi_+(-\omega), & \omega \in [-\omega_0/2, 0] \end{cases} \tag{7.4.12}$$

is a solution of problem $\overline{(MVP)}$.

Lemma 7.8. *Let* $\frac{\gamma_+}{a_+} = \frac{\gamma_-}{a_-}$ *and* $\psi_+(\omega) \not\equiv 0$ *be a solution of the problem (MiP). Suppose, in addition, that* $m \geq 2$ *and the inequality*

$$\lambda^m(q + m - 1 + \mu) + \lambda^{m-1}(2 - m + \tau) > a_0 \tag{7.4.13}$$

holds. Then $\psi_+(\omega) > 0$ *in* $(0, \omega_0/2]$ *and* $\psi''_+(\omega) < 0$ *in* $(0, \omega_0/2)$.

Proof. We rewrite the (MiP) equation in the form (7.3.6). By setting $\psi'_+(\omega)/\psi_+(\omega) = \overline{y}(\omega)$, we arrive at the Cauchy problem for $\overline{y}(\omega)$, $\overline{\lambda}$:

$$\begin{cases} \left[(m-1)\overline{y}^2 + \overline{\lambda}^2\right](\overline{y}^2 + \overline{\lambda}^2)^{\frac{m-4}{2}}\overline{y}' + (m - 1 + \overline{\mu})(\overline{y}^2 + \overline{\lambda}^2)^{\frac{m}{2}} \\ \qquad + \overline{\lambda}(2 - m + \tau)(\overline{y}^2 + \overline{\lambda}^2)^{\frac{m-2}{2}} = \overline{a}_0, \quad \omega \in (0, \omega_0/2); \\ \\ \overline{y}(0) = +\infty; \quad \overline{a}_+\left(\overline{y}^2(\omega_0/2) + \overline{\lambda}^2\right)^{\frac{m-2}{2}} \cdot \overline{y}(\omega_0/2) = -\gamma_+. \end{cases} \tag{CPE}$$

From the equation of (CPE) we get:

$$\begin{aligned} &-\left[(m-1)\overline{y}^2 + \overline{\lambda}^2\right](\overline{y}^2 + \overline{\lambda}^2)^{\frac{m-4}{2}}\overline{y}' \\ &= (m - 1 + \overline{\mu})(\overline{y}^2 + \overline{\lambda}^2)^{\frac{m}{2}} + \lambda(2 - m + \tau)(\overline{y}^2 + \overline{\lambda}^2)^{\frac{m-2}{2}} - \overline{a}_0 \\ &= (\overline{y}^2 + \overline{\lambda}^2)^{\frac{m-2}{2}}\left[(m - 1 + \overline{\mu})(\overline{y}^2 + \overline{\lambda}^2) + \overline{\lambda}(2 - m + \tau)\right] - \overline{a}_0 \\ &\geq (\overline{y}^2 + \overline{\lambda}^2)^{\frac{m-2}{2}}[\overline{\lambda}^2(m - 1 + \overline{\mu}) + \overline{\lambda}(2 - m + \tau)] - \overline{a}_0 \end{aligned}$$

$$\geq \overline{\lambda}^m(m - 1 + \overline{\mu}) + \overline{\lambda}^{m-1}(2 - m + \tau) - \overline{a}_0 > 0$$

by virtue of (7.4.13) and $m \geq 2$. Thus, it is proved that $\overline{y}'(\omega) < 0$, $\omega \in (0, \omega_0/2)$. Therefore $\overline{y}(\omega)$ *decreases* on the interval $(0, \omega_0/2)$. Hence it follows that $\overline{y}(\omega_0/2) < \overline{y}(\omega) < \overline{y}(0) = \infty$ and therefore $\psi_+(\omega) \neq 0$, $\forall \omega \in (0, \omega_0/2)$. By the continuity, this means that the function $\psi_+(\omega)$ retains its sign on the interval $(0, \omega_0/2)$. In addition, we note that the solutions of (CPE) are determined uniquely up to a scalar multiple. Thus we can assume $\psi_+(\omega) > 0$ in $(0, \omega_0/2]$ and then, by Lemma 7.4, $\psi_+''(\omega) < 0$ in $(0, \omega_0/2)$. □

7.4.2 Properties of the eigenvalue λ for (CPE)

Let $y\left(\frac{\omega_0}{2}\right) = y_+$. Taking into account that $\overline{a} = a\varsigma^{m-1}$, $\overline{a}_0 = a_0\varsigma^{1-m}$, $\overline{\mu} = \mu\varsigma$, $\overline{\lambda} = \frac{\lambda}{\varsigma}$ and performing the change of the variable $\overline{y}(\omega) = \frac{y(\omega)}{\varsigma}$ we integrate (CPE) and obtain the equation for the determination of λ:

$$\begin{cases} \int\limits_{y_+}^{\infty} \dfrac{\left[(m-1)y^2 + \lambda^2\right](y^2+\lambda^2)^{\frac{m-4}{2}}dy}{(m-1+q+\mu)(y^2+\lambda^2)^{\frac{m}{2}} + \lambda(2-m+\tau)(y^2+\lambda^2)^{\frac{m-2}{2}} - a_0} = \frac{\omega_0}{2}, \\ a_+\left(\lambda^2 + y_+^2\right)^{\frac{m-2}{2}} \cdot y_+ = -\gamma_+, \\ \lambda^m(q + m - 1 + \mu) + \lambda^{m-1}(2 - m + \tau) > a_0. \end{cases} \qquad (\Lambda)_+$$

We set

$$\mathcal{F}(\lambda, a_0, \omega_0) \qquad (7.4.14)$$
$$= -\frac{\omega_0}{2} + \int\limits_{y_+}^{+\infty} \frac{[(m-1)y^2+\lambda^2](y^2+\lambda^2)^{\frac{m-4}{2}}dy}{(m-1+q+\mu)(y^2+\lambda^2)^{\frac{m}{2}} + \lambda(2-m+\tau)(y^2+\lambda^2)^{\frac{m-2}{2}} - a_0}.$$

By performing the substitution: $y = \lambda t$, $y_+ = \lambda t_+$, $t \in (t_+, +\infty)$ we obtain

$$\mathcal{F}(\lambda, a_0, \omega_0) = -\frac{\omega_0}{2} + \int\limits_{t_+}^{+\infty} \Lambda(\lambda, a_0, t)dt, \qquad (7.4.15)$$

where

$$\Lambda(\lambda, a_0, t) \equiv \frac{[(m-1)t^2+1](t^2+1)^{\frac{m-4}{2}}}{\lambda(m-1+q+\mu)(t^2+1)^{\frac{m}{2}} + (2-m+\tau)(t^2+1)^{\frac{m-2}{2}} - a_0\lambda^{1-m}} > 0$$
$$\text{(by the third inequality in } (\Lambda)_+\text{);} \qquad (7.4.16)$$

$$a_+\lambda^{m-1}\left(1+t_+^2\right)^{\frac{m-2}{2}} \cdot t_+ = -\gamma_+ \qquad (7.4.17)$$

(we note that (7.4.17) follows from the second equation in $(\Lambda)_+$). Then the first equation of $(\Lambda)_+$ takes the form

$$\mathcal{F}(\lambda, a_0, \omega_0) = 0. \tag{7.4.18}$$

Considering λ as a function $\lambda(a_0, \omega_0)$ we set $\lambda_0 = \lambda(0, \omega_0)$. Then we have

$$\mathcal{F}(\lambda_0, 0, \omega_0) = 0. \tag{7.4.19}$$

The direct calculations yield

$$\frac{\partial \Lambda}{\partial \lambda} = -[(m-1)t^2 + 1](t^2+1)^{\frac{m-4}{2}} \times \frac{(m-1+q+\mu)(t^2+1)^{\frac{m}{2}} + a_0(m-1)/\lambda^m}{\left[\lambda(m-1+q+\mu)(t^2+1)^{\frac{m}{2}} + (2-m+\tau)(t^2+1)^{\frac{m-2}{2}} - a_0\lambda^{1-m}\right]^2} < 0, \tag{7.4.20}$$

$$\frac{\partial \Lambda}{\partial a_0} = \frac{\lambda^{1-m}[(m-1)t^2+1](t^2+1)^{\frac{m-4}{2}}}{\left[\lambda(m-1+q+\mu)(t^2+1)^{\frac{m}{2}} + (2-m+\tau)(t^2+1)^{\frac{m-2}{2}} - a_0\lambda^{1-m}\right]^2} > 0, \tag{7.4.21}$$

for all t, λ, a_0. Therefore, we can apply the theorem about implicit functions. In a certain neighborhood of the point $(\lambda_0, 0)$ the equation (7.4.18) determines $\lambda = \lambda(a_0, \omega_0)$ as a single-valued *continuous* function of a_0, depending continuously on the parameter ω_0 and having continuous partial derivatives $\frac{\partial \lambda}{\partial a_0}, \frac{\partial \lambda}{\partial \omega_0}$. Now, we analyze the properties of λ as the function $\lambda(a_0, \omega_0)$. First, from (7.4.18) we get:

$$\frac{\partial \mathcal{F}}{\partial \lambda}\frac{\partial \lambda}{\partial a_0} + \frac{\partial \mathcal{F}}{\partial a_0} = 0 \quad \text{and} \quad \frac{\partial \mathcal{F}}{\partial \lambda}\frac{\partial \lambda}{\partial \omega_0} + \frac{\partial \mathcal{F}}{\partial \omega_0} = 0.$$

Hence it follows that

$$\frac{\partial \lambda}{\partial a_0} = -\frac{\left(\frac{\partial \mathcal{F}}{\partial a_0}\right)}{\left(\frac{\partial \mathcal{F}}{\partial \lambda}\right)} \quad \text{and} \quad \frac{\partial \lambda}{\partial \omega_0} = -\frac{\left(\frac{\partial \mathcal{F}}{\partial \omega_0}\right)}{\left(\frac{\partial \mathcal{F}}{\partial \lambda}\right)}. \tag{7.4.22}$$

Now, from (7.4.15) and in virtue of (7.4.21), we have

$$\frac{\partial \mathcal{F}}{\partial \omega_0} = -\frac{1}{2}, \qquad \frac{\partial \mathcal{F}}{\partial a_0} = \int_{t_+}^{+\infty} \frac{\partial \Lambda}{\partial a_0} dt > 0, \quad \forall (\lambda, a_0). \tag{7.4.23}$$

The equation (7.4.17) means that t_+ is a negative function of λ, i.e., $t_+ = t_+(\lambda)$. Differentiating this equation we obtain

$$\frac{dt_+}{d\lambda} = \frac{(m-1)\gamma_+}{a_+\lambda^m \left\langle 1 + (m-1)t_+^2 \right\rangle (1+t_+^2)^{\frac{m-4}{2}}} > 0.$$

Hence and from (7.4.15) it follows that

$$\frac{\partial \mathcal{F}}{\partial \lambda} = \int_{t_+}^{+\infty} \frac{\partial \Lambda}{\partial \lambda} dt - \Lambda(\lambda, a_0, t_+) \frac{dt_+}{d\lambda} < 0, \tag{7.4.24}$$

by virtue of (7.4.16) and (7.4.20).

From (7.4.22)–(7.4.24) we get:

$$\frac{\partial \lambda}{\partial a_0} > 0 \text{ and } \frac{\partial \lambda}{\partial \omega_0} < 0 \quad \text{for any } a_0 \geq 0. \tag{7.4.25}$$

So, *the function* $\lambda(a_0, \omega_0)$ *increases with respect to* a_0 *and decreases with respect to* ω_0. Applying the analytic continuation method, we obtain the solvability of $(\Lambda)_+$, $\forall a_0 \geq 0$.

Corollary 7.9.

$$\lambda = \lambda(a_0, \omega_0) \geq \lambda_0 > 0 \quad \textit{for any } a_0 \geq 0.$$

Remark 5. **The eigenfunction existence.** It follows from Corollary 7.7 that it is sufficient to verify the existence of the (MiP) non-trivial solution $\psi_+(\omega)$. In this connection, by the Lemma 7.8 proof, we must integrate the (CPE) equation. As a result, according to (7.3.3) we get

$$\begin{cases} \Phi_+(\omega) = \exp\left(\int y(\omega) d\omega\right); \\ \\ \displaystyle\int_{y(\omega)}^{+\infty} \frac{\left[(m-1)y^2 + \lambda^2\right](y^2+\lambda^2)^{\frac{m-4}{2}} dy}{(m-1+q+\mu)(y^2+\lambda^2)^{\frac{m}{2}} + \lambda(2-m+\tau)(y^2+\lambda^2)^{\frac{m-2}{2}} - a_0} = \omega, \end{cases}$$

where λ is the $(\Lambda)_+$ solution.

7.4.3 Perturbation of problem (MiP)

We consider the perturbation of problem (MiP). Namely, for $\forall \varepsilon \in (0, \frac{\omega_0}{2}]$ on the segment $[-\varepsilon, \frac{\omega_0}{2}]$, we define the problem for $(\lambda_\varepsilon, \psi_\varepsilon)$:

$$\begin{cases} \bar{a}_+ \dfrac{d}{d\omega}\left[\left(\lambda_\varepsilon^2 \psi_\varepsilon^2 + \psi_\varepsilon'^2\right)^{\frac{m-2}{2}} \psi_\varepsilon'\right] \\ +\bar{a}_+ \lambda_\varepsilon [\lambda_\varepsilon(m-1) - m + 2 + \tau] \psi_\varepsilon \left(\lambda_\varepsilon^2 \psi_\varepsilon^2 + \psi_\varepsilon'^2\right)^{\frac{m-2}{2}} \\ \qquad +\bar{a}_+ \frac{\mu}{\psi_\varepsilon} \left(\lambda_\varepsilon^2 \psi_\varepsilon^2 + \psi_\varepsilon'^2\right)^{\frac{m}{2}} = a_+(a_0 - \varepsilon) \psi_\varepsilon |\psi_\varepsilon|^{m-2}, \quad \omega \in (-\varepsilon, \frac{\omega_0}{2}); \\ \\ \psi_\varepsilon(-\varepsilon) = 0, \quad \bar{a}_+ \left(\lambda_\varepsilon^2 \psi_\varepsilon^2 + \psi_\varepsilon'^2\right)^{\frac{m-2}{2}} \psi_\varepsilon' \Big|_{\omega=\omega_0/2} + \gamma_+ \psi_\varepsilon |\psi_\varepsilon|^{m-2} \Big|_{\omega=\omega_0/2} = 0. \end{cases} \tag{$(MiP)_\varepsilon$}$$

Let us define function $y_\varepsilon(\omega) = \frac{\psi'_\varepsilon(\omega)}{\psi_\varepsilon(\omega)}$. From $(MiP)_\varepsilon$, by integrating, we get the equation for λ_ε:

$$\begin{cases} \displaystyle\int_{y_\varepsilon(\omega_0/2)}^{+\infty} \frac{\left[(m-1)y^2+\lambda_\varepsilon^2\right](y^2+\lambda_\varepsilon^2)^{\frac{m-4}{2}}dy}{(m-1+\overline{\mu})(y^2+\lambda_\varepsilon^2)^{\frac{m}{2}}+\lambda_\varepsilon(2-m+\tau)(y^2+\lambda_\varepsilon^2)^{\frac{m-2}{2}}+\varepsilon\varsigma^{1-m}-\overline{a}_0} \\ = \frac{\omega_0}{2}+\varepsilon, \\ \overline{a}_+\left(\lambda_\varepsilon^2+y_\varepsilon^2(\omega_0/2)\right)^{\frac{m-2}{2}}\cdot y_\varepsilon(\omega_0/2) = -\gamma_+. \end{cases} \qquad (\Lambda)_\varepsilon$$

Problem $(MiP)_\varepsilon$ is obtained from problem (MiP) by replacing in the latter a_0 by $a_0-\varepsilon$. By virtue of the monotonicity of the function $\overline{\lambda}(\omega_0,\overline{a}_0)$ established above, we get

$$0<\lambda_\varepsilon<\overline{\lambda}, \quad \lim_{\varepsilon\to+0}\lambda_\varepsilon=\overline{\lambda}. \qquad (7.4.26)$$

We denote by $\overline{\lambda}_0$ the value of $\overline{\lambda}$ for $\overline{a}_0=0$. It easily follows from (7.4.15) that $\overline{\lambda}_0=\lambda_0\Big|_{q=0;\ \mu=\overline{\mu}}$. By virtue of Corollary 7.9, we have $\overline{\lambda}>\overline{\lambda}_0$. From (7.4.26) we draw that

$$0<\frac{1}{2}\overline{\lambda}_0<\lambda_\varepsilon<\overline{\lambda} \qquad (7.4.27)$$

for a sufficiently small $\varepsilon>0$.

Lemma 7.10. *Let assumptions of Lemma* 7.8 *be satisfied. There exists* $\varepsilon^*>0$ *such that*

$$\psi_\varepsilon(0)\geq\frac{\varepsilon}{\omega_0}\psi_\varepsilon\left(\frac{\omega_0}{2}\right), \quad \forall\varepsilon\in(0,\varepsilon^*). \qquad (7.4.28)$$

Proof. We turn back to the inequality $P_m(\lambda)\equiv\lambda^m(q+m-1+\mu)+\lambda^{m-1}(2-m+\tau)-a_0>0$ from $(\Lambda)_+$, or $\overline{P}_m(\overline{\lambda})\equiv\overline{\lambda}^m(m-1+\overline{\mu})+\overline{\lambda}^{m-1}(2-m+\tau)-\overline{a}_0>0$, because of $\overline{\lambda}=\frac{\lambda}{\varsigma}$, $\overline{\mu}=\varsigma\mu$, $\overline{a}_0=a_0\varsigma^{1-m}$, $\varsigma=\frac{m-1}{q+m-1}$. Since $\overline{P}_m(\overline{\lambda})$ is an exponential function with positive exponent, then, by the continuity, there exists a δ^*-neighborhood of the point $\overline{\lambda}$, in which the inequality $\overline{P}_m(\overline{\lambda})>0$ remains valid, that is there exists $\delta^*>0$ such that $\overline{P}_m(\lambda)>0$, $\forall\lambda \mid |\lambda-\overline{\lambda}|<\delta^*$. We choose the number $\delta^*>0$ to guarantee this. In particular, the inequality

$$\overline{P}_m(\overline{\lambda}-\delta)>0, \quad \forall\delta\in(0,\delta^*) \qquad (7.4.29)$$

holds. Let us recall that $\overline{\lambda}$ is a solution of (CPE). By (7.4.26) we can now put, for every $\delta\in(0,\delta^*)$,

$$\lambda_\varepsilon=\overline{\lambda}-\delta \qquad (7.4.30)$$

and solve $(\Lambda)_\varepsilon$ together with this λ_ε with respect to ε. Let $\varepsilon(\delta)>0$ be the obtained solution. Since (7.4.26) holds, $\lim_{\delta\to+0}\varepsilon(\delta)=+0$. Thus, we get the sequence of problems $(MiP)_\varepsilon$–$(\Lambda)_\varepsilon$ with respect to

$$(\lambda_\varepsilon,\psi_\varepsilon(\omega)) \quad \forall\varepsilon \text{ such that } 0<\varepsilon<\min\{\varepsilon(\delta),\frac{\omega_0}{2}\}=\varepsilon^*(\delta),\ \forall\delta\in(0,\delta^*). \quad (7.4.31)$$

We consider $\psi_\varepsilon(\omega) \geq 0$ from (7.4.31). In the same way as in Lemma 7.4 we verify that $\psi''_\varepsilon(\omega) < 0 \; \forall \omega \in \left(-\varepsilon, \frac{\omega_0}{2}\right)$. This inequality means that the function $\psi_\varepsilon(\omega)$ is convex (concave downward) on the segment $[-\varepsilon, \frac{\omega_0}{2}]$, that is

$$\psi_\varepsilon(\alpha_1\omega_1 + \alpha_2\omega_2) \geq \alpha_1\psi_\varepsilon(\omega_1) + \alpha_2\psi_\varepsilon(\omega_2) \quad \forall \omega_1, \omega_2 \in \left[-\varepsilon, \frac{\omega_0}{2}\right]$$
$$\text{for } \alpha_1 \geq 0, \; \alpha_2 \geq 0 \; \big| \alpha_1 + \alpha_2 = 1.$$

We put $\alpha_1 = \frac{\omega_0}{2\varepsilon+\omega_0}, \quad \alpha_2 = \frac{2\varepsilon}{2\varepsilon+\omega_0}$ and $\omega_1 = -\varepsilon, \quad \omega_2 = \frac{\omega_0}{2}$. By $(MiP)_\varepsilon$ we get

$$\psi_\varepsilon(0) \geq \frac{2\varepsilon}{\omega_0 + 2\varepsilon}\psi_\varepsilon\left(\frac{\omega_0}{2}\right) \geq \frac{\varepsilon}{\omega_0}\psi_\varepsilon\left(\frac{\omega_0}{2}\right).$$

Thus, the lemma is proved. □

Hence, because solutions of $(MiP)_\varepsilon$ are determined uniquely up to a scalar factor, we have the following

Corollary 7.11.

$$\varepsilon \leq \psi_\varepsilon(\omega) \leq 1 \quad \forall \omega \subset [0, \omega_0/2], \quad \forall \varepsilon \subset (0, \varepsilon^*). \tag{7.4.32}$$

Lemma 7.12. *Let $\tau \geq m - 2$ and assumptions of Lemma 7.8 be satisfied. Then there exist positive constants C_1, C_2 depending only on $m, q, \mu, \tau, \lambda, \varepsilon$ such that*

$$|\psi'_\varepsilon(\omega)| \leq C_1, \quad |\psi''_\varepsilon(\omega)| \leq C_2, \quad \forall \omega \in [0, \omega_0/2].$$

Proof. If $a_0 > 0$ we will assume that $\varepsilon < a_0$. From the $(MiP)_\varepsilon$ equation we have

$$\psi''_\varepsilon(\omega) = F(\omega, \psi_\varepsilon, \psi'_\varepsilon), \quad \omega \in (-\varepsilon, \frac{\omega_0}{2}),$$

where

$$|F(\omega, \psi_\varepsilon, \psi'_\varepsilon)| = \Big\langle \overline{\mu}\psi_\varepsilon^{-1}(\lambda_\varepsilon^2\psi_\varepsilon^2 + \psi_\varepsilon'^2)^{\frac{m}{2}} + (m-2)\lambda_\varepsilon^2\psi_\varepsilon\psi_\varepsilon'^2(\lambda_\varepsilon^2\psi_\varepsilon^2 + \psi_\varepsilon'^2)^{\frac{m-4}{2}}$$
$$+ \lambda_\varepsilon \left\langle \lambda_\varepsilon(m-1) + 2 - m + \tau \right\rangle \psi_\varepsilon(\lambda_\varepsilon^2\psi_\varepsilon^2 + \psi_\varepsilon'^2)^{\frac{m-2}{2}} - \varsigma^{1-m}(a_0 - \varepsilon)\psi_\varepsilon^{m-1}\Big\rangle$$
$$\times \left\langle \lambda_\varepsilon^2\psi_\varepsilon^2 + (m-1)\psi_\varepsilon'^2 \right\rangle^{-1} (\lambda_\varepsilon^2\psi_\varepsilon^2 + \psi_\varepsilon'^2)^{\frac{4-m}{2}}.$$

We can estimate from above the right-hand side of this equality. Taking into account $m \geq 2$, $\tau \geq m - 2$ and (7.4.32) we derive

$$|F(\omega, \psi_\varepsilon, \psi'_\varepsilon)| \leq \varepsilon^{-1}\overline{\mu}\psi_\varepsilon'^2 + (2m - 3 + \overline{\mu})\lambda_\varepsilon^2 + (2 - m + \tau)\lambda_\varepsilon, \; \omega \in [0, \omega_0/2].$$

Now we apply the S. Bernstein Theorem 1.25. In our case we have for this theorem:

$$A = \varepsilon^{-1}\overline{\mu}, \quad B = (2m - 3 + \overline{\mu})\lambda_\varepsilon^2 + (2 - m + \tau)\lambda_\varepsilon, \ M = 1.$$

Therefore we obtain

$$|\psi_\varepsilon'(\omega)| \le C_1 = \sqrt{\varepsilon \frac{(2m - 3 + \overline{\mu})\lambda_\varepsilon^2 + (2 - m + \tau)\lambda_\varepsilon}{2\overline{\mu}}} \cdot \exp(4\overline{\mu}\varepsilon^{-1}).$$

Next, from the equation for ψ_ε we have

$$|\psi_\varepsilon''(\omega)| \le C_2 = \langle (2m - 3 + \overline{\mu})\lambda_\varepsilon^2 + (2 - m + \tau)\lambda_\varepsilon \rangle \cdot \left(\frac{1}{2}\exp(8\overline{\mu}\varepsilon^{-1}) + 1 \right). \qquad \square$$

7.4.4 Estimates of the $(TDQL)$ solution modulus

In this Subsection we denote, for $x_2 > 0$,

$$x_\pm \equiv x\Big|_{\overline{G_\pm}} = (x_1, \pm x_2, x').$$

Theorem 7.13. *Let u be a weak solution of the problem $(TDQL)$ and assumptions 1)–10) are satisfied with (7.3.1), $m \ge 2$, $\frac{\gamma_+}{a_+} = \frac{\gamma_-}{a_-}$. Let λ be the least positive solution of $(\Lambda)_+$ and assume that $M_0 = \max_{x \in \overline{G}} |u(x)|$ is known . Suppose, in addition, that the following compatibility conditions are satisfied:*

$$\begin{aligned} a_- a_-^i(x_-, -\varkappa u_+, -\varkappa \nabla u_+) &= -a_+ a_+^i(x_+, u_+, \nabla u_+), \ \forall i \ne 2; \\ a_- a_-^2(x_-, -\varkappa u_+, -\varkappa \nabla u_+) &= +a_+ a_+^2(x_+, u_+, \nabla u_+); \\ a_- b_-(x_-, -\varkappa u_+, -\varkappa \nabla u_+) &= -a_+ b_+(x_+, u_+, \nabla u_+); \\ a_- c_-(x_-, -\varkappa u_+) &= -a_+ c_+(x_+, u_+); \\ g_-(x_-, -\varkappa u_+) &= -g_+(x_+, u_+); \qquad\qquad (CC) \\ \text{where} \quad \varkappa &= \left(\frac{a_+}{a_-}\right)^{\frac{1}{q+m-1}}; \end{aligned}$$

$$a_- f_-(x_-) = -a_+ f_+(x_+); \quad a_+\gamma_-\left(-\tfrac{\omega_0}{2}\right) = a_-\gamma_+\left(\tfrac{\omega_0}{2}\right).$$

Then for every $\varepsilon > 0$ there are a $c_\varepsilon > 0$ and a $d \in (0,1)$ depending only on the parameters and the norms of functions occuring in the assumptions and independent of u such that

$$|u(x)| \le c_\varepsilon r^{\lambda - \varepsilon}, \quad \forall x \in G_0^d. \tag{7.4.33}$$

Proof. At first, we consider an auxiliary function $v_+(x)$ as a strong solution of the mixed problem

$$\begin{cases} -\frac{d}{dx_i}\mathcal{A}_+^i(x, \nabla v_+) + \mathcal{B}_+(x, v_+, \nabla v_+) + a_0\mathcal{C}_+(x, v_+) = f_+(x), & x \in G_+^d; \\ v_+\Big|_{\Sigma_0} = 0; & \\ a_+\mathcal{A}_+^i(x, \nabla v_+)n_i + \gamma_+(\omega)r^{\tau - m + 1}v_+|v_+|^{m-2} = \mathcal{G}_+(x, v_+), & x \in \Gamma_+^d. \end{cases} \qquad (MP)_+$$

Now we define the operator

$$
\begin{aligned}
&Q_+(v_+,\eta)\\
&\equiv a_+\cdot \int\limits_{(G_0^d)_+} \Big\langle \mathcal{A}_+^i(x,\nabla v_+)\frac{\partial \eta}{\partial x_i} + \mathcal{B}_+(x,v_+,\nabla v_+)\eta + a_0\mathcal{C}_+(x,v_+)\eta - f_+(x)\eta\Big\rangle dx\\
&\quad + \gamma_+\left(\frac{\omega_0}{2}\right)\cdot \int\limits_{\Gamma_+^d} r^{\tau-m+1}v_+|v_+|^{m-2}\eta(x)ds - \int\limits_{\Gamma_+^d}\mathcal{G}_+(x,v_+)\eta(x)ds\\
&\quad - a_+\cdot \int\limits_{\Omega_d^+}\mathcal{A}_+^i(x,\nabla v_+)\cos(r,x_i)\eta(x)d\Omega_d
\end{aligned}
$$

for any $\eta(x)\in C^0(\overline{(G_0^d)_+})\cap\mathfrak{N}^1_{m,0,\tau}((G_0^d)_+)$. It is obvious that every strong solution $v_+(x)$ of problem $(MP)_+$ satisfies the integral identity $Q_+(v_+,\eta)=0$ for any non-negative $\eta(x)\in C^0(\overline{(G_0^d)_+})\cap\mathfrak{N}^1_{m,0,\tau}((G_0^d)_+)$ with $\eta\Big|_{\Sigma_0^d}=0$.

We consider $z_\varepsilon(x)=r^{\lambda_\varepsilon}\psi_\varepsilon(\omega)$, where $(\lambda_\varepsilon,\psi_\varepsilon)$ is the solution of perturbed eigenvalue problem $(MiP)_\varepsilon$–$(\Lambda)_\varepsilon$, as a barrier function and apply the comparison principle to $v_+(x)$ and $z_\varepsilon(r,\omega)$. We will show that $Q_+(Az_\varepsilon,\eta)\geq 0 = Q_+(v_+,\eta)$ for all non-negative $\eta\in C^0(\overline{(G_0^d)_+})\cap\mathfrak{N}^1_{m,0,\tau}((G_0^d)_+)$ with $\eta\Big|_{\Sigma_0^d}=0$ and some $A>0$. Integrating by parts in the first integral, we have:

$$
\begin{aligned}
&Q_+(Az_\varepsilon,\eta)\\
&\equiv a_+\cdot \int\limits_{(G_0^d)_+}\Big\langle -\frac{d}{dx_i}\mathcal{A}_+^i(x,A\nabla z_\varepsilon)+\mathcal{B}_+(x,Az_\varepsilon,A\nabla z_\varepsilon)+a_0\mathcal{C}_+(x,Az_\varepsilon)-f_+(x)\Big\rangle\eta(x)dx\\
&\quad+\int\limits_{\Gamma_+^d}\Big\langle a_+\mathcal{A}_+^i(x,A\nabla z_\varepsilon)n_i+\gamma_+\left(\frac{\omega_0}{2}\right)A^{m-1}r^{\tau-m+1}z_\varepsilon|z_\varepsilon|^{m-2}\Big\rangle\eta(x)ds\\
&\quad-\int\limits_{\Gamma_+^d}\mathcal{G}_+(x,Az_\varepsilon)\eta(x)ds.
\end{aligned}
$$

We verify that $z_\varepsilon(r,\omega)$ satisfies the problem

$$
\begin{cases}
\mathcal{L}_0^+z_\varepsilon\equiv -\overline{a}_+\frac{d}{dx_i}(r^\tau|\nabla z_\varepsilon|^{m-2}\frac{\partial z_\varepsilon}{\partial x_i})+\overline{a}_+\overline{a}_0r^{\tau-m}z_\varepsilon|z_\varepsilon|^{m-2}-\overline{a}_+\overline{\mu}r^\tau z_\varepsilon^{-1}|\nabla z_\varepsilon|^m\\
\qquad\qquad = \overline{a}_+\varepsilon r^{\tau-m}z_\varepsilon|z_\varepsilon|^{m-2},\quad r\in(0,d),\ \omega\in\left(-\varepsilon,\frac{\omega_0}{2}\right);\\
z_\varepsilon\Big|_{\omega=-\varepsilon}=0,\ \overline{a}_+|\nabla z_\varepsilon|^{m-2}\frac{\partial z_\varepsilon}{\partial\vec{n}}+\gamma_+r^{1-m}z_\varepsilon|z_\varepsilon|^{m-2}=0,\quad r\in(0,d),\ \omega=\frac{\omega_0}{2}
\end{cases}
\qquad (MP)_\varepsilon
$$

and therefore we get

$$
\begin{aligned}
&Q_+(Az_\varepsilon, \eta) \\
&\equiv \int\limits_{(G_0^d)_+} \Big\langle \mathcal{L}_0^+(Az_\varepsilon) - a_+ \frac{d}{dx_i}\left(\mathcal{A}_+^i(x, A\nabla z_\varepsilon) - (A\varsigma)^{m-1} r^\tau |\nabla z_\varepsilon|^{m-2}\frac{\partial z_\varepsilon}{\partial x_i}\right) \\
&\quad + \overline{a}_+ A^{m-1}\overline{\mu} r^\tau z_\varepsilon^{-1}|\nabla z_\varepsilon|^m - a_+ f_+(x) + a_+\mathcal{B}_+(x, Az_\varepsilon, A\nabla z_\varepsilon) \\
&\quad + a_+ a_0\left(\mathcal{C}_+(x, Az_\varepsilon) - A^{m-1} r^{\tau-m} z_\varepsilon |z_\varepsilon|^{m-2}\right)\Big\rangle \eta(x)dx \\
&\quad + \int\limits_{\Gamma_+^d} \Big\langle a_+\left(\mathcal{A}_+^i(x, A\nabla z_\varepsilon)n_i - (A\varsigma)^{m-1} r^\tau |\nabla z_\varepsilon|^{m-2}\frac{\partial z_\varepsilon}{\partial \vec{n}}\right) \qquad (7.4.34) \\
&\quad + \left\langle \gamma_+\left(\frac{\omega_0}{2}\right) - \gamma_+\right\rangle A^{m-1} r^{\tau-m+1} z_\varepsilon |z_\varepsilon|^{m-2}\Big\rangle \eta(x)ds - \int\limits_{\Gamma_+^d} \mathcal{G}_+(x, Az_\varepsilon)\eta(x)ds.
\end{aligned}
$$

Further, from the definition of $z_\varepsilon(x)$, by Corollary 7.11 and Lemma 7.12, we calculate

$$
\begin{aligned}
&\varepsilon r^{\lambda_\varepsilon} \le z_\varepsilon \le r^{\lambda_\varepsilon}; \\
&|\nabla z_\varepsilon| = \sqrt{\lambda_\varepsilon^2\psi_\varepsilon^2 + \psi_\varepsilon'^2}\cdot r^{\lambda_\varepsilon - 1} \le \sqrt{\lambda_\varepsilon^2 + C_1^2}\cdot r^{\lambda_\varepsilon-1} \equiv \widetilde{C_1} r^{\lambda_\varepsilon - 1}; \qquad (7.4.35) \\
&|D^2 z_\varepsilon| \le c(\lambda_\varepsilon)\left(|\psi_\varepsilon''| + |\psi_\varepsilon'| + |\psi_\varepsilon|\right)\cdot r^{\lambda_\varepsilon-2} \le (1 + C_1 + C_2)\cdot r^{\lambda_\varepsilon - 2} \equiv \widetilde{C_2} r^{\lambda_\varepsilon-2}.
\end{aligned}
$$

From the equation of $(MP)_\varepsilon$ by (7.4.32) we have

$$
\mathcal{L}_0^+(Az_\varepsilon) \ge a_+(A\varsigma)^{m-1}\varepsilon^m r^{\tau - m + (m-1)\lambda_\varepsilon}. \qquad (7.4.36)
$$

By direct calculation,

$$
\begin{aligned}
&\frac{d}{dx_i}\left(\mathcal{A}_+^i(x, A\nabla z_\varepsilon) - (A\varsigma)^{m-1} r^\tau |\nabla z_\varepsilon|^{m-2}\frac{\partial z_\varepsilon}{\partial x_i}\right) \\
&= \left(\frac{\partial \mathcal{A}_+^i(x, A\nabla z_\varepsilon)}{\partial x_i} - \tau (A\varsigma)^{m-1} r^{\tau-2}|\nabla z_\varepsilon|^{m-2}\sum_{i=1}^2 x_i z_{\varepsilon x_i}\right) \qquad (7.4.37) \\
&\quad + z_{\varepsilon x_i x_j}\Big\langle A\frac{\partial \mathcal{A}_+^i(x, A\nabla z_\varepsilon)}{\partial (A z_{\varepsilon x_j})} \\
&\qquad - (A\varsigma)^{m-1} r^\tau |\nabla z_\varepsilon|^{m-4}\left(\delta_i^j |\nabla z_\varepsilon|^2 + (m-2) z_{\varepsilon x_i} z_{\varepsilon x_j}\right)\Big\rangle.
\end{aligned}
$$

Therefore, by assumptions 3)′, 4)′, 5)′, 6)′, 7)′ and estimates (7.4.35),

$$
\begin{aligned}
&a_+ \left| \frac{d}{dx_i} \left(\mathcal{A}^i_+(x, A\nabla z_\varepsilon) - (A\varsigma)^{m-1} r^\tau |\nabla z_\varepsilon|^{m-2} \frac{\partial z_\varepsilon}{\partial x_i} \right) \right| \\
&\le \overline{a}_+ A^{m-1} \mathcal{A}(r) r^\tau |\nabla z_\varepsilon|^{m-2} \left(r^{-1} |\nabla z_\varepsilon| + |D^2 z_\varepsilon| \right) \\
&\le C_3 \overline{a}_+ A^{m-1} \mathcal{A}(r) r^{\tau+(m-1)\lambda_\varepsilon - m}, \qquad x \in (G_0^d)_+; \\
&\overline{a}_+ A^{m-1} \overline{\mu} r^\tau z_\varepsilon^{-1} |\nabla z_\varepsilon|^m + a_+ \mathcal{B}_+(x, Az_\varepsilon, A\nabla z_\varepsilon) \ge (\mu - \mu_0) \overline{a}_+ A^{m-1} \varsigma r^\tau z_\varepsilon^{-1} |\nabla z_\varepsilon|^m \\
&= (\mu - \mu_0) \overline{a}_+ A^{m-1} \varsigma r^{\tau+(m-1)\lambda_\varepsilon - m}, \qquad x \in (G_0^d)_+; \\
&\mathcal{C}_+(x, Az_\varepsilon) - A^{m-1} r^{\tau-m} z_\varepsilon |z_\varepsilon|^{m-2} \ge 0, \qquad x \in (G_0^d)_+; \\
&a_+ \left| \mathcal{A}^i_+(x, A\nabla z_\varepsilon) n_i - (A\varsigma)^{m-1} r^\tau |\nabla z_\varepsilon|^{m-2} \frac{\partial z_\varepsilon}{\partial \vec{n}} \right| \le \overline{a}_+ A^{m-1} \mathcal{A}(r) r^\tau |\nabla z_\varepsilon|^{m-1} \\
&\le C_4 \overline{a}_+ A^{m-1} \mathcal{A}(r) r^{\tau+(m-1)\lambda_\varepsilon - m+1}, \qquad x \in \Gamma_+^d.
\end{aligned}
$$

Therefore from (7.4.34) it follows that

$$
\begin{aligned}
&Q_+(Az_\varepsilon, \eta) \\
&\ge \int\limits_{(G_0^d)_+} \left\{ \overline{a}_+ A^{m-1} r^{\tau+(m-1)\lambda_\varepsilon - m} \left\langle \varepsilon^m + \varsigma(\mu - \mu_0) - C_3 \mathcal{A}(r) \right\rangle - a_+ f_+(x) \right\} \eta(x) dx \\
&\quad + \int\limits_{\Gamma_+^d} \left\{ \left(\varepsilon^{m-1} \left\langle \gamma_+ \left(\frac{\omega_0}{2} \right) - \gamma_+ \right\rangle - C_4 \overline{a}_+ \mathcal{A}(r) \right) \cdot A^{m-1} r^{\tau+(m-1)\lambda_\varepsilon - m+1} \right. \\
&\qquad \left. - \mathcal{G}_+(x, Az_\varepsilon) \right\} \eta(x) ds. \qquad (7.4.38)
\end{aligned}
$$

Next, by (7.4.27) and (7.3.3), $\lambda_\varepsilon < \overline{\lambda} = \frac{q+m-1}{m-1}\lambda$; consequently, assumption 10) takes the form

$$
|f(x)| \le f_1 r^{\tau-m+(m-1)\lambda_\varepsilon}, \quad |g(x,0)| \le g_1 r^{\tau-m+1+(m-1)\lambda_\varepsilon}.
$$

Since

$$
\mathcal{G}(x,v) = \mathcal{G}(x,0) + v \cdot \int_0^1 \frac{\partial \mathcal{G}(x, \tau v)}{\partial(\tau v)} d\tau = g(x,0) + v \cdot \int_0^1 \frac{\partial \mathcal{G}(x, \tau v)}{\partial(\tau v)} d\tau, \qquad (7.4.39)
$$

by assumption 8)$'$, we have

$$
-\mathcal{G}_+(x, Az_\varepsilon) \ge -g(x,0) \ge -g_1 r^{\tau-m+1+(m-1)\lambda_\varepsilon}.
$$

Now, from (7.4.38) we derive

$$
\begin{aligned}
Q_+(Az_\varepsilon, \eta) &\ge a_+ (A\varsigma)^{m-1} \left\langle \varepsilon^m + \varsigma(\mu - \mu_0) - C_3 \mathcal{A}(d) - f_1 (A\varsigma)^{1-m} \right\rangle \\
&\quad \times \int\limits_{(G_0^d)_+} r^{\tau+(m-1)\lambda_\varepsilon - m} \eta(x) dx +
\end{aligned}
$$

$$+\left\{\left(\varepsilon^{m-1}\left\langle\gamma_+\left(\frac{\omega_0}{2}\right)-\gamma_+\right\rangle - C_4\overline{a}_+\mathcal{A}(d)\right)\cdot A^{m-1}-g_1\right\}$$
$$\times\int\limits_{\Gamma_+^d} r^{\tau+(m-1)\lambda_\varepsilon-m+1}\eta(x)ds.$$

Now we fix $\varepsilon>0$ and choose at first $d>0$, by the continuity of $\mathcal{A}(r)$ at zero, so small that

$$\mathcal{A}(d)\le\frac{\gamma_+\left(\frac{\omega_0}{2}\right)-\gamma_+}{2\overline{a}_+C_4}\varepsilon^{m-1}$$

and next

$$A\ge\max\left\{\left(\frac{q+m-1}{m-1}\right)^{\frac{m}{m-1}}\cdot\left(\frac{2f_1}{\mu-\mu_0}\right)^{\frac{1}{m-1}};\quad\frac{1}{\varepsilon}\left\langle\frac{2g_1}{\gamma_+\left(\frac{\omega_0}{2}\right)-\gamma_+}\right\rangle^{\frac{1}{m-1}}\right\}.$$

Then we get $Q_+(Az_\varepsilon,\eta)\ge0$.

Now, by the continuity of $v_+(x)$, we have

$$v_+(x)\Big|_{\Omega_d}\le\max_{\overline{G_+}}|v_+(x)|\le M_0^{1/\varsigma}.$$

On the other hand, by virtue of Corollary 7.11,

$$Az_\varepsilon\Big|_{\Omega_d}\ge A\varepsilon d^{\lambda_\varepsilon}\ge M_0^{1/\varsigma}\ge v_+(x)\Big|_{\Omega_d}$$

provided that $A>0$ is chosen sufficiently large, namely $A\ge\frac{M_0^{1/\varsigma}}{\varepsilon d^{\lambda_\varepsilon}}$. Thus, from above we get

$$\begin{cases}Q_+(Az_\varepsilon,\eta)\ge0=Q_+(v_+,\eta), & x\in(G_0^d)_+;\\ Az_\varepsilon\ge v_+(x), & x\in\partial(G_0^d)_+\setminus\Gamma_+^d\end{cases}$$

for all non-negative $\eta\in C^0(\overline{(G_0^d)}_+)\cap\mathfrak{N}^1_{m,0,\tau}((G_0^d)_+)$, $\eta\Big|_{\Sigma_0^d}=0$. Besides that, one can readily verify that all conditions of the comparison principle (Theorem 9.6 in [14]) are fulfilled. By this principle we get

$$v_+(x)\le Az_\varepsilon(r,\omega),\quad\forall x\in\overline{(G_0^d)}_+.$$

Similarly, one can prove that

$$v_+(x)\ge -Az_\varepsilon(r,\omega),\quad\forall x\in\overline{(G_0^d)}_+.$$

Thus, finally, we obtain

$$|v_+(x)|\le Az_\varepsilon(r,\omega)\le Ar^{\lambda_\varepsilon},\quad\forall x\in\overline{(G_0^d)}_+. \tag{7.4.40}$$

□

Lemma 7.14. *Let $v_+(x)$ be a strong solution of the mixed problem $(MP)_+$. Let*

$$v_-(x_1, -x_2, x') = -kv_+(x),\ x_2 \geq 0; \quad k = \left(\frac{a_+}{a_-}\right)^{\frac{1}{m-1}}. \tag{7.4.41}$$

Suppose that conditions (CC) are satisfied. Then function $v_-(x)$, $x \in \overline{G_-}$ satisfies the following problem:

$$\begin{cases} -\frac{d}{dx_i}\mathcal{A}^i_-(x, \nabla v_-) + \mathcal{B}_-(x, v_-, \nabla v_-) + a_0\mathcal{C}_-(x, v_-) = f_-(x), & x \in G^d_-; \\ v_-\Big|_{\Sigma_0} = 0; \\ a_-\mathcal{A}^i_-(x, \nabla v_-)n_i + \gamma_-(\omega)r^{\tau-m+1}v_-|v_-|^{m-2} = \mathcal{G}_-(x, v_-), & x \in \Gamma^d_- \end{cases} \tag{MP$_-$}$$

and thus for function $v(x) = \begin{cases} v_+(x), & x \in \overline{G_+}, \\ v_-(x), & x \in \overline{G_-} \end{cases}$ the conjunction conditions

$$[v]_{\Sigma_0} = 0, \quad \left[a\mathcal{A}^i(x, \nabla v)n_i\right]_{\Sigma_0} = 0 \tag{7.4.42}$$

are satisfied.

Proof. At first, we observe that for the function $v_-(x)$, $x \in \overline{G_-}$ conditions (CC) have the form

$$a_-\mathcal{A}^i_-(x, \nabla v_-)\Big|_{\overline{G_-}} = -a_+\mathcal{A}^i_+(x, \nabla v_+)\Big|_{\overline{G_+}}, \ \forall i \neq 2;$$

$$a_-\mathcal{A}^2_-(x, \nabla v_-)\Big|_{\overline{G_-}} = +a_+\mathcal{A}^2_+(x, \nabla v_+)\Big|_{\overline{G_+}};$$

$$a_-\mathcal{B}_-(x, v_-, \nabla v_-)\Big|_{\overline{G_-}} = -a_+\mathcal{B}_+(x, v_+, \nabla v_+)\Big|_{\overline{G_+}};$$

$$a_-\mathcal{C}_-(x, v_-)\Big|_{\overline{G_-}} = -a_+\mathcal{C}_+(x, v_+)\Big|_{\overline{G_+}}; \tag{CC$'$}$$

$$\mathcal{G}_-(x, v_-)\Big|_{\overline{G_-}} = -\mathcal{G}_+(x, v_+)\Big|_{\overline{G_+}}; \tag{7.4.43}$$

$$a_-f_-(x_-) = -a_+f_+(x_+); \quad a_+\gamma_-\left(-\tfrac{\omega_0}{2}\right) = a_-\gamma_+\left(\tfrac{\omega_0}{2}\right).$$

Because of $x_2\Big|_{G_-} \leq 0$ and $x_2\Big|_{G_+} \geq 0$, we calculate for $x_2 \geq 0$,

$$\frac{\partial v_-}{\partial x_i}\bigg|_{(x_1, -x_2, x')} = -k\frac{\partial v_+}{\partial x_i}\bigg|_{(x_1, x_2, x')}, \ \forall i \neq 2; \quad \frac{\partial v_-}{\partial x_2}\bigg|_{(x_1, -x_2, x')} = k\frac{\partial v_+}{\partial x_2}\bigg|_{(x_1, x_2, x')}. \tag{7.4.44}$$

Next, by $(CC)'$,

$$-a_-\frac{d\mathcal{A}^i_-(x, \nabla v_-)}{dx_i}\bigg|_{(x_1, -x_2, x')} = a_+\frac{d\mathcal{A}^i_+(x, \nabla v_+)}{dx_i}\bigg|_{(x_1, x_2, x')}$$

and therefore, again by $(CC)'$, from the equation of $(MP)_+$ it follows the equation of $(MP)_-$.

Now, $v_-\Big|_{\Sigma_0} = -kv_+\Big|_{\Sigma_0} = 0$. Further, we calculate

$$\cos(\vec{n}, x_1)\Big|_{\Gamma^d_\pm} = -\sin\left(\frac{\omega_0}{2}\right), \ \cos(\vec{n}, x_2)\Big|_{\Gamma^d_\pm} = \pm\cos\left(\frac{\omega_0}{2}\right),$$

$$\cos(\vec{n}, x_i)\Big|_{\Gamma^d_\pm} = 0 \ \forall i \geq 3 \quad \Longrightarrow$$

$$a_+\mathcal{A}^i_+(x, \nabla v_+)n_i\Big|_{\Gamma^d_+} = -a_-\mathcal{A}^i_-(x, \nabla v_-)n_i\Big|_{\Gamma^d_-},$$

by virtue of $(CC)'$. Next, because of $a_+\gamma_-\left(-\frac{\omega_0}{2}\right) = a_-\gamma_+\left(\frac{\omega_0}{2}\right)$ we derive

$$\gamma_+(\omega)v_+|v_+|^{m-2}\Big|_{\Gamma_+} = -\gamma_+\left(\frac{\omega_0}{2}\right)\frac{a_-}{a_+}v_-|v_-|^{m-2}\Big|_{\Gamma_-} = -\gamma_-\left(-\frac{\omega_0}{2}\right)v_-|v_-|^{m-2}\Big|_{\Gamma_-}$$
$$= -\gamma_-(\omega)v_-|v_-|^{m-2}\Big|_{\Gamma_-}.$$

Thus, we have proved that $v_-(x)$ solves problem $(MP)_-$.

Similarly, by $(CC)'$ and $\cos(\overrightarrow{n}, x_i)\Big|_{\Sigma_0} = \begin{cases} -1, & i = 2, \\ 0, & \forall i \neq 2, \end{cases}$ we obtain that

$$a_+\mathcal{A}^i_+(x, \nabla v_+)n_i\Big|_{\Sigma^d_0} = a_-\mathcal{A}^i_-(x, \nabla v_-)n_i\Big|_{\Sigma^d_0}.$$

Hence it follows that the conjunction conditions (7.4.42) are satisfied.

Corollary 7.15. *Let the function $v(x)$ is defined by Lemma* 7.14 *and compatibility conditions $(CC)'$ be satisfied. Then $v(x)$ is a strong solution of the transmission problem*

$$\begin{cases} -\varsigma^{m-1}\frac{d}{dx_i}(r^\tau|\nabla v|^{m-2}v_{x_i}) + \mathcal{B}(x, v, \nabla v) + a_0\mathcal{C}(x, v) = f(x), & x \in G^d_0 \setminus \Sigma^d_0; \\ [v]_{\Sigma^d_0} = 0, \quad [\overline{a}|\nabla v|^{m-2}\frac{\partial v}{\partial \vec{n}}]_{\Sigma^d_0} = 0; & \\ \overline{a}r^\tau|\nabla v|^{m-2}\frac{\partial v}{\partial \vec{n}} + \gamma r^{\tau-m+1}v|v|^{m-2} = \mathcal{G}(x, v), & x \in \Gamma^d. \end{cases} \tag{TrP}$$

Finally, from (7.4.40)–(7.4.41) we get

$$|v(x)| \leq c_0 r^{\lambda_\varepsilon}.$$

On returning to the old variables by virtue of (6.1.1) according to (7.4.26)–(7.4.30) we get the required estimate (7.4.33). ☐

7.4.5 Example

Here we consider the two-dimensional transmission problem for the Laplace operator with absorbtion term in an angular domain. Suppose $n = 2$ and the domain G lies inside the corner

$$G_0 = \{(r, \omega) \,|r > 0; \; -\frac{\omega_0}{2} < \omega < \frac{\omega_0}{2}\}, \quad \omega_0 \in]0, 2\pi[;$$

$\mathcal{O} \in \partial G$ and in some neighborhood of $\mathcal{O}$ the boundary ∂G coincides with the sides of the corner $\omega = -\frac{\omega_0}{2}$ and $\omega = \frac{\omega_0}{2}$. We write

$$\Gamma_\pm = \{(r, \omega) \mid r > 0; \; \omega = \pm\frac{\omega_0}{2}\}, \quad \Sigma_0 = \{(r, \omega) \mid r > 0; \; \omega = 0\}.$$

We consider problem (MVP) for $m = 2$, $\beta = \gamma = 0$:

$$\begin{cases} -\frac{d}{dx_i}(r^\tau |w|^q w_{x_i}) + a_0 r^{\tau-2} w|w|^q - \mu r^\tau w|w|^{q-2}|\nabla w|^2 = 0, & x \in G_0 \setminus \Sigma_0, \\ [w]_{\Sigma_0} = 0, \; \left[a|w|^q \frac{\partial w}{\partial \vec{n}}\right]_{\Sigma_0} = 0, & x \in \Sigma_0, \\ a_\pm |w_\pm|^q \frac{\partial w_\pm}{\partial \vec{n}} = 0, & x \in \Gamma_\pm, \end{cases} \tag{$(MVP)_0$}$$

where

$$a = \begin{cases} a_+, & x \in G_+, \\ a_-, & x \in G_-, \end{cases} \qquad a_\pm > 0, \; q \geq 0, \; \tau \geq 0, \; a_0 \geq 0, \; 0 \leq \mu < 1 + q$$

are constants.

By solving this problem we get (cf. Example 4.6 in [7] or Example 9.18, p. 393 in [14])

$$w(r, \omega) = r^\lambda \cdot \begin{cases} a_-^{\frac{1}{1+q}} \sin^{\frac{1}{1+q+\mu}}\left(\frac{\pi\omega}{\omega_0}\right), & \omega \in \left[0, \frac{\omega_0}{2}\right], \\ -a_+^{\frac{1}{1+q}} \left|\sin\left(\frac{\pi\omega}{\omega_0}\right)\right|^{\frac{1}{1+q+\mu}}, & \omega \in \left[-\frac{\omega_0}{2}, 0\right], \end{cases}$$

where

$$\lambda = \frac{\sqrt{\tau^2 + 4(\pi/\omega_0)^2 + 4a_0(1 + q + \mu)} - \tau}{2(1 + q + \mu)}.$$

The principal part in the $(MVP)_0$ equation is weak nonlinear. Therefore we obtained the precise exponent λ of the solution decreasing rate. This fact is completely compatible with the results of Section 5. In the case of the linear Neumann transmission problem for the Laplace equation ($\tau = q = a_0 = \mu = 0$) we obtain also a well-known result

$$w(r,\omega) = r^{\frac{\pi}{\omega_0}} \cdot \begin{cases} a_- \sin\left(\frac{\pi\omega}{\omega_0}\right), & \omega \in \left[0, \frac{\omega_0}{2}\right], \\ a_+ \sin\left(\frac{\pi\omega}{\omega_0}\right), & \omega \in \left[-\frac{\omega_0}{2}, 0\right] \end{cases}$$

(see Subsection 3.6).

7.5 The case $\frac{\gamma_+}{a_+} \neq \frac{\gamma_-}{a_-}$

Now we turn our attention to the details of the properties of the barrier function $z(x) = r^{\overline{\lambda}}\psi(\omega)$ defined by (7.3.3), that is a solution of $(\overline{MVP})$. We remind that $\psi(\omega)$ is an eigenfunction of the Sturm-Liouville boundary problem $\overline{(StL)}$.

7.5.1 Properties of solutions to the Sturm-Liouville boundary problem $\overline{(StL)}$

First of all, we observe that in our case $\psi(0) \neq 0$ and that solutions of $(\overline{StL})$ are determined uniquely up to a scalar multiple. Therefore we can consider the solution $\psi(\omega)$ normed by the condition

$$\psi_-(0) = \psi_+(0) = 1. \tag{7.5.1}$$

Lemma 7.16. *Let $\psi(\omega)$ be an eigenfunction of the Sturm-Liouville boundary problem $\overline{(StL)}$ and $\psi(0) \neq 0$. Suppose, in addition, that $m \geq 2$ and the inequality (7.4.13) holds. Then $\psi(\omega) > 0$ in $[-\omega_0/2, \omega_0/2]$ and $\psi''(\omega) < 0$ in $(-\omega_0/2, \omega_0/2)$.*

Proof. We rewrite the $\overline{(StL)}$ equation in the form (7.3.6). By setting $\psi'(\omega)/\psi(\omega) = \overline{y}(\omega)$, we arrive at the problem for $\overline{y}(\omega)$, $\overline{\lambda}$:

$$\begin{cases} \left[(m-1)\overline{y}^2 + \overline{\lambda}^2\right](\overline{y}^2 + \overline{\lambda}^2)^{\frac{m-4}{2}}\overline{y}' + (m-1+\overline{\mu})(\overline{y}^2 + \overline{\lambda}^2)^{\frac{m}{2}} \\ +\overline{\lambda}(2-m+\tau)(\overline{y}^2 + \overline{\lambda}^2)^{\frac{m-2}{2}} = \overline{a}_0, \quad \omega \in (-\omega_0/2, 0) \cup (0, \omega_0/2); \\ \overline{a}_+ \left(\overline{\lambda}^2 + \overline{y}_+^2(0)\right)^{\frac{m-2}{2}} \cdot \overline{y}_+(0) - \overline{a}_- \left(\overline{\lambda}^2 + \overline{y}_-^2(0)\right)^{\frac{m-2}{2}} \cdot \overline{y}_-(0) = \beta; \\ \overline{a}_- \left(\overline{\lambda}^2 + \overline{y}_-^2(-\omega_0/2)\right)^{\frac{m-2}{2}} \cdot \overline{y}_-(-\omega_0/2) = \gamma_-; \\ \overline{a}_+ \left(\overline{y}_+^2(\omega_0/2) + \overline{\lambda}^2\right)^{\frac{m-2}{2}} \cdot \overline{y}_+(\omega_0/2) = -\gamma_+. \end{cases} \tag{7.5.2}$$

From the equation of (7.5.2) we get:

$$\begin{aligned}
&-\left[(m-1)\overline{y}^2+\overline{\lambda}^2\right](\overline{y}^2+\overline{\lambda}^2)^{\frac{m-4}{2}}\overline{y}' \\
&=(m-1+\overline{\mu})(\overline{y}^2+\overline{\lambda}^2)^{\frac{m}{2}}+\lambda(2-m+\tau)(y^2+\overline{\lambda}^2)^{\frac{m-2}{2}}-\overline{a}_0 \\
&=(\overline{y}^2+\overline{\lambda}^2)^{\frac{m-2}{2}}\left[(m-1+\overline{\mu})(\overline{y}^2+\overline{\lambda}^2)+\overline{\lambda}(2-m+\tau)\right]-\overline{a}_0 \\
&\geq(\overline{y}^2+\overline{\lambda}^2)^{\frac{m-2}{2}}[\overline{\lambda}^2(m-1+\overline{\mu})+\overline{\lambda}(2-m+\tau)]-\overline{a}_0 \\
&\geq\overline{\lambda}^m(m-1+\overline{\mu})+\overline{\lambda}^{m-1}(2-m+\tau)-\overline{a}_0>0
\end{aligned}$$

by virtue of (7.4.13) and $m \geq 2$. Thus, it is proved that $\overline{y}'(\omega) < 0$, $\omega \in (-\omega_0/2, 0) \cup (0, \omega_0/2)$. Therefore $\overline{y}(\omega)$ *decreases* on the interval $(-\omega_0/2, 0) \cup (0, \omega_0/2)$. Hence it follows that

$$\begin{aligned}
&\overline{y}_-(0) \leq \overline{y}_-(\omega) \leq \overline{y}_-\left(-\frac{\omega_0}{2}\right), \omega \in \left[-\frac{\omega_0}{2}, 0\right]; \\
&\overline{y}_+(0) \geq \overline{y}_+(\omega) \geq \overline{y}_+\left(\frac{\omega_0}{2}\right), \ \omega \in \left[0, \frac{\omega_0}{2}\right]
\end{aligned} \tag{7.5.3}$$

and therefore $\psi(\omega) \neq 0$, $\forall\omega \in [-\omega_0/2, \omega_0/2]$. By the continuity, it means that function $\psi(\omega)$ retains its sign on the interval $(-\omega_0/2, \omega_0/2)$ and, by (7.5.1), $\psi(\omega) > 0$ in $[-\omega_0/2, \omega_0/2]$. Finally, by Lemma 7.4, $\psi''_+(\omega) < 0$ in $(-\omega_0/2, \omega_0/2)$. □

Now we can estimate $\psi(\omega)$ from below.

Lemma 7.17. *Let $\psi(\omega)$ be an eigenfunction of the Sturm-Liouville boundary problem $\overline{(StL)}$ and $m \geq 2$. Then*

$$\psi_\pm(\omega) \geq \exp\left(-\frac{\omega_0}{2}\cdot\left(\frac{\gamma_\pm}{\overline{a}_\pm}\right)^{\frac{1}{m-1}}\right). \tag{7.5.4}$$

Proof. First of all, by the definition of $\overline{y}(\omega)$, we have

$$\begin{aligned}
&\psi_-(\omega)=\exp\left(-\int_\omega^0\overline{y}_-(\xi)d\xi\right), \ \omega \in \left(-\frac{\omega_0}{2}, 0\right); \\
&\psi_+(\omega)=\exp\left(\int_0^\omega\overline{y}_+(\xi)d\xi\right), \ \omega \in \left(0, \frac{\omega_0}{2}\right).
\end{aligned} \tag{7.5.5}$$

By the boundary condition of (7.5.2), we have

$$\frac{\gamma_\pm}{\overline{a}_\pm}=\left|\overline{y}_\pm\left(\pm\frac{\omega_0}{2}\right)\right|\cdot\left(\overline{y}_\pm^2\left(\pm\frac{\omega_0}{2}\right)+\overline{\lambda}^2\right)^{\frac{m-2}{2}}\geq\left|\overline{y}_\pm\left(\pm\frac{\omega_0}{2}\right)\right|^{m-1}\Longrightarrow$$

$$\left|\overline{y}_\pm\left(\pm\frac{\omega_0}{2}\right)\right|\leq\left(\frac{\gamma_\pm}{\overline{a}_\pm}\right)^{\frac{1}{m-1}}.$$

Hence, again by the boundary conditions according to $\gamma_\pm > 0$, it follows that

$$-\left(\frac{\gamma_+}{\overline{a}_+}\right)^{\frac{1}{m-1}} \le \overline{y}_+\left(\frac{\omega_0}{2}\right) < 0 \quad \text{and} \quad 0 < \overline{y}_-\left(-\frac{\omega_0}{2}\right) \le \left(\frac{\gamma_-}{\overline{a}_-}\right)^{\frac{1}{m-1}}, \tag{7.5.6}$$

or, by virtue of the decrease of $y(\omega)$,

$$\overline{y}_-(\omega) \le \left(\frac{\gamma_-}{\overline{a}_-}\right)^{\frac{1}{m-1}}, \ \omega \in \left[-\frac{\omega_0}{2}, 0\right]; \qquad \overline{y}_+(\omega) \ge -\left(\frac{\gamma_+}{\overline{a}_+}\right)^{\frac{1}{m-1}}, \ \omega \in \left[0, \frac{\omega_0}{2}\right]. \tag{7.5.7}$$

From (7.5.5)–(7.5.7) the required (7.5.4) follows. □

By integrating (7.5.2) with regard to $\overline{a} = a\varsigma^{m-1}$, $\overline{a}_0 = a_0\varsigma^{1-m}$, $\overline{\mu} = \mu\varsigma$, $\overline{\lambda} = \frac{\lambda}{\varsigma}$, $\overline{y}(\omega) = \frac{y(\omega)}{\varsigma}$, we find that the eigenvalue λ satisfies the system

$$\begin{cases} a_+\left(\lambda^2 + y_+^2(\omega_0/2)\right)^{\frac{m-2}{2}} \cdot y_+(\omega_0/2) = -\gamma_+, \\ a_-\left(\lambda^2 + y_-^2(-\omega_0/2)\right)^{\frac{m-2}{2}} \cdot y_-(-\omega_0/2) = \gamma_-, \\ \displaystyle\int\limits_{y_+(\omega_0/2)}^{y_+(0)} \frac{\left[(m-1)y^2 + \lambda^2\right](y^2+\lambda^2)^{\frac{m-4}{2}}dy}{(m-1+q+\mu)(y^2+\lambda^2)^{\frac{m}{2}} + \lambda(2-m+\tau)(y^2+\lambda^2)^{\frac{m-2}{2}} - a_0} = \frac{\omega_0}{2}, \\ \displaystyle\int\limits_{y_-(0)}^{y_-(-\omega_0/2)} \frac{\left[(m-1)y^2 + \lambda^2\right](y^2+\lambda^2)^{\frac{m-4}{2}}dy}{(m-1+q+\mu)(y^2+\lambda^2)^{\frac{m}{2}} + \lambda(2-m+\tau)(y^2+\lambda^2)^{\frac{m-2}{2}} - a_0} = \frac{\omega_0}{2}, \\ a_+\left(\lambda^2 + y_+^2(0)\right)^{\frac{m-2}{2}} \cdot y_+(0) - a_-\left(\lambda^2 + y_-^2(0)\right)^{\frac{m-2}{2}} \cdot y_-(0) = \beta, \\ \lambda^m(q+m-1+\mu) + \lambda^{m-1}(2-m+\tau) > a_0, \end{cases} \tag{Λ}$$

where $a_\pm > 0$, $q \ge 0$, $\tau \ge m-2$, $a_0 \ge 0$, $\mu > \mu_0$, $0 < \beta < \beta_0$, $0 < \gamma_\pm < \gamma_\pm\left(\pm\frac{\omega_0}{2}\right)$ (see (7.3.1)) and $\frac{\gamma_+}{a_+} \ne \frac{\gamma_-}{a_-}$. We may integrate system (Λ) for $m = 2$ or $a_0 = 0$. In fact, by integrating of this system we obtain the following.

$\boxed{m = 2}$

Let us define the value $\Upsilon = \sqrt{\lambda^2 + \frac{\tau\lambda - a_0}{1+q+\mu}}$. Then λ is the least positive root of the transcendental equation

$$a_+\frac{a_+\Upsilon\tan\left((1+q+\mu)\Upsilon\frac{\omega_0}{2}\right) - \gamma_+}{a_+\Upsilon + \gamma_+\tan\left((1+q+\mu)\Upsilon\frac{\omega_0}{2}\right)} + a_-\frac{a_-\Upsilon\tan\left((1+q+\mu)\Upsilon\frac{\omega_0}{2}\right) - \gamma_-}{a_-\Upsilon + \gamma_-\tan\left((1+q+\mu)\Upsilon\frac{\omega_0}{2}\right)} = \frac{\beta}{\Upsilon}. \tag{7.5.8}$$

Then according to (7.3.2) and (StL) we find the corresponding solution of problem (MVP):

$$w(r,\omega) = r^{\lambda} \begin{cases} \dfrac{\cos^{\frac{1}{1+q+\mu}} \left\langle (1+q+\mu)\Upsilon\left(\omega - \frac{\omega_0}{2}\right) + \arctan\left(\frac{1}{\Upsilon} \cdot \frac{\gamma_+}{a_+}\right)\right\rangle}{\cos^{\frac{1}{1+q+\mu}} \left\langle (1+q+\mu)\Upsilon \frac{\omega_0}{2} - \arctan\left(\frac{1}{\Upsilon} \cdot \frac{\gamma_+}{a_+}\right)\right\rangle}, & \omega \in \left[0, \frac{\omega_0}{2}\right], \\ \\ \dfrac{\cos^{\frac{1}{1+q+\mu}} \left\langle (1+q+\mu)\Upsilon\left(\omega + \frac{\omega_0}{2}\right) - \arctan\left(\frac{1}{\Upsilon} \cdot \frac{\gamma_-}{a_-}\right)\right\rangle}{\cos^{\frac{1}{1+q+\mu}} \left\langle (1+q+\mu)\Upsilon \frac{\omega_0}{2} - \arctan\left(\frac{1}{\Upsilon} \cdot \frac{\gamma_-}{a_-}\right)\right\rangle}, & \omega \in \left[-\frac{\omega_0}{2}, 0\right]. \end{cases} \tag{7.5.9}$$

The principal part in the (MVP) equation for $m = 2$ is weak nonlinear. Therefore we have obtained the precise exponent λ of the solution decreasing rate. This fact is completely compatible with the results of Section 5.

$$\boxed{a_0 = 0}$$

Let us define the value

$$\Lambda = \sqrt{\lambda^2 + \frac{(\tau + 2 - m)\lambda}{m - 1 + q + \mu}}. \tag{7.5.10}$$

Then our system (Λ) takes the form

$$\begin{cases} a_+ \left(\lambda^2 + y_+^2(\omega_0/2)\right)^{\frac{m-2}{2}} \cdot y_+(\omega_0/2) = -\gamma_+, \\ \\ a_- \left(\lambda^2 + y_-^2(-\omega_0/2)\right)^{\frac{m-2}{2}} \cdot y_-(-\omega_0/2) = \gamma_-, \\ \\ A\left(\arctan \frac{y_+(0)}{\lambda} - \arctan \frac{y_+(\omega_0/2)}{\lambda}\right) + \frac{B}{\Lambda}\left(\arctan \frac{y_+(0)}{\Lambda} - \arctan \frac{y_+(\omega_0/2)}{\Lambda}\right) - \frac{\omega_0}{2}, \\ \\ A\left(\arctan \frac{y_-(0)}{\lambda} - \arctan \frac{y_-(-\omega_0/2)}{\lambda}\right) + \frac{B}{\Lambda}\left(\arctan \frac{y_-(0)}{\Lambda} - \arctan \frac{y_-(-\omega_0/2)}{\Lambda}\right) \\ \hfill = -\frac{\omega_0}{2}, \\ \\ a_+ \left(\lambda^2 + y_+^2(0)\right)^{\frac{m-2}{2}} \cdot y_+(0) - a_- \left(\lambda^2 + y_-^2(0)\right)^{\frac{m-2}{2}} \cdot y_-(0) = \beta, \\ \\ A = \frac{2-m}{2-m+\tau}, \quad B = \frac{m-1}{m-1+q+\mu} - \lambda \cdot \frac{2-m}{2-m+\tau}. \end{cases} \tag{$(\Lambda)_0$}$$

If $m = 2$, then (7.5.8) follows for $a_0 = 0$.

Lemma 7.18. *Let $\psi(\omega)$ be an eigenfunction of the Sturm-Liouville boundary problem $\overline{(StL)}$ and $m \geq 2$. Then*

$$\psi_\pm(\omega) \leq \exp\left\{\frac{q + m - 1}{m - 1} \cdot \frac{\omega_0}{2}\Lambda \cdot \tan\left\langle (m - 1 + q + \mu)\Lambda \frac{\omega_0}{2}\right\rangle\right\}. \tag{7.5.11}$$

Proof. Because $m \geq 2$, $a_0 \geq 0$ and (7.5.10) from system (Λ) it follows that

$$\begin{aligned}
\frac{\omega_0}{2} &\geq \int\limits_{y_+(\omega_0/2)}^{y_+(0)} \frac{dy}{(m-1+q+\mu)(y^2+\lambda^2)+\lambda(2-m+\tau)} \\
&= \frac{1}{m-1+q+\mu} \cdot \int\limits_{y_+(\omega_0/2)}^{y_+(0)} \frac{dy}{y^2+\Lambda^2} \\
&= \frac{1}{(m-1+q+\mu)\Lambda}\left(\arctan\frac{y_+(0)}{\Lambda} - \arctan\frac{y_+(\omega_0/2)}{\Lambda}\right) \\
&\geq \frac{1}{(m-1+q+\mu)\Lambda}\arctan\frac{y_+(0)}{\Lambda}; \qquad (7.5.12) \\
\frac{\omega_0}{2} &\geq \int\limits_{y_-(0)}^{y_-(-\omega_0/2)} \frac{dy}{(m-1+q+\mu)(y^2+\lambda^2)+\lambda(2-m+\tau)} \\
&= \frac{1}{m-1+q+\mu} \cdot \int\limits_{y_-(0)}^{y_-(-\omega_0/2)} \frac{dy}{y^2+\Lambda^2} \\
&= \frac{1}{(m-1+q+\mu)\Lambda}\left(\arctan\frac{y_-(-\omega_0/2)}{\Lambda} - \arctan\frac{y_-(0)}{\Lambda}\right) \\
&\geq -\frac{1}{(m-1+q+\mu)\Lambda}\arctan\frac{y_-(0)}{\Lambda}.
\end{aligned}$$

Hence we get

$$\begin{aligned}
y_+(0) &\leq \Lambda \tan\left\langle (m-1+q+\mu)\Lambda\frac{\omega_0}{2}\right\rangle; \\
y_-(0) &\geq -\Lambda \tan\left\langle (m-1+q+\mu)\Lambda\frac{\omega_0}{2}\right\rangle. \qquad (7.5.13)
\end{aligned}$$

Now, from (7.5.5), (7.5.3) and (7.5.13) we derive the required (7.5.11). □

From Lemmas 7.17 and 7.18 follows

Corollary 7.19. *Let $\psi(\omega)$ be an eigenfunction of the Sturm-Liouville boundary problem $\overline{(StL)}$ and $m \geq 2$. Then*

$$\begin{aligned}
&\exp\left(-\frac{\omega_0}{2}\cdot\left(\frac{\gamma_\pm}{\overline{a}_\pm}\right)^{\frac{1}{m-1}}\right) \leq \psi_\pm(\omega) \\
&\leq \exp\left\{\frac{q+m-1}{m-1}\cdot\frac{\omega_0}{2}\Lambda\cdot\tan\left\langle (m-1+q+\mu)\Lambda\frac{\omega_0}{2}\right\rangle\right\} \equiv \Psi_0.
\end{aligned}$$

Further, we can estimate $\psi'_\pm(\omega)$.

Lemma 7.20. *Let $\psi(\omega)$ be an eigenfunction of the Sturm-Liouville boundary problem $\overline{(StL)}$ and $m \geq 2$. Then*

$$-\frac{q+m-1}{m-1}\Lambda \cdot \tan\left\langle (m-1+q+\mu)\Lambda\frac{\omega_0}{2}\right\rangle \Psi_0 \leq \psi'_-(\omega)$$

$$\leq \frac{q+m-1}{m-1}\left(\frac{\gamma_-}{a_-}\right)^{\frac{1}{m-1}}\Psi_0, \qquad \omega \in \left[-\frac{\omega_0}{2}, 0\right];$$

$$-\frac{q+m-1}{m-1}\left(\frac{\gamma_+}{a_+}\right)^{\frac{1}{m-1}}\Psi_0 \leq \psi'_+(\omega)$$

$$\leq \frac{q+m-1}{m-1}\Lambda \cdot \tan\left\langle (m-1+q+\mu)\Lambda\frac{\omega_0}{2}\right\rangle \Psi_0, \qquad \omega \in \left[0, \frac{\omega_0}{2}\right];$$

$$|\psi''(\omega)| \leq \psi_0^{-1}\overline{\mu}\Psi_1^2 + (2m-3+\overline{\mu})\overline{\lambda}^2 + (2-m+\tau)\overline{\lambda}, \qquad \omega \in \left(-\frac{\omega_0}{2}, \frac{\omega_0}{2}\right).$$

Proof. In fact, because $\psi'(\omega) = \overline{y}(\omega)\psi(\omega)$, $\psi(\omega) > 0$ and (7.5.3) we have

$$\frac{q+m-1}{m-1}y_-(0)\psi_-(\omega) \leq \psi'_-(\omega) \leq \overline{y}_-\left(-\frac{\omega_0}{2}\right)\psi_-(\omega), \qquad \omega \in \left[-\frac{\omega_0}{2}, 0\right];$$

$$\overline{y}_+\left(\frac{\omega_0}{2}\right)\psi_+(\omega) \leq \psi'_+(\omega) \leq \frac{q+m-1}{m-1}y_+(0)\psi_+(\omega), \qquad \omega \in \left[0, \frac{\omega_0}{2}\right].$$

Next we use inequalities (7.5.6), (7.5.13) and Corollary 7.19. For the estimating of $|\psi''(\omega)|$ we refer to the proof of Lemma 7.12. Thus, we obtain the required statements. □

Remark 6. **The eigenfunction existence.** By the proof of Lemma 7.16, we must integrate the equation in (7.5.2). As a result, according to (7.3.3) we get

$$\begin{cases} \Phi(\omega) = \exp\left(\int\limits_0^\omega y(\xi)d\xi\right); \\ \int\limits_{y(\omega)}^{y(0)} \frac{\left[(m-1)y^2+\lambda^2\right](y^2+\lambda^2)^{\frac{m-4}{2}}dy}{(m-1+q+\mu)(y^2+\lambda^2)^{\frac{m}{2}} + \lambda(2-m+\tau)(y^2+\lambda^2)^{\frac{m-2}{2}} - a_0} = \omega, \end{cases}$$

where λ is the solution of (Λ).

7.5.2 Estimates of the $(TDQL)$ solution modulus

Now we can estimate $|u(x)|$ for the problem $(TDQL)$ in a neighborhood of the edge.

Theorem 7.21. *Let u be a weak solution of the problem $(TDQL)$ with $\gamma\left(\pm\frac{\omega_0}{2}\right) > \gamma_\pm > 0$ and $\frac{\gamma_+}{a_+} \neq \frac{\gamma_-}{a_-}$. Let us assume that $M_0 = \max\limits_{x\in\overline{G}}|u(x)|$ is known. Let assumptions 1)–10) with $m \geq 2$ be satisfied and λ be the least positive number satisfying* *the system* (Λ). *Then there exist* $d \in (0,1)$ *and a constant* $C_0 > 0$ *independent of* u *such that*

$$|u(x)| \le C_0 r^{\lambda}, \quad \forall x \in G_0^d. \tag{7.5.14}$$

Proof. At first, we perform the function change (6.1.1) and will consider the function $v(x)$. For the proof we use the above constructed barrier function $z(x)$ (as a solution of problem $(\overline{MVP})$) and apply the comparison principle to $v(x)$ and $z(x)$. Taking into account $(\widehat{II})_{loc}$ we define the operator

$$\begin{aligned} Q(v,\eta) \equiv & \int\limits_{G_0^d} \Big\langle a\mathcal{A}^i(x,v_x)\eta_{x_i} + a\mathcal{B}(x,v,v_x)\eta + aa_0\mathcal{C}(x,v)\eta - af(x)\eta(x)\Big\rangle dx \\ & + \int\limits_{\Gamma^d} \gamma(\omega) r^{\tau-m+1} v|v|^{m-2}\eta(x)ds + \beta_0 \int\limits_{\Sigma_0^d} r^{\tau-m+1}v|v|^{m-2}\eta(x)ds \\ & - \int\limits_{\Omega_d} a\mathcal{A}^i(x,v_x)\cos(r,x_i)\eta(x)d\Omega_\varrho - \int\limits_{\Gamma^d} \mathcal{G}(x,v)\eta(x)ds - \int\limits_{\Sigma_0^d} \mathcal{H}(x,v)\eta(x)ds. \end{aligned} \tag{7.5.15}$$

We will show that $Q(Az,\eta) \ge 0 = Q(v,\eta)$ for all non-negative $\eta \in \mathbf{C}^0(\overline{G_0^d}) \cap \mathfrak{N}^1_{m,0,\tau}(G_0^d)$ and some $A > 0$. By definition (7.5.15) and integrating by parts in the first integral, we have:

$$\begin{aligned} & Q(Az,\eta) \\ & \equiv \int\limits_{G_0^d} \Big\langle -a\frac{d\mathcal{A}^i(x,A\nabla z)}{dx_i} + a\mathcal{B}(x,Az,A\nabla z) + a_0 a\mathcal{C}(x,Az) - af(x)\Big\rangle \eta(x)dx \\ & \quad + \int\limits_{\Gamma^d} \Big\langle a\mathcal{A}^i(x,A\nabla z)n_i + A^{m-1}\gamma(\omega)r^{\tau-m+1}z|z|^{m-2} - \mathcal{G}(x,Az)\Big\rangle \eta(x)ds \\ & \quad + \int\limits_{\Sigma_0^d} \Big\langle \big[a\mathcal{A}^i(x,A\nabla z)n_i\big] + \beta_0 A^{m-1} r^{\tau-m+1}z|z|^{m-2} - \mathcal{H}(x,Az)\Big\rangle \eta(x)ds \end{aligned}$$

or, by virtue of $z(x)$ is a solution of problem $(\overline{MVP})$,

$$\begin{aligned} Q(Az,\eta) \equiv \int\limits_{G_0^d} a\Big\{ & -\frac{d}{dx_i}\Big(\mathcal{A}^i(x,A\nabla z) - (A\varsigma)^{m-1} r^{\tau}|\nabla z|^{m-2}\frac{\partial z}{\partial x_i}\Big) - f(x) \\ & + \big\langle \mathcal{B}(x,Az,A\nabla z) + A^{m-1}\varsigma^m \mu r^{\tau} z^{-1}|\nabla z|^m\big\rangle \\ & + a_0 \big\langle \mathcal{C}(x,Az) - A^{m-1} r^{\tau-m} z|z|^{m-2}\big\rangle \Big\}\eta(x)dx \end{aligned} \tag{7.5.16}$$

$$
\begin{aligned}
&+\int_{\Gamma^d}\Big\{a\left\langle \mathcal{A}^i(x,A\nabla z)-(A\varsigma)^{m-1}r^\tau|\nabla z|^{m-2}\frac{\partial z}{\partial x_i}\right\rangle n_i\\
&\quad+(\gamma(\omega)-\gamma)\,A^{m-1}r^{\tau-m+1}z|z|^{m-2}-\mathcal{G}(x,Az)\Big\}\eta(x)ds\\
&+\int_{\Sigma_0^d}\Big\langle\left[a\left\langle \mathcal{A}^i(x,A\nabla z)-(A\varsigma)^{m-1}r^\tau|\nabla z|^{m-2}\frac{\partial z}{\partial x_i}\right\rangle n_i\right]\\
&\quad+(\beta_0-\beta)A^{m-1}r^{\tau-m+1}z|z|^{m-2}-\mathcal{H}(x,Az)\Big\rangle\eta(x)ds.
\end{aligned}
$$

Now we estimate all terms in (7.5.16). We note that $z(x)\geq 0$, by Lemma 7.17. Because of (7.4.37) and our assumptions, we derive:

- $\left|\frac{d}{dx_i}\left(\mathcal{A}^i(x,A\nabla z)-(A\varsigma)^{m-1}r^\tau|\nabla z|^{m-2}\frac{\partial z}{\partial x_i}\right)\right|$ $\leq (A\varsigma)^{m-1}\mathcal{A}(r)\left(r^{\tau-1}|\nabla z|^{m-1}+r^\tau|\nabla z|^{m-2}|D^2z|\right)$;
- $\mathcal{B}(x,Az,A\nabla z)+A^{m-1}\varsigma^m\mu r^\tau z^{-1}|\nabla z|^m\geq(\mu-\mu_0)\varsigma(A\varsigma)^{m-1}r^\tau z^{-1}|\nabla z|^m>0$, since $\mu>\mu_0$;
- $\mathcal{C}(x,Az)-A^{m-1}r^{\tau-m}z|z|^{m-2}\geq 0$;
- $\left|\left\langle\mathcal{A}^i(x,A\nabla z)-(A\varsigma)^{m-1}r^\tau|\nabla z|^{m-2}\frac{\partial z}{\partial x_i}\right\rangle n_i\right|\leq(A\varsigma)^{m-1}\mathcal{A}(r)r^\tau|\nabla z|^{m-1}$;
- by (7.4.39) and assumptions 8)′, 10),

$$
\begin{aligned}
-\mathcal{G}(x,Az)&\geq -g(x,0)\geq -g_1r^{\tau-m+1+(q+m-1)\lambda};\\
-\mathcal{H}(x,Az)&\geq -h(x,0)\geq -h_1r^{\tau-m+1+(q+m-1)\lambda};\\
-f(x)&\geq -f_1r^{\tau-m+(q+m-1)\lambda}.
\end{aligned}
$$

Further, by (7.3.3), Corollary 7.19 and Lemma 7.20, similarly (7.4.35) we obtain

$$
\begin{aligned}
&z(x)=r^{\overline{\lambda}}\psi(\omega)\implies\psi_0r^{\overline{\lambda}}\leq z(x)\leq\Psi_0r^{\overline{\lambda}};\\
&|\nabla z|^2=z_r^2+r^{-2}z_\omega^2=r^{2\overline{\lambda}-2}(\psi'^2+\overline{\lambda}^2\psi^2)\implies\left|\frac{\nabla z}{z}\right|\geq\overline{\lambda}r^{-1};\quad|\nabla z|\leq\widehat{C}_1r^{\overline{\lambda}-1};\\
&|D^2z|\leq c(\overline{\lambda})\left(|\psi''|+|\psi'|+|\psi|\right)\cdot r^{\overline{\lambda}-2}\implies|D^2z|\leq\widehat{C}_2r^{\overline{\lambda}-2},
\end{aligned}
\tag{7.5.17}
$$

where $\psi_0=\exp\left(-\frac{\omega_0}{2}\gamma_0\right)$, $\gamma_0=\max\left(\frac{\gamma_-}{a_-};\frac{\gamma_+}{a_+}\right)$ and $\widehat{C}_1$, $\widehat{C}_2$ are some positive constants. Therefore from (7.5.16) according to the above inequalities it follows that

$$
Q(Az,\eta)\geq\int_{G_0^d}a\Big\{(\mu-\mu_0)\varsigma(A\varsigma)^{m-1}r^\tau z^{-1}|\nabla z|^m-(A\varsigma)^{m-1}\mathcal{A}(r)r^{\tau-1}|\nabla z|^{m-1}
$$

$$
\begin{aligned}
&- (A\varsigma)^{m-1}\mathcal{A}(r)r^{\tau}|\nabla z|^{m-2}|D^2 z| - f_1 r^{\tau-m+\overline{\lambda}(m-1)}\Big\}\eta(x)dx \\
&+ \int\limits_{\Gamma^d}\Big\{\left(\gamma(\omega)-\gamma\right)A^{m-1}r^{\tau-m+1}z^{m-1} - a(A\varsigma)^{m-1}\mathcal{A}(r)r^{\tau}|\nabla z|^{m-1} \\
&- g_1 r^{\tau-m+1+\overline{\lambda}(m-1)}\Big\}\eta(x)ds \\
&+ \int\limits_{\Sigma_0^d}\Big\{(\beta_0-\beta)A^{m-1}r^{\tau-m+1}z^{m-1} - a(A\varsigma)^{m-1}\mathcal{A}(r)r^{\tau}|\nabla z|^{m-1} \\
&- h_1 r^{\tau-m+1+\overline{\lambda}(m-1)}\Big\}\eta(x)ds \qquad (7.5.18)\\
\geq &\Big\{\lambda(A\lambda\psi_0)^{m-1}\Big\langle(\mu-\mu_0) - C_3\mathcal{A}(d)\Big\rangle - f_1\Big\}\cdot\int\limits_{G_0^d} a r^{\tau-m+\overline{\lambda}(m-1)}\eta(x)dx \\
&+ \int\limits_{\Gamma^d} r^{\tau-m+1+\overline{\lambda}(m-1)}\Big\{(A\psi_0)^{m-1}\Big\langle(\gamma(\omega)-\gamma) - C_4\mathcal{A}(d)\Big\rangle - g_1\Big\}\eta(x)ds \\
&+ \Big\{(A\psi_0)^{m-1}\Big\langle(\beta_0-\beta) - C_5\mathcal{A}(d)\Big\rangle - h_1\Big\}\cdot\int\limits_{\Sigma_0^d} r^{\tau-m+1+\overline{\lambda}(m-1)}\eta(x)ds.
\end{aligned}
$$

Now, we take into consideration that $\mu > \mu_0$, $\beta < \beta_0$, $\gamma < \gamma\left(\pm\frac{\omega_0}{2}\right)$ (see (7.3.1)) and we can choose $d > 0$ so small that, by the continuity of $\mathcal{A}(r)$ at zero,

$$
\mathcal{A}(d) \leq \min\left\{\frac{\mu-\mu_0}{2C_3};\ \frac{\gamma\left(-\frac{\omega_0}{2}\right)-\gamma}{2C_4};\ \frac{\gamma\left(\frac{\omega_0}{2}\right)-\gamma}{2C_4};\ \frac{\beta_0-\beta}{2C_5}\right\}. \qquad (7.5.19)
$$

Therefore from (7.5.18)–(7.5.19) we get

$$
\begin{aligned}
Q(Az,\eta) \geq &\left\langle\lambda(A\lambda\psi_0)^{m-1}\frac{\mu-\mu_0}{2} - f_1\right\rangle\cdot\int\limits_{G_0^d} a r^{\tau-m+\overline{\lambda}(m-1)}\eta(x)dx \\
&+ \int\limits_{\Gamma^d} r^{\tau-m+1+\overline{\lambda}(m-1)}\left\langle(A\psi_0)^{m-1}\frac{\gamma(\omega)-\gamma}{2} - g_1\right\rangle\eta(x)ds \\
&+ \left\langle(A\psi_0)^{m-1}\frac{\beta_0-\beta}{2} - h_1\right\rangle\cdot\int\limits_{\Sigma_0^d} r^{\tau-m+1+\overline{\lambda}(m-1)}\eta(x)ds. \qquad (7.5.20)
\end{aligned}
$$

If we choose $A > 0$ sufficiently large, namely

$$A \geq \frac{1}{\psi_0} \max\left\{ \frac{1}{\lambda^{\frac{m}{m-1}}} \cdot \left(\frac{2f_1}{\mu - \mu_0}\right)^{\frac{1}{m-1}}, \left(\frac{2g_1}{\gamma(-\omega/2) - \gamma}\right)^{\frac{1}{m-1}}, \right.$$
$$\left. \left(\frac{2g_1}{\gamma(\omega/2) - \gamma}\right)^{\frac{1}{m-1}}, \left(\frac{2h_1}{\beta_0 - \beta}\right)^{\frac{1}{m-1}} \right\}, \quad (7.5.21)$$

then we provide the required inequality $Q(Az, \eta) \geq 0 = Q(v, \eta)$.

Now, by the continuity of $v(x)$, we have

$$v(x)\Big|_{\Omega_d} \leq M_0^{1/\varsigma} = \max_{\overline{G}} |v(x)|.$$

On the other hand, by virtue of Lemma 7.17 according to (7.5.17),

$$Az\Big|_{\Omega_d} \geq Ad^{\overline{\lambda}}\psi_0 \geq M_0^{1/\varsigma} \geq v\Big|_{\Omega_d}$$

provided that $A > 0$ is chosen sufficiently large, namely $A \geq \left(d^{\overline{\lambda}}\psi_0\right)^{-1} \cdot M_0^{1/\varsigma}$. Thus from above we get

$$\begin{cases} Q(Az, \eta) \geq Q(v, \eta), & \forall \eta \in \mathbf{C}^0(\overline{G_0^d}) \cap \mathfrak{N}^1_{m,0,\tau}(G_0^d); \\ Az\Big|_{\Omega_d} \geq v\Big|_{\Omega_d}. \end{cases}$$

Besides that, one can readily verify that all conditions of the comparison principle (Proposition 7.3) are fulfilled. By this principle we get

$$v(x) \leq Az(r, \omega), \quad \forall x \in \overline{G_0^d}.$$

Similarly one can prove that

$$v(x) \geq -Az(r, \omega), \quad \forall x \in \overline{G_0^d}.$$

Thus, finally, we obtain

$$|v| \leq Az(r, \omega) \leq c_0 r^{\overline{\lambda}}, \quad \forall x \in \overline{G_0^d}; \qquad c_0 = A\Psi_0, \quad (7.5.22)$$

by (7.5.17). On returning to the old variables, by virtue of (6.1.1), we get the required estimate (7.5.14). □

Bibliography

[1] S. Agmon, A. Douglis and L. Nirenberg, Estimates near the boundary for solutions of elliptic partial differential equations satisfying general boundary conditions, Comm. Pure Appl. Math. **12** (1959), 623–727.

[2] E. Beckenbach and R. Bellman, *Inequalities,* Springer-Verlag (1961).

[3] Ben M'Barek, M. Merigot, Régularité de la solution d'un problème de transmission, C.R. Acad. Sci. Paris Sér. I Math. **280** (1975), 1591–1593.

[4] S. Bernstein, Sur les équations du calcul des variations, Ann. Ec. Norm. **29** (1912), 431–485.

[5] H. Blumenfeld, Eigenlösungen von gemischten interface Problemen auf Gebieten mit Ecken – ihre Regularität und Approximation mittels Finiter Elemente, Dissertation. Freie Universität Berlin, 1983.

[6] M.V. Borsuk, A priori estimates and solvability of second order quasilinear elliptic equations in a composite domain with nonlinear boundary conditions and conjunction condition. Proc. Steklov Inst. of Math. **103** (1970), 13–51.

[7] M. Borsuk, The behavior of weak solutions to the boundary value problems for elliptic quasilinear equations with triple degeneration in a neighborhood of a boundary edge, Nonlinear Analysis: Theory, Methods and Applications **56**, no 3 (2004), 347–384.

[8] M. Borsuk, Best possible estimates of solutions to the interface problem for linear elliptic divergence second-order equations in a conical domain. // International Conference on Numerical Analysis and Applied Mathematics 2006, Wiley-VCH Verlag, pp. 56–60.

[9] M. Borsuk, The transmission problem for quasi-linear elliptic second order equations in a conical domain. // Ukrainer Mathematical Bulleten, **4**, no. 4 (2007), 485–525.

[10] M. Borsuk, Best possible estimates of solutions to the transmission problem for the Laplace operator with N different media in a conical domain. // Applicationes Mathematicae, **34**, no. 4 (2007), 445–491. 12.

[11] M. Borsuk, The transmission problem for elliptic second order equations in a conical domain. // Annales Academiae Paedagogicae Cracoviensis. Studia Mathematica, **7** (2008), 61–89.

[12] M. Borsuk, The transmission problem for elliptic second order equations in a domain with conical boundary points. // Lectures Notes in Computer Sciences, Springer. **5434** (2009), pp. 1–12.

[13] M. Borsuk, The transmission problem for quasi-linear elliptic second order equations in a conical domain. I, II. // Nonlinear Analysis: Theory, Methods and Applications . **71**, no 10 (2009), pp. 5032–5083.

[14] M. Borsuk, V. Kondratiev, Elliptic Boundary Value Problems of Second Order in Piecewise Smooth Domains. North-Holland Mathematical Library, **69**, ELSEVIER (2006), 531 p.

[15] M. Carriero, Problemi discontinui e di transmissione per equazioni ellitiche a coefficienti costanti relativi ad angoli consecutivi, Le Matematische (Catania), **29** (1974), pp. 444–484; **31** (1976), pp. 228–245.

[16] M. Carriero, Problemi discontinui e di transmissione per due equazioni ellitiche di ordine diverso a coefficienti variabli relativi ad angoli consecutivi, Annali di Mat. Pura e appl,, **109** (1976), pp. 247–271.

[17] M. Carriero, Problemi ellitici di transmissione in un poligono, Rend. Circ. Mat. Palermo, **28** (1979), pp. 411–444.

[18] Ya-Zhe Chen, Lan-Cheng Wu, Second order elliptic equations and elliptic systems. Transl. of Math. Monographs, **174**. AMS, Providence, Rhode Island, 246 p.

[19] W. Chikouche, D. Mercier and S. Nicaise, Regularity of the solution of some unilateral boundary value problems in polygonal and polyhedral domains, Communications in partial differential equations, **29**, no 1&2 (2004), 43–70.

[20] G.R. Cirmi and M.M. Porzio, $L^{\infty}-$ *solutions for some nonlinear degenerate elliptic and parabolic equations,* Ann. mat. pura ed appl. (IV) **169** (1995), 67–86.

[21] M. Costabel, E.P. Stephan, A direct boundary integral equation method for transmission problems, J. Math. Anal. Appl. **106**, (1985), 367–413.

[22] M. Dauge, Oblique derivative problems on polyhedral domains and polygonal interface problems, Comm. PDE **14**, (1989), 1193–1227.

[23] M. Dauge, S. Nicaise, Oblique derivative and interface problems on polygonal domains and networks, Comm. PDE **14**, (1989), 1147–1192.

[24] M. Dobrowolski, Numerical approximation of elliptic interface and corner problems, Habilitationschrift, Bonn, 1981.

[25] N. Dunford and J.T. Schwartz, *Linear operators, Part I: General Theory.* Interscience Publishers, New York, London, 1958.

[26] C. Ebmeyer, J. Frehse, M. Kassmann, Boundary regularity for nonlinear elliptic systems: applications to the transmission problem. Geometric analysis and nonlinear partial differential equations, Springer Verlag, 2002, pp. 505–517.

[27] L. Escauriaza, E. Fabes, G. Verchota, On a regularity theorem for weak solutions to transmission problems with internal Lipschitz boundaries, Proc. Amer. Math. Soc. **115**, no 4 (1992), 1069–1076.

[28] I.M. Gel'fand, G.E. Shilov, Generalized functions, v. I. Academic Press, 1964.

[29] D. Gilbarg and N.S. Trudinger, Elliptic Partial Differential Equations of Second Order, Springer-Verlag, Berlin/Heidelberg/New York, 1977. Revised Third Printing, 1998.

[30] G.H. Hardy, J.E. Littlewood and G. Pólya, Inequalities, 1934; University Press, Cambridge, 1952.

[31] V.A. Il'in, On the solvability of the Dirichlet and Neumann problems for linear elliptic operator with discontinuous coefficients, Doklady AN USSR. **137**, no 1 (1961), 28–30. – Soviet Math. Dokl. **2** (1961), 228–230.

[32] V.A. Il'in, I.A. Šišmarev, The method of potentials for the problems of Dirichlet and Neumann in the case of equations with discontinuous coefficients, Sibirsk. Math. Ž. **2**, no 1 (1961), 46–58.

[33] D. Kapanadze, B.-W. Schulze, Boundary-contact problems for domains with conical singularities. Journal of Differential Equations, **217**, no 2 (2005), 456–500.

[34] D. Kapanadze, B.-W. Schulze, Boundary-contact problems for domains with edge singularities. Journal of Differential Equations, **234**, (2007), 26–53.

[35] R.G. Kellogg, Singularities in interface problems, Symposium on Numerical Solutions of Partial Differential Equations, II, Academic Press, (1971), 351-400.

[36] R.G. Kellogg, Higher order singularities for interface problems, SThe Math. foundations of finite elements method with application to Partial Differential Equations, Ed. Aziz, Academic Press, (1972), 589–602.

[37] R.G. Kellogg, On the Poisson equation with intersecting interfaces, Applicable Analysis, **4**, (1975), 101–129.

[38] R.G. Kellogg, Notes on piecewise smooth elliptic boundary value problems. Industrial Mathematics Institute, University of South Carolina, Columbia, USA. Preprint, (2005), 97 pp.

[39] D. Knees, On the regularity of weak solutions of nonlinear transmission problems on polyhedral domains, Zeitschrift fur Analysis und ihre Anwendungen, vol. 23, no 3, pp. 509-546, 2004. (Preprint)

[40] D. Knees, A.-M. Sändig, Regularity of elastic fields in composites, Multifield problems in solid and fluid mechanics, Lecture Notes Appl. Comput. Mech., **28** (2006), 331–360. Springer, Berlin.

[41] N. Kutev, P.L. Lions, Nonlinear second-order elliptic equations with jump discontinuous coefficients. Part I: Quasilinear equations. Differential and Integral Equations, **5**, no 6 (1992), 1201–1217.

[42] O.A. Ladyzhenskaya, *Boundary value problems of mathematical physics,* Nauka, Moscow, 1973 (in Russian).

[43] O.A. Ladyzhenskaya, N.N. Ural'tseva, Linear and Quasilinear Elliptic Equations, Academic Press, New York, 1968.

[44] O.A. Ladyzhenskaya, V.Ja. Rivkind, N.N. Ural'tseva, On the classical solvability of diffraction problems for equations of elliptic and parabolic types, Proc. Steklov. Inst. Math, **92** (1966), 132–166.

[45] K. Lemrabet, Régularité de la solution d'un problème de transmission, J. Math. Pures Appl. **56** (1977), 1–38.

[46] K. Lemrabet, An interface problem in a domain of $\mathbb{R}^3$, J. Math. Anal. Appl. **63** (1978), 549–562.

[47] Gary M. Lieberman, Boundary regularity for solutions of degenerate elliptic equations, Nonlinear Analysis. **12**, no 11 (1988), 1203-1219.

[48] G.M. Lieberman, Local estimates for subsolutions and supersolutions of oblique derivative problems for general second order elliptic equations. Trans. of AMS., **304**, 1 (1987), 343-353.

[49] Gary M. Lieberman, *Second order parabolic differential equations,* World Scientific, Singapore – New Jersey – London – Hong Kong, 1996, 439 p.

[50] A. Lorenzi, On elliptic equations with piecewise constant coefficients. II, Ann. Scuola Norm. Pisa, **26** (1972), pp. 839–870.

[51] A. Maghnouji, S. Nicaise, Interface problems with operators of different order on polygons, Annales de la faculté des Sciences de Toulouse, Série 6, **I**, (1992), pp. 187–209.

[52] V.G. Maz'ya, *Sobolev spaces,* Springer Verlag, Springer Series in Soviet Mathematics, 1985.

[53] E. Meister, Integral equations for the Fourier-transformed boundary values for the transmission problems for fight-angles wedges and octants, Math. Meth. in the Appl. Sci., **8**, (1986), pp. 182–205.

[54] E. Meister and F.O. Speck, A contribution to the quarter-plane problem in diffraction, J. Math. Anal. Appl., **130**, (1988), pp. 223–236.

[55] E. Meister, F. Penzel, F.O. Speck and F.S. Texeira, Two media scattering problems in a half space, Preprint, Technische Hochschule Darmstadt, (1991).

[56] Mercier D., Minimal regularity of the solutions of some transmission problems, Math. Meth. in Appl. Sci. **26** (2003), pp. 321–348.

[57] M.K.V. Murthy and G. Stampacchia, *Boundary value problem for some degenerate elliptic operators* Ann. Math. Pura Appl. Ser.IV, **80** (1968), 1-122.

[58] S. Nicaise, Polygonal Interface Problems, Peter Lang (1993), 250 p. (Methoden und Verfahren der mathematischen Physik; Bd. 39).

[59] S. Nicaise, A.-M. Sändig, General interface problems I, II. Math. Meth. Appl. Sci, **17**, no 6 (1994), 395–450.

[60] S. Nicaise, A.-M. Sändig, Transmission problems for the Laplace and elasticity operators: Regularity and boundary integral formulation, Math. Models Meth. Appl. Sci, **9** (1999), 855–898.

[61] O.A. Oleinik, Boundary value problems for linear elliptic and parabolic equations with discontinuous coefficients, Amer. Math. Soc. Transl., **42** (1964), 175–194.

[62] T. von Petersdorff, Boundary integral equations for mixed Dirichlet, Neumann and transmission conditions, Math. Meth. Appl. Sci, **11** (1989), 185–213.

[63] M. Petzoldt, Regularity results for Laplace interface problems in two dimensions, Zeitschrift für Analysis und ihre Anwendungen, **20**, no. 2 (2001), 431–455.

[64] V.Ja. Rivkind, N.N. Ural'tseva, The classical solvability and linear schemes of the approximate solution of diffraction problems for quasilinear equations of parabolic and elliptic types. Problems of Math. Analysis, Leningrad. Univ., **3** (1972), 69–110.

[65] M. Schechter, A generalization of the problem of transmission. Ann. Scuola Norm. Sup. Pisa (3) **14** (1960), 207–236.

[66] Z.G. Sheftel, Estimates in L_p of solutions of elliptic equations with discontinuous coefficients and satisfying general boundary conditions and conjugacy conditions, Soviet Math. Dokl. **4** (1963), 321–324.

[67] Z.G. Sheftel, Energy inequalities and general boundary problems for elliptic equations with discontinuous coefficients, Sibirsk. Math. Zh. **6** (1965), 636–668. (Russian).

[68] S.L. Sobolev, *Some applications of functional analysis in mathematical physics,* Trans. of Math. Monographs, **90** (1991). AMS, Providence, Rhode Island.

[69] Van Tun, About estimates of solutions of boundary value problems for general elliptic second-order equation with discontinuous coefficients, Journal of computing mathematics. **4**, no 3 (1964), 577–580.

[70] W.P. Ziemer, *Weakly Differentiable Functions,* Springer-Verlag, New York, 1989.

Index

Notation Index

Frontiers in Mathematics

This series is designed to be a repository for up-to-date research results which have been prepared for a wider audience. Graduates and postgraduates as well as scientists will benefit from the latest developments at the research frontiers in mathematics and at the "frontiers" between mathematics and other fields like computer science, physics, biology, economics, finance, etc.

BIRKHÄUSER

■ **Aleman, A. / Feldman, N.S. / Ross, W.T.**, The Hardy Space of a Slit Domain (2009). ISBN 978-3-0346-0097-2

The book begins with an exposition of Hardy spaces of slit domains and then proceeds to several descriptions of the invariant subspaces of the operator multiplication by z. Along the way, we discuss and characterize the nearly invariant subspaces of these Hardy spaces and examine conditions for z-invariant subspaces to be cyclic. This work also makes important connections to model spaces for the standard backward shift operator as well as the de Branges spaces of entire functions.

■ **Avkhadiev, F.G. / Wirths, K.-J.**, Schwarz-Pick Type Inequalities (2009). ISBN 978-3-7643-9999-3

This book discusses in detail the extension of the Schwarz-Pick inequality to higher order derivatives of analytic functions with given images. It is the first systematic account of the main results in this area. Recent results in geometric function theory presented here include the attractive steps on coefficient problems from Bieberbach to de Branges, applications of some hyperbolic characteristics of domains via Beardon-Pommerenke's theorem, a new interpretation of coefficient estimates as certain properties of the Poincaré metric, and a successful combination of the classical ideas of Littlewood, Löwner and Teichmüller with modern approaches.

■ **Duggal, K.L. / Sahin, B.**, Differential Geometry of Lightlike Submanifolds (2010). ISBN 978-3-0346-0250-1

■ **Huber, M.**, Flag-transitive Steiner Designs (2009). ISBN 978-3-0346-0001-9

The monograph provides the first full discussion of flag-transitive Steiner designs. This is a central part of the study of highly symmetric combinatorial configurations at the interface of several mathematical disciplines, like finite or incidence geometry, group theory, combinatorics, coding theory, and cryptography. In a sufficiently self-contained and unified manner the classification of all flag-transitive Steiner designs is presented. This recent result settles interesting and challenging questions that have been object of research for more than 40 years. Its proof combines methods from finite group theory, incidence geometry, combinatorics, and number theory. The book contains a broad introduction to the topic, along with many illustrative examples. Moreover, a census of some of the most general results on highly symmetric Steiner designs is given in a survey chapter.

■ **Kasch, F. / Mader, A.**, Regularity and Substructures of Hom (2009). ISBN 978-3-7643-9989-4

■ **Kravchenko, V.V.**, Applied Pseudoanalytic Function Theory (2009). ISBN 978-3-0346-0003-3

Pseudoanalytic function theory generalizes and preserves many crucial features of complex analytic function theory. The Cauchy-Riemann system is replaced by a much more general first-order system with variable coefficients which turns out to be closely related to important equations of mathematical physics. This relation supplies powerful tools for studying and solving Schrödinger, Dirac, Maxwell, Klein-Gordon and other equations with the aid of complex-analytic methods. The book is dedicated to these recent developments in pseudoanalytic function theory and their applications as well as to multidimensional generalizations.